“十四五”普通高等教育本科部委级规划教材

纺织科学与工程一流学科建设教材

纺织工程一流本科专业建设教材

U0747216

机织物组织与结构
Weaves and Woven Structure
（双语）

郑天勇　赵　健　编著

中国纺织出版社有限公司

内容提要 / Abstract

本书以中英文对照的形式介绍了机织物的基本特点、织物上机图的表达、各种经典机织物组织的特点、结构与外观形成原理、组织设计方法、织物上机生产要点和机织物几何结构的基本原理。书中关于机织物结构形成和外观效果的浮长线原理为最新研究成果的总结，有利于读者理解机织物设计的基本原理。

本书可作为纺织工程本科生和留学生的教材，也可作为纺织工程技术人员和研究人员的参考用书。

This book introduces the basic characteristics of woven fabrics, the expressions and drawings of looming plans, the characteristics of various classic weaves, the forming principles of structure and appearance of woven fabrics, the methods of weave design, the key points of looming and the geometry of woven fabrics. In the book, the principles of forming the geometric structure and the appearance of woven fabrics by anaylizing the floats in weaves are summarized from the latest research results, which is helpful for readers to understand the basic principles of woven fabric design.

The book can be used as a teaching textbook for undergraduates and international students majoring in textile engineering, as well as a reference for textile engineers, technicians and researchers.

图书在版编目（CIP）数据

机织物组织与结构 ＝ Weaves and Woven Structure：双语：汉文、英文 / 郑天勇，赵健编著. —北京：中国纺织出版社有限公司，2022.11

"十四五"普通高等教育本科部委级规划教材　纺织科学与工程一流学科建设教材　纺织工程一流本科专业建设教材

ISBN 978-7-5180-9695-4

Ⅰ．①机…　Ⅱ．①郑…②赵…　Ⅲ．①机织物—织物组织—双语教学—高等学校—教材—汉、英　Ⅳ．①TS105.1

中国版本图书馆CIP数据核字（2022）第125339号

责任编辑：孔会云　沈　靖　责任校对：楼旭红
责任印制：王艳丽

中国纺织出版社有限公司出版发行
地址：北京市朝阳区百子湾东里A407号楼　邮政编码：100124
销售电话：010—67004422　传真：010—87155801
http://www.c-textilep.com
中国纺织出版社天猫旗舰店
官方微博 http://weibo.com/2119887771
三河市宏盛印务有限公司印刷　各地新华书店经销
2022年11月第1版第1次印刷
开本：787×1092　1/16　印张：21
字数：500千字　定价：58.00元

凡购本书，如有缺页、倒页、脱页，由本社图书营销中心调换

前　言

为了适应我国纺织教育国际化进程，促进纺织工业在"一带一路"沿线国家的发展，天津工业大学组织具有丰富教学经验的教师，编写了纺织专业核心课程双语教学的规划教材，为纺织工程专业的留学生提供中文和英文教学两种教学模式，以适应不同语言背景的留学生的学习需要。

对机织物组织与结构的认识与正确理解，是学生掌握纺织工程专业核心知识体系和能力培养的重要环节。本教材传承了国内外关于"机织物结构与设计"的经典理论，涵盖了常见的组织。编著者在多年的研究基础上，增补了浮长线对织物结构的影响机理，形成了机织物组织与结构的形成与外观效应之间的关系的统一理论体系，方便读者理解经典织物组织与结构之间的关系，为后续相关专业课程的学习提供了坚实的基础。限于篇幅，同类教材中的多重多层组织、纱罗织物、大提花织物、织物设计等内容没有列入本教材。

本书由郑天勇、赵健编著。编写分工如下：第一章、第二章、第五章、第六章、第七章由郑天勇编写，第三章、第四章由赵健编写。全书由郑天勇统稿，郑懿配音。

本书在编写与出版过程中，得到了天津工业大学国际教育学院姜亚明教授的大力支持。同时，对书中借鉴与引用的文献的作者表示崇高的敬意！

由于作者水平有限，书中难免有不成熟的见解，不当之处在所难免，热忱欢迎读者批评指正，以便将来修正。

编著者
2022年4月

Preface

In order to apply textile education internationalization in our country, and promote the textile industry in the countries along the Belt and Road initiative, Tiangong University organized the plan to edit the bilingual textbooks for specialized core curriculum in textile engineering, which is suitable for both domestic students and international students with different language backgrounds.

It is an important step for students in major of textile engineering to master the core knowledge system and cultivate ability by understanding the weaves and the structures of woven fabrics. This book inherits the classic theories of all versions of the textbook *Woven Structure and Design* at home and abroad, covering classical weaves. Based on the editor's many years of research, the theory of floats' effects on the woven structure is introduced in the book, and a unified theory system of forming the woven structure and the appearance is elaborated, which makes it easier for readers to understand the classic weaves and fabric structures, and provides a solid foundation for the related professional courses in subsequent study. Constrained by the space, the textbook excludes the contents of multilayer weave, leno fabric, jacquard fabric and fabric design in similar textbooks.

The book is written by ZHENG Tianyong and ZHAO Jian. The division of labor is as follows: Chapters 1, 2, 5, 6 and 7 are written by ZHENG Tianyong, while chapters 3 and 4 are written by ZHAO Jian. The book is finally edited by ZHENG Tianyong and voicd by ZHENG Yi.

In the process of writing and publishing this book, professor JIANG Yaming, School of International Education, Tiangong University, has given great support. At the same time, we would like to express great respect to the authors of the references and references in the book.

Due to the authors' limited level, it is inevitable that there are immature opinions in the book, and mistakes and improper places are inevitable. Readers' criticisms and corrections are warmly welcome for future correction.

Authors
April, 2022

目　录

第一章　机织物组织与上机图 /Woven Weaves and Looming Plans ·················· 1

第一节　机织物概述 /A Survey of Woven Fabrics ·················· 1

一、机织物的分类 /Classification of Woven Fabrics ·················· 1

二、机织物的组成 /Compositions of Woven Fabrics ·················· 2

三、机织物的形成原理 /Principle of Forming the Woven Fabrics ·················· 3

四、机织物的量度 /Fabric Measurement ·················· 4

第二节　织物组织 /Weave ·················· 6

一、织物组织的概念 /Concept of Weave ·················· 6

二、织物组织的表示方法 /Representation of Weave ·················· 7

三、飞数 /Step Number ·················· 11

四、同面组织和异面组织 /Even Sided Weave and Unbalanced Weave ·················· 12

五、规则组织图的绘制 /Drawing of Regular Weave Patterns ·················· 13

第三节　上机图 /Looming Plans ·················· 16

一、上机图的组成 /Components of Looming Plans ·················· 16

二、穿综图 /Draft ·················· 16

三、穿筘图 /Denting Plan ·················· 23

四、纹板图 /Lifting Plan ·················· 26

五、组织图、穿综图和纹板图之间的关系 /The Relationship Among Design, Draft and Lifting Plan ·················· 34

习题 /Questions ·················· 35

第二章　机织物几何结构 /Geometric Structure of Woven Fabrics ·················· 39

第一节　机织物几何结构参数 /Parameters of Woven Geometric Structure ·················· 39

一、机织物几何结构的概念 /Concept of Geometric Structure ·················· 39

二、织物中纱线的几何形态 /Contour and Crimp of Yarns ·················· 40

1

三、屈曲波高与几何结构相 /Crimp Height and Geometry Phase·················43

四、影响织物屈曲波高的因素 /Factors Affecting Crimp Height··············47

五、织物其他主要参数 /Other Parameters of Fabric·····················48

第二节　织物组织对织物几何结构的影响 /Influence of Woven Structure on
　　　　Woven Geometric Structure·····························52

一、交错与浮长 /Intersections and Float Length·····················53

二、交织对纱线形态和织物性能的影响 /Effects of Interlacing on the Yarn
　　Contour and the Properties of Fabrics··························53

三、浮长对织物结构和外观的影响 /Effects of Floats on the Structure
　　and Appearance of Fabrics··································55

四、紧密织物与紧密度 /Tight Fabrics and Tightness·················62

五、紧密织物的结构相 /Geometric Phase of Tight Fabrics···············65

六、紧密织物的织缩率 /Contraction Rate of Tight Fabrics···············67

第三节　几何结构对织物外观性能的影响 /Influence of Woven Geometric
　　　　Structure on the Properties of Fabrics·····················68

一、外观效应 /Appearance···································69

二、力学性能 /Mechanical Properties····························69

三、与孔隙率、厚度相关的特性 /Properties Related to Porosity and Thickness········73

习题 /Questions···75

第三章　原组织 /Elementary Weaves·····························77

第一节　原组织概述 /A Survey of Elementary Weaves·················77

一、定义 /Definition······································77

二、原组织的特征 /Characteristics of Elementary Weaves···············77

三、原组织的类型 /Types of Elementary Weaves·····················77

第二节　平纹组织 /Plain Weaves·······························78

一、平纹组织表示方法 /Representation of Plain Weaves··················78

二、平纹组织的特征 /Characteristics of Plain Weaves·················79

三、平纹织物上机 /Looming for Plain Weaves·······················79

四、平纹织物结构与外观性能特点 /Appearance and Properties of Plain Weaves·······80

五、平纹组织的应用 /The Application of Plain Weaves·················80

六、平纹织物的特殊效应 /Special Effects of Plain Fabrics···············81

第三节　斜纹组织 /Basic Twill Weaves·························85

一、斜纹组织斜向 /Direction of Basic Twill Weaves·················86

二、斜纹组织参数与表示方法 /Parameters and Representation of
　　Basic Twill Weaves·····································86

三、斜纹织物结构特征及外观形成原理 /Appearance and Characteristics
of Basic Twill Weaves ··· 87

四、斜纹组织图绘制 /Drawing of Basic Twill Weaves ············· 90

五、斜纹织物上机 /Looming for Basic Twill Weaves ············· 91

六、斜纹组织的应用 /The Application of Basic Twill Weaves ··········· 92

第四节 缎纹组织 /Satin/Sateen Weaves ··········· 92

一、缎纹组织形成条件 /Prerequisites of Satin/Sateen Weaves ········· 93

二、缎纹组织表示方法 /Representation of Satin/Sateen Weaves ········· 93

三、缎纹组织的特征 /Characteristics of Satin/Sateen Weaves ········· 93

四、缎纹织物外观与性能 /Appearance and Properties of Satin/Sateen Weaves ········· 94

五、缎纹组织图绘制 /Drawing of Satins/Sateen Weaves ············· 96

六、缎纹织物上机 /Looming for Satin/Sateen Weaves ············· 97

七、纱线捻向对缎纹织物风格的影响 /Effects of Yarn Twist on
Satin/Sateen Weaves ··· 97

八、缎纹组织的应用 /The Application of Satin/Sateen Weaves ········· 99

第五节 原组织对比 /Comparison Among Elementary Weaves ········· 99

一、平均浮长与飞数 /Average Length of Float and Step Number ········· 99

二、织物正反面外观特征 /Appearance of Both Sides ············· 100

三、织物性能 /Properties of Elementary Weave Fabrics ············· 101

习题 /Questions ··· 102

第四章 变化组织 /Derivatives of Elementary Weaves ············· 103

第一节 平纹变化组织 /Plain Weave Derivatives ············· 103

一、经重平组织 /Warp Rib Weaves ··· 104

二、纬重平组织 /Weft Rib Weaves ··· 105

三、方平组织 /Basket Weaves ··· 108

第二节 斜纹变化组织 /Twill Weave Derivatives ············· 112

一、加强斜纹组织 /Double Twills ··· 112

二、复合斜纹组织 /Composed Twills ············· 114

三、角度斜纹组织 /Elongated Twills ············· 116

四、曲线斜纹组织 /Curved Twills ············· 119

五、山形斜纹组织 /Waved Twills ············· 123

六、锯齿斜纹组织 /Zigzag Twills ············· 126

七、破斜纹组织 /Broken Twills ············· 128

八、菱形斜纹组织 /Diamond Twills ············· 134

九、芦席斜纹组织 /Entwining Twills ············· 137

第三节　缎纹变化组织 /Satin/Sateen Weave Derivatives ················· 139

　　一、加强缎纹组织 /Double Satins/Sateens ······················· 139

　　二、变则缎纹组织 /Irregular Satins/Sateens ····················· 141

　　三、重缎纹组织 /Extended Satins/Sateens ······················· 143

　　四、阴影缎纹组织 /Shaded Satins/Sateens ······················· 144

第四节　传统布边 /Traditional Selvage ································ 145

　　一、传统布边设计原则 /Principle of Designing Traditional Selvage ···· 146

　　二、布边织疵及解决方案 /Faults in Selvage and Solutions ·········· 147

　　三、布边设计实例 /Applications ······························· 149

习题 /Questions ·· 152

第五章　联合组织 /Combined Weaves ································ 155

第一节　条格组织 /Stripe and Check Weaves ························· 155

　　一、纵条纹组织 /Longitudinal Stripe Weaves ···················· 156

　　二、方格组织 /Check Weaves ································· 159

　　三、格子组织 /Two-way Stripe Weaves ························· 162

第二节　绉组织 /Crepe Weaves ····································· 163

　　一、绉效应 /Crepe Effects ···································· 163

　　二、起绉方法 /Methods of Making Crepe Weaves ················· 163

　　三、起绉原理与设计要点 /Principles and Key Points of Designing Crepe Weaves ······ 164

　　四、绉组织设计 /Designing of Crepe Weaves ···················· 165

第三节　透孔组织 /Open Gauze Weaves ····························· 171

　　一、外观特点 /Description ···································· 171

　　二、组织特点与透孔的形成原理 /Construction and Formation of the Structure ······ 172

　　三、简单透孔组织的组织图绘制 /Designing of Simple Open Gauze Weaves ·········· 177

　　四、透孔织物上机要点 /Looming for Open Gauze Weaves ·········· 177

　　五、变化透孔组织 /Variations of Open Gauze Weaves ·············· 178

　　六、透孔组织的应用 /Applications of Open Gauze Weaves ·········· 179

第四节　蜂巢组织 /Honeycomb Weaves ······························ 179

　　一、外观特点 /Description ···································· 179

　　二、组织特点与蜂巢的形成原理 /Construction and Formation of the Structure ······ 180

　　三、简单蜂巢组织的组织图绘制 /Designing of Ordinary Honeycomb Weaves ········· 182

　　四、变化蜂巢组织 /Variations of Honeycomb Weaves ·············· 183

　　五、蜂巢组织上机与应用 /Looming and Applications of Honeycomb Weaves ········· 185

第五节　凸条组织 /Bedford Cord Weaves ····························· 187

　　一、外观特点 /Description ···································· 187

　　二、组织特点与凸条的形成原理 /Construction and Formation of the Structure ……… 187

　　三、简单凸条组织的组织图绘制 /Designing of Simple Bedford Cord Weaves ……… 189

　　四、增加凸条隆起效果的方法 /Measures to Enhance the Cords ……… 191

　　五、凸条织物上机要点 /Looming for Bedford Cord Weaves ……… 192

　　六、凸条组织的应用 /Applications of Bedford Cord Weaves ……… 193

第六节　网目组织 /Spider Weaves ……… 194

　　一、外观特点 /Description ……… 194

　　二、组织特点与网目的形成原理 /Construction and Formation of
　　　　the Structure ……… 196

　　三、简单网目组织的组织图绘制 /Designing of Simple Spider Weaves ……… 197

　　四、网目织物上机要点 /Looming for Spider Weaves ……… 198

　　五、增加网目效果的方法 /Measures to Enhance Distorted Effects ……… 199

　　六、变化网目组织 /Variations of Spider Weaves ……… 200

第七节　小提花组织 /Dobby Spot Weaves ……… 201

　　一、小提花组织的构成 /Construction of Dobby Spot Weaves ……… 201

　　二、小提花织物设计原则 /Principles of Designing Dobby Spot Weaves ……… 202

　　三、小提花织物设计步骤 /Procedures of Designing Dobby Spot Weaves ……… 203

　　四、小提花组织设计实例 /Examples of Designing Dobby Spot Weaves ……… 203

第八节　配色模纹效果 /Color and Weave Effects ……… 205

　　一、配色模纹图 /Color and Weave Effect Diagram ……… 206

　　二、配色模纹图的绘制 /Drafting the Color and Weave Effects ……… 208

　　三、根据配色模纹图反求色纱排列与组织 /Determine the Weave and the
　　　　Orders of the Coloring by the Effects ……… 209

　　四、配色模纹类型 /Variations of Color and Weave Effects ……… 211

习题 /Questions ……… 211

第六章　二重及双层组织 /Backed Weaves and Multi-Layer Weaves ……… 216

第一节　经二重组织 /Warp-Backed Weaves ……… 216

　　一、经二重组织的构成 /Construction of Warp-Backed Weaves ……… 217

　　二、经二重组织的判断 /Judgement of Warp-Backed Weaves ……… 218

　　三、经二重组织的设计原则 /Principles of Designing Warp-Backed Weaves ……… 219

　　四、经二重组织的组织图绘制 /Drawing of Warp-Backed Weaves ……… 220

　　五、经二重组织的上机要点 /Looming for Warp-Backed Weaves ……… 221

　　六、经起花组织 /Extra Warp Figured Weaves ……… 222

　　七、经二重组织的其他应用 /Other Applications ……… 225

第二节 纬二重组织 /Weft-Backed Weaves ·················· 226

一、纬二重组织的构成 /Construction of Weft-Backed Weaves ·············· 226

二、纬二重组织的判断 /Judgement of Weft-Backed Weaves ·············· 227

三、纬二重组织的设计原则 /Principles of Designing Weft-Backed Weaves ·········· 228

四、纬二重组织的组织图绘制 /Drawing of Weft-Backed Weaves ·········· 229

五、纬二重组织的上机要点 /Looming for Weft-Backed Weaves ·········· 229

六、表里换纬二重组织 /Interchanging Weft-Backed Weaves ·········· 230

七、纬起花组织 /Extra Weft Figured Weaves ·············· 232

八、重经重纬组织的比较 /Comparison Between Warp-Backed Weaves and

Weft-Backed Weaves ················· 233

第三节 双层组织 /Double-Layer Weaves ·················· 234

一、双层组织的结构与形成原理 /Construction and Formation of

Double-Layer Weaves ··············· 235

二、管状织物 /Tubular Weaves ················· 239

三、双幅织物 /Double-Width Weaves ·············· 246

四、表里换层织物 /Double Interchanging Weaves ·········· 249

五、接结双层织物 /Stitched Double Weaves ·········· 253

习题 /Questions ···················· 264

第七章 起毛起圈织物 /Fleecy Fabrics ·················· 269

第一节 起毛起圈组织的类型 /Classification of Fleecy Fabrics ·········· 269

一、割绒起圈织物 /Pile Fabrics ··············· 269

二、拉绒织物 /Napped Fabrics ··············· 272

第二节 灯芯绒 /Corduroy Weaves ·················· 273

一、灯芯绒组织的构成 /Construction of Corduroy Weaves ·········· 273

二、割绒工序 /Fustian Cutting ··············· 276

三、灯芯绒织物设计要点 /Key Points of Designing Corduroy Weaves ·········· 277

四、灯芯绒组织设计 /Designing of Corduroy Weaves ·········· 283

五、灯芯绒上机要点 /Looming for Corduroy Weaves ·········· 284

六、花式灯芯绒 /Figured Corduroy Weaves ·········· 285

第三节 经起毛组织 /Warp Pile Weaves ·················· 287

一、杆织法 /With the Aid of Wires ·············· 288

二、经浮长通割法 /By Cutting the Warp Floats ·········· 289

三、双层织制法 /On the Principle of Face to Face ·········· 290

第四节 毛巾组织 /Terry Weaves ·················· 307

一、毛巾组织的结构 /Construction of Terry Weaves ·········· 308

二、毛圈的形成原理 /Formation of Terry Weaves ················· 309

三、毛圈成形良好的关键 /The Conditions Required for Terry Loops ············· 311

四、毛巾组织设计 /Designing of Terry Weaves ················· 314

五、花式毛巾组织 /Fancy Terry Weaves ················· 316

六、毛巾织物上机 /Looming for Terry Fabrics ················· 318

习题 /Questions ················· 319

参考文献 /References ················· 322

第一章　机织物组织与上机图 /Woven Weaves and Looming Plans

第一节　机织物概述 /A Survey of Woven Fabrics

一、机织物的分类 /Classification of Woven Fabrics

织物是由相互垂直的两个系统纱线按照一定规律交织成的片状纤维集合体，使用最广泛。机织物有多种分类方法，具体见表 1-1-1。一些特殊的织物如图 1-1-1 ～图 1-1-4 所示。

Woven fabrics are generally used to refer to the fiber assembly composed of two series of threads, warp and filling, which are interlaced at right angle based on a certain pattern. The woven fabrics are most widely used and can be classified in different ways as shown in Table 1-1-1. Figures from Fig. 1-1-1 to Fig. 1-1-4 show some special fabrics.

表 1-1-1　机织物分类

分类方法	类别	描述/品种
按原料分	纯纺织物	由同种纤维纺成的纱线构成
	混纺织物	由两种及以上不同纤维混纺成的纱线织成
	交织物	经纱、纬纱分别由不同纤维纺成
按用途分	衣着用织物	内衣、外衣、衬里织物
	装饰用织物	挂毯、床上用品、家具布等织物
	产业用织物	经专门设计、具有工程结构特点和特定应用领域的纺织品，包括农业栽培用、渔业和水产养殖用、土工用、交通运输用、过滤分离用、隔层及绝缘用、包装用、安全防护用、文体休闲用、医疗卫生用、结构增强用、国防军事用织物及其他
按组织复杂程度分	原组织织物	平纹组织、斜纹组织、缎纹组织织物
	变化组织织物	平纹变化组织、斜纹变化组织、缎纹变化组织织物
	联合组织/小提花织物	绉组织、条格组织、蜂巢组织、透孔组织等织物
	复杂组织织物	多重组织、多层组织、毛巾组织、起毛组织、纱罗组织织物
	大提花组织织物	织锦缎、古香缎等
其他分类	三向织物	三组纱线相互成60°交织而成
	三维织物	由三个方向（长度、宽度和高度）而不是两个方向的纱线构成，分四类:异型织物、可展开织物、三维正交织物、异型三维正交织物
	机织针织联合织物	机织物与针织物联合形成

Table 1-1-1　Classification of woven fabrics

Methods of Classification	Species	Descriptions/Items
By materials	Pure fabrics	Comprising threads of same fibers
	Blended fabrics	Comprising blended yarns that are mixed by not less than two species of fibers
	Union fabrics	Fabrics with a type of warp thread and another type of weft thread
By usage	Clothing/apparel fabrics	Underwear, jacket, lining fabric, etc.
	Ornamental/ decorative fabrics	Tapestries, bedding, furniture fabrics, etc.
	Industrial fabric/ technical fabrics	Fabrics or fibrous components are selected principally (but not exclusively) for its performance and properties. Applications include for agriculture, fisheries and aquaculture, civil engineering, transportation, sails, filtering, safety and protection, polishing and absorption, insulation, package, entertainment, medical and hygiene, composites, aerospace, defense and military, etc.
By complexity of the weaves	Basic weaves	Fabrics of plain weave, twill weave, satin and sateen weave
	Derivatives of basic weaves	Plain weave derivatives, twill kindred weaves, satin weave derivatives
	Combined weaves	Fabrics of crepe, stripe, honeycomb, open gauze, etc.
	Compound weaves	Fabrics of backed weaves, multi-layer weaves, terry, piled weaves, lenos, etc.
	Jacquard fabrics	Brocade, Damask, etc.
Others	Tri-axial fabrics	Comprises of three series of yarns that are interlaced at 60°
	3D fabrics	The component threads are simultaneously woven in three directions (length, width, and thickness) rather than in the conventional two. The types of structures that can be produced into four broad classes: contoured fabrics, expandable fabrics, interwoven fabrics, and contoured interwoven fabrics
	Woven-knit combined fabrics	Combination of woven fabrics and knitted fabrics

图1-1-1　三向织物
Tri-axial fabric

图1-1-2　三维织物
3D fabric

图1-1-3　异型三维织物
3D special fabric

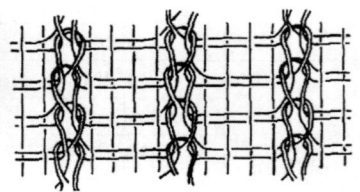

图1-1-4　机织针织联合织物
Woven-knitted combined fabric

二、机织物的组成 /Compositions of Woven Fabrics

机织物由经纱与纬纱组成。经纱方向从织机后方到前方排列，与织物布边平行；纬纱与织机幅宽方向平行，与布边垂直。

Woven fabrics consist of warps and wefts. The warp direction is from the back of the loom to the front, parallel to the selvedge of the fabric; the weft crosses the width of the loom and is perpendicular to the selvedge.

机织物由布身与布边组成。布身是织物的主体部分，布边有光边与毛边之分。

织物有正反面之分，有的织物正面与反面相差不大，有的织物正反面区别明显。一般来说，外观效应好的一边为织物的正面。

A woven fabric may also be considered to consist of the body and the selvedge. The body is the main part of the fabric, and the selvedge includes the clear selvedge and fringe selvedge.

Fabrics have the face (obverse, front or right) side and back (reverse or wrong) side. There is a tiny difference in two sides of some fabrics while others may be totally different. Generally, the side with good appearance effect is the front side of the fabric.

三、机织物的形成原理 /Principle of Forming the Woven Fabrics

机织物从织机上织造形成。经纱从织轴上退绕，经过后梁、分绞棒、经停片、综丝眼、钢筘，在织口处与纬纱交织成织物，再经过胸梁，被卷取到卷布辊上（图1-1-5）。

The woven fabrics are formed on a loom. The warps are unwound from weaver's beam, then pass on the back rest roller, leasing rods, drop wires, eyes of heald wire (or mail eyes), split of reed, and interlace with the filling yarns at fell. Then, the newly formed fabric passes over the breast beam and is wound on the clothing roller (Fig. 1–1–5).

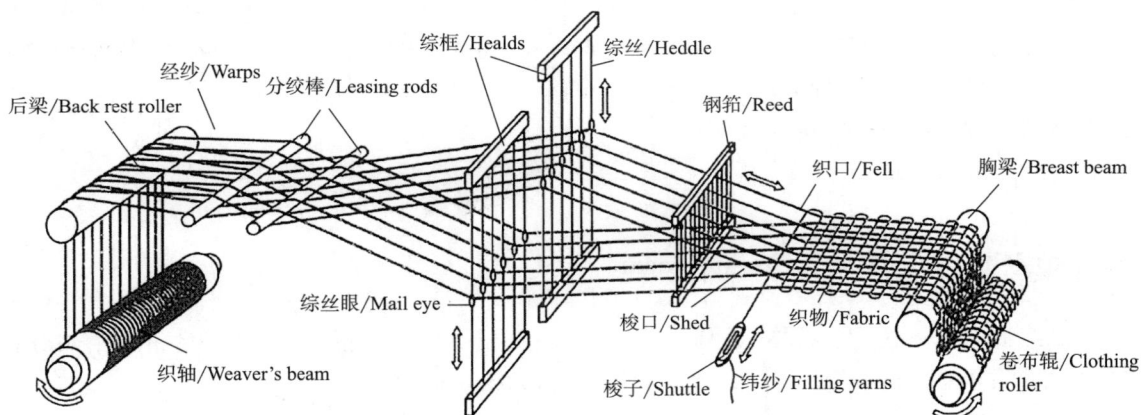

图1-1-5　机织物的形成原理
The principle of weaving

织机在织造织物时，至少需要开口运动、引纬运动和打纬运动。

开口运动就是经纱由综框控制，分成上下两层，形成梭口，以便梭子或者引纬器进入，织入纬纱。开口时，必须按照织物交织规

To form a woven fabric, at least three movements are needed, i.e., shedding, picking and beating-up.

In shedding, the warps are controlled by heald shafts to form two layers which form an insertion path for the shuttle or carrier to carry the filling yarn to pass through the passage. The woven structure is formed by controlling the

3

律所决定的提综顺序，通过控制综框来控制经纱的升降顺序，形成所需的织物组织结构。如使用凸轮开口机构，综框数量在8页（片）以内；若采用多臂开口机构，综框数量在16页以内，但最多可达32页；大提花开口机构可以直接控制每根经纱独立运动，经纱规律可以更加复杂。

引纬运动将纬纱引入梭口中，不同的引纬方式会影响开口尺寸、引纬效率、布边形态和能源消耗。打纬运动则是通过钢筘的往复运动将刚引入的纬纱推向织口，与经纱交织成织物。钢筘还具有控制经纱密度和导引纬纱的作用。

为了使织造能够连续不断地进行，还需要卷取运动和送经运动将已经织造完的织物引离织口，同时在织轴上放送相应长度的经纱。卷取运动和送经运动共同控制纬纱密度。若需要织入不同品种的纬纱，织机需要配备选色机构或多梭箱装置。

order of raising the heddle frames which are determined by the interlacing of the warps and the wefts. If a cam shedding mechanism is used, the maximum number of the healds is 8. While in dobby shedding, the number of healds is usually limited in 16 and never exceeds 32. For Jacquard shedding, since each end is individually controlled by heddles, more complex interlacing can be achieved.

In picking, the filling yarn is inserted into the shed. The way of inserting affects the size of shedding passage, efficiency, forms of selvedge and energy consumption. In beating-up, the reciprocating sley pushes the newly inserted filling yarns to the fell to interlace with the warps. The reed can also be used to control the warp density and the sley helps to guide the weft carrier or shuttle.

To make the weaving continuously, the taking-up movement forwards the formed fabric and the letting-off movement supplies the warps at given speed. Both the two movements control the tension in weaving and the weft density of the fabric. If different filling yarns are required to produce a transverse stripe, a box motion or filling-select mechanism is fixed on a loom.

四、机织物的量度 /Fabric Measurement

机织物可以从长度、幅宽、厚度、重量等方面进行度量。

1. 长度 /Piece Length

织物的长度用米表示，商业上常使用匹长，即从一个接缝到另一个接缝的织物长度，匹长随着织物的重量、厚度和幅宽的变化而变化，一般长度为40m，厚重织物为10m左右，特别轻薄的织物可以超过100m。

2. 幅宽 /Width

机织物的幅宽由织物用途与

Woven fabrics are measured in length, width, thickness, weight, etc.

Woven fabrics are commercially produced and sold in length units, commonly called pieces, the standard (full) length of cloth from seam to seam, which varies according to the weights and widths of the fabrics. The piece length of fabrics is usually 40m. Very heavy of special types of fabrics may be in pieces as short as 10 meters, whilst light fabrics may be in pieces of 100 meters in length or more.

The width of the woven fabric is based on the

生产设备决定。一般来说，机织物幅宽在 30cm 以上，不包括带状等特殊机织物。对于一些传统织机来说，制作服装的织物幅宽为 90～114cm，粗纺毛织物的幅宽为 130～140cm，精纺毛织物的幅宽为 150cm。对于服装生产来说，幅宽越宽，生产成本越低。目前织物有宽幅化的趋势，片梭织机的幅宽可以达到 5.4m，喷气织机的幅宽为 4.2m，剑杆织机的幅宽为 4.6m，喷水织机的幅宽为 2.3m，而一些地毯织机的幅宽可以达到 7m 以上。

application and production equipment. The width of the woven fabric varies from 30cm upwards, but this excludes very narrow fabrics as ribbons, tapes, and braids which are made by a special section of the textile industry. There are examples of widths of some traditional fabrics, for shirtings and dress fabrics, they range from 90cm to 114cm, woolen fabrics may range from 130cm to 140cm, and the worsted suiting is as wide as 150cm. It is more economical to weave the fabric in a wider width for cutting in garment. Therefore, there is a tendency for fabrics to widen at present. The width of fabric woven by a gripper loom can be as wide as 5.4m; 4.2m by air-jet loom, 4.6m by rapier loom and 2.3m by water-jet loom. The width of some carpets can reach 7m or more.

3. 厚度 /Thickness

织物的厚度一般定义为在一定压力下织物正面与反面的距离（mm）。

In textiles, thickness is the distance (mm) between the upper and lower surfaces of the material, measured under a specified pressure.

4. 重量 /Weight

织物的重量是指每平方米无浆干燥织物的克重（g）。机织物的克重范围在 $15g/m^2$（雪纺绸）到 $600g/m^2$（大衣呢），地毯织物则更加厚重。

Weights are often expressed in grammes per square meters of the dried, unsized fabric. The range of the weights in woven fabric varies from as little as $15g/m^2$ to $600g/m^2$ or more for heavy carpet fabrics.

5. 织物密度 /Density

织物密度是指 10cm 内纱线的根数，分为经纱密度和纬纱密度。英制的织物密度指 1 英寸内的纱线根数，也有的表示方法指每平方英寸中经纬纱根数之和。粗纺麦尔登和苎麻织物的经纱密度可以低至 150 根 /10cm，丝织物的经纱密度可以高达 1200 根 /10cm。

Fabric density is the number of threads in 10cm. It includes warp density and weft density. For Imperial system, the density is the number of threads in one inch. Thread /cloth count is the number of warp threads and filling threads in a square inch of fabric. For the ramie fabric or woolen fabric like Melton, the warp density can be as low as 150 threads/10cm, however, the warp density of some silk fabrics can be as high as 1200 threads/10cm.

6. 织物规格表示方法 /Specification of the Woven Fabric

机织物规格一般由四部分组成，包括幅宽、纱线线密度、织物密度和织物组织。表示方法为：织物幅宽，经纱（纺纱方法，原料，线密度）× 纬纱（纺纱方法，原

The specifications of the woven fabric are usually represented in 4 parts: width, yarn fineness: density and weave pattern. A full representation of the specification is written as "width of the fabric, warp yarn (manufacturing system, material, linear density) × weft yarn (manufacturing

料，线密度），经密 × 纬密，织物组织。如：160cm，JC 14.5 tex × JC 14.5tex，547 × 283，府绸。表示某府绸织物幅宽为 160cm，经纱与纬纱都采用精梳棉纱，线密度均为 14.5tex，经纱密度为 547 根 /10cm，纬纱密度为 283 根 /10cm，采用平纹组织织造。

部分织物规格采用英制表示，例如：72 英寸，（JC 40 英支竹 + T48 英支）（1：1）×（JC 40 英支竹 + T48 英支）（1：1），133×78，小提花。表示某小提花织物，织物幅宽为 72 英寸，经纬纱都是 2 合股股线，单纱分别是 40 英支精梳棉与竹纤维混纺纱和 48 英支涤纶纱，经纱密度为 133 根 / 英寸，纬纱密度为 78 根 / 英寸，采用小提花组织织造。

system, material, linear density), warp density × weft density, weave pattern".

For example, "160cm, JC 14.5tex × JC 14.5tex, 547 × 283, combed poplin" means that the width of the poplin is 160cm. Both the warp thread and weft thread are 14.5 tex combed yarn, and there are 547 ends per10cm and 283 picks per 10cm. The warp thread and weft thread are interwoven in plain weave.

Some fabrics are expressed by Imperial System. For example, "72″, (JC 40s bamboo + T48s)(1:1) × (JC 40s bamboo + T48s)(1:1), 133 × 78, dobby" means that the fabric is 72 inches in width. Both the warp thread and the weft thread are the 2-plied yarn composed of the combed 40s blended yarn of cotton and baboo fibers and 48s yarn of polyester fiber. The fabric is woven in a dobby weave with 133 warps per inch and 78 picks per inch.

第二节　织物组织 /Weave

Concept of Weave

一、织物组织的概念 /Concept of Weave

织物中，经纱与纬纱相互交错（交织）或者沉浮的规律称为织物组织。交错使得纱线弯曲，固定织物中经纱和纬纱的位置，在空间形成特定的纱线形态，即织物的几何结构，它影响织物的各项性能。图 1-2-1 和图 1-2-2 是两种不同织物的经纱与纬纱交织规律，经纱与纬纱交叉并重叠之处称为组织点，若经纱在纬纱之上称为经组织点或经浮点，若纬纱在经纱之上称为纬组织点或纬浮点。当所有的经纬纱交织规律达到重复的最小单元，称

Weave is the system or pattern of intersecting warps and wefts. Intersection bends the threads and anchors the warps and the filling threads to form a special geometric structure, which affects physical properties of the fabric. Figs. 1-1-1 and 1-1-2 show two interlacing diagrams of two different weaves, in which the point where a warp thread and a weft thread overlap or cross is called an interlacing point. If the warp is over the weft, the point is called warp over weft (raiser or warp-up point), otherwise, is called weft over warp (sinker or weft-up point). A weave design repeats on a definite number of ends and picks. Weave repeat is the smallest number of ends and picks on which a weave-interlacing pattern can be represented. In this book, the

为一个组织循环或一个完全组织。此时的经纱根数称为经纱循环根数，用 R_j 表示；纬纱根数称为纬纱循环根数，用 R_w 表示。本书中，R 表示循环根数，下标 j 和 w 分别表示经向、纬向。图 1-2-1 所示机织物的 $R_j=R_w=2$，A 点是经组织点，B 点是纬组织点；图 1-2-2 所示机织物的 $R_j=R_w=6$，A 点是经组织点，B 点是纬组织点。

number of the ends in a weave repeat is represented as R_j, and R_w for picks, where R for size of repeat and subscript j and w denoting warp-wise and weft-wise respectively. In Fig. 1-2-1, $R_j=R_w=2$, and A is the interlacing of warp over weft while B is weft over warp. In Fig. 1-2-2, $R_j=R_w=6$, and A is a raiser while B is a sinker.

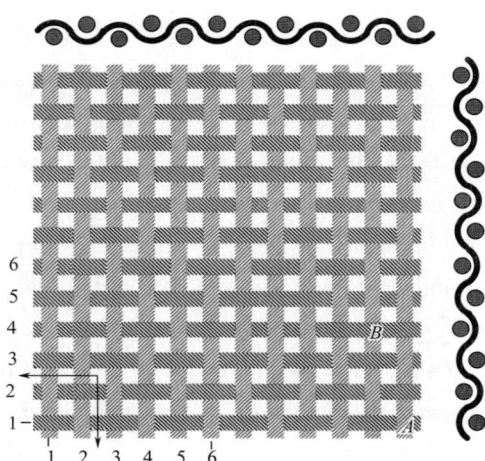

图1-2-1　机织物1的交织规律示意图
Interlacing schematic diagram of weave 1

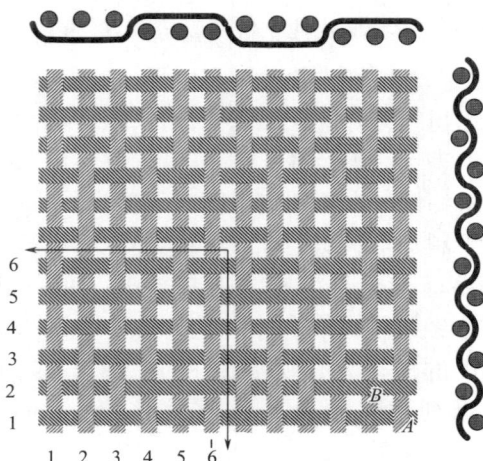

图1-2-2　机织物2的交织规律示意图
Interlacing schematic diagram of weave 2

二、织物组织的表示方法 /Representation of Weave

织物组织可用多种方法表示，如交织图、组织图、分数式+说明、截面图（剖面图）等。

A weave can be represented by various ways, such as interlacing diagram, weave pattern, fraction + verbal statement, and sectional diagram.

Representation of Weave

1. 交织图 /Interlacing Diagram

图 1-2-1 和图 1-2-2 的主体部分是表示织物交织规律的交织图。交织图中，经纱的顺序是自左向右，纬纱的顺序是自下而上。从图 1-2-1 可以看出，经纱与纬纱交织规律彼此都是一沉一浮，这是最

The main parts of Fig. 1-2-1 and Fig. 1-2-2 are the interlacing diagrams where the ends are ordered from left to right and the picks are ordered from bottom to top. Fig. 1-2-1 shows the simplest weave—plain weave where the warps float over and below the wefts alternately. In a more complicate weave shown in Fig. 1-2-2, some warps

简单的织物组织，称为平纹组织。图 1-2-2 稍微复杂，部分经纱和纬纱是彼此一沉一浮，另外一些经纱和纬纱则是连续浮在 3 根纱线之上，然后连续沉在 3 根纱线之下。对组织循环较大的组织来说，交织图绘制较为费时。

2. 组织图 /Weave Pattern

在方格纸（也称意匠纸）上用 ■、×、▲ 等符号表示经组织点、用空白格表示纬组织点的方式表示织物经纬纱交织规律的图解称为组织图。意匠纸一般每 8 个格子一组，用粗线分开。在组织图上，每一竖列表示一根经纱，每一横行表示一根纬纱。在绘制组织图时，至少要绘制一个组织循环，组织循环内每一横行与每一竖列上都至少有一个经组织点和一个纬组织点。组织图的横行数必须是 R_w 的整数倍，纵列数必须是 R_j 的整数倍，图 1-2-3 和图 1-2-4 分别为对应图 1-2-2 和图 1-2-1 的组织图。这种表示方法绘图快捷、简单。

如不特殊说明，只要方格图上有符号，即说明是经组织点。有时不同的符号和颜色表示该组织点对织物结构的形成具有不同的作用。

float over and below the wefts alternately, and the other ends float over and below 3 successive wefts. Obviously, it is a time-consuming work to draw the interlacing diagram of a weave upon large repeat number.

In a point paper, the weave pattern is a plan where the symbols or marks like " ■ " " × " " ▲ " are used to indicate the interlacing point "warp over weft" while the blank is used to indicate the interlacing point "weft over warp" unless otherwise stated. The standard textile design paper is ruled in groups of 8×8, these being separated by thicker bar lines. In a weave pattern, each vertical space (column) is taken to represent a warp end and each horizontal space (row) a weft pick. Since the weave repeats on a definite number of ends and picks (or of vertical and horizontal spaces), one repeat is needed and enough on design paper. In each full repeat of the weave, every vertical space and horizontal space must have at least one mark and at least one blank, otherwise, the threads do not interlace but merely form loose floats which do not become woven into the cloth. The number of the rows and the number of the columns in a weave pattern must be the multiple of R_w and R_j respectively. Fig. 1-2-3 and Fig. 1-2-4 are the weave patterns corresponding to Fig. 1-2-2 and Fig. 1-2-1 respectively. It is more convenient to represent the interlacings by a design paper (point paper, or square paper).

Without special explanation, any symbol or mark in a weave means that the weaving point is a riser. Several types of marks may be used in one design simultaneously to indicate that some ends differently marked vary in thickness, color or function from the others.

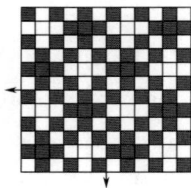

图1-2-3　织物2组织图（2个循环）
The weave 2 pattern (2 repeats)

图 1-2-4　织物1组织图（1个循环）
The weave 1 pattern (Only one repeat)

3. 分数式 + 说明 /Fraction + Verbal Statement

织物组织中，若每根纱线的交织规律相同，可以用"分数式＋说明"表示。这里的分数式通常是第一根纱线（可能是经纱，也可能是纬纱）交织方式，分子表示连续的经组织点个数，分母则是连续的纬组织点个数。说明部分则是将第一根纱线推广到整个循环的方式，包括参数。

图 1–2–3 所示的组织可以表述为 $\dfrac{1}{1}$ 平纹，读作一上一下平纹，表示第一根经纱上，经纱在与纬纱交织时，经纱的位置次序是连续 1 个在上方（经组织点）、连续 1 个在下方（纬组织点）。图 1–2–5 所示的各种组织分别是 $\dfrac{2}{2}$ 方平、$\dfrac{2}{2}\nearrow$ 斜纹、$\dfrac{2}{2}$ 经重平、$\dfrac{2}{2}$ 纬重平、$\dfrac{3}{2}\nearrow$ 斜纹为基础、经向飞数（后面介绍）S_j=3 的急斜纹，$\dfrac{1\ 2\ 2}{1\ 1\ 1}$（一上一下、二上一下、二上一下）斜纹为基础、K_w=8 的纬山形斜纹。如果说明中没有涉及"纬向"，则默认是第一根经纱的交织规律，否则，则是第一根纬纱上经纱的交织规律。如纬山形斜纹组织中，第一纬上经纱的位置交织次序是连续 1 个经组织点（上）、1 个纬组织点（下），连续 2 个经组织点（上）、1 个纬组织点（下），连续 2 个经组织点（上）、1 个纬组织点（下）。

Some regular weaves can be represented by "fraction + verbal statement", where "fraction" usually indicates the interlacings of the first thread, and the numerator is the number of the successive marks while the denominator is the number of the successive blanks. The "verbal statement (perhaps including the parameters)" explains how the intersection is extended to the other threads.

The weave shown in Fig. 1–2–3 is designated as $\dfrac{1}{1}$ plain, and read as "one up one down". In interweaving with the weft threads, the first end floats over one weft to form a riser and under the next weft to form a sinker. The weaves in Fig. 1–2–5 are designated as $\dfrac{2}{2}$ hopsack, $\dfrac{2}{2}\nearrow$ twill, $\dfrac{2}{2}$ warp rib, $\dfrac{2}{2}$ weft rib, $\dfrac{3}{2}\nearrow$ steep twill with S_j=3, and $\dfrac{1\ 2\ 2}{1\ 1\ 1}$ composed transverse waved twill based on K_w=8 respectively. The order of interlacing of the first thread is usually meant the direction of warp unless otherwise stated. For the transvers waved twill, the interlacing order of the first pick is 1 riser, 1 sinker, 2 continuous risers, 1 sinker, 2 continuous risers and 1 sinker.

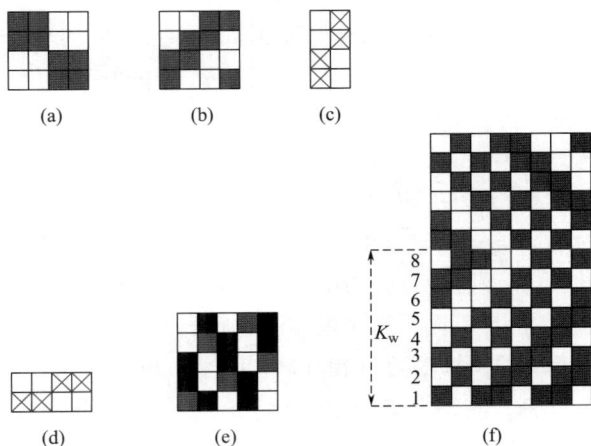

图1-2-5　可以用分数式+说明表示的织物组织
The weaves can be represented by a fraction expression

不同的组织有不同的说明方式，有不同的参数要求，导致计算 R_j 和 R_w 不同，后面其他纱线的交织顺序不同。如上面的表达式中，出现了斜向、经向飞数 S_j、K_w 等参数，将在以后的章节中讲解。

不是所有的组织都可以用"分数式＋说明"的方法表示。图 1-2-4 的组织，每根纱线交织规律不同，就不能用这种方式表示。

4. 织物截面（剖面）示意图 /Sectional Diagram

以上三种表示方法只能看到经纬纱交织的平面状态，织物截面图（剖面图）能更清楚地反映经纱与纬纱交织后在空间的几何形态。一般来说，经纱截面图仅显示一根经纱与所有纬纱的交织情况，纬纱截面图仅显示一根纬纱与所有经纱的交织情况，如图 1-2-1、图 1-2-2、图 1-2-6 与图 1-2-7 所示。有的截面图能显示所有的纱线交织情况，如图 1-2-8 所示。纬纱截面图相当于正视图，经纱截面图如果在组织图左侧，则相当于左视图，左侧为织物正面；若在组织图右侧，则相当于右视图，右侧为织物正面。图 1-2-6 中经纱截面图（a）与（c）显示的是第 1 根经纱与纬纱的交织情况，而图（b）显示的是第 1 根纬纱与所有经纱的交织情况。同一组织的不同位置，经纬纱的截面图很可能不同，图 1-2-2 上方显示的是第 2 根纬纱的截面图，第 1 根纬纱的截面图则与图 1-2-1 的第 1 根纬纱相同。

根据截面图，可以看出经纬纱交错时的沉浮情况，容易绘制织物组织图。图 1-2-8 的组织图就是根据截面图绘制的。

Different weaves may require various statements and parameters and resulting in a different calculating for R_j and R_w. Various symbols and parameters such as S_j, K_w have been given in above designations, which will be discussed in the following chapters.

However, not all the weaves can be represented by fraction and verbal statement. The weave in which the threads have the different interlacings can not be designated by the way as shown in Fig. 1-2-4.

Sectional diagram is used to indicate the configuration of the threads in the cloth as well as the interlacing. Many compound structures cannot be properly understood without the use of the sectional diagrams. Generally speaking, a vertical sectional diagram (or warp profile) only shows the interlacing of one end and all wefts, and a horizontal sectional diagram (or weft profile) only shows the interlacing of one pick and all warps, as shown in Figs. 1-2-1, 1-2-2, 1-2-6 and 1-2-7. Some sectional diagrams show the interlacings of all the threads, as shown in Fig. 1-2-8. The horizontal sectional diagram is equivalent to the front view, while the vertical sectional diagram is equivalent to the left view if it is on the left side of the weave pattern, and the left side is the front view of the fabric. If on the right side of the weave pattern, the right side is the front of the fabric relative to the right view. (a) and (c) of Fig. 1-2-6 show the interlacing of the first warp and the weft, while (b) shows the interlacing of the first weft and all the warps. The sectional diagrams of the warp and weft are likely to be different at different positions of the same weave pattern. The transverse sectional diagrams of the second weft thread shown at the top of Fig. 1-2-2, for the first weft, just the same as those of the first weft in Fig. 1-2-1.

The weave pattern can be easily drawn based on the sectional diagram as shown in Fig. 1-2-8 as the interlacing of the threads is clear at a glance.

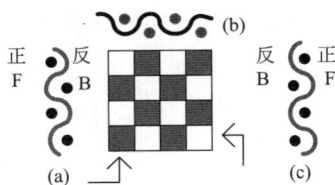

图1-2-6　平布截面图
Sectional diagram of habotai

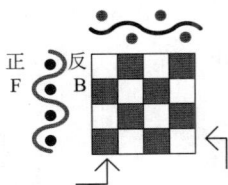

图1-2-7　府绸截面图
Sectional diagram of poplin

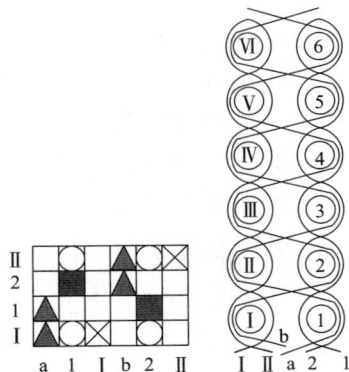

图1-2-8　显示全部交织情况的截面图
Sectional diagram showing total threads

需要说明的是，截面图的绘制，不能想当然，必须要遵照实际情况。组织相同、其他织造条件不同的织物，其截面图可能不同。图1-2-6所示平布（平纹组织）的经纬纱弯曲程度大致相同；图1-2-7所示府绸（平纹组织）的经纱弯曲程度大，但纬纱弯曲程度小。一些复杂组织中，部分组织点仅仅显示经纱与纬纱的上下关系，并没有真正的交织，图1-2-8所示的织物组织经向截面图中，纬纱的放置方式为两列，说明是双层织物。如果像其他图示一样单层排列，就不能得到正确的截面图。因此，仅仅根据织物组织图，不一定能正确绘制截面图。

It needs to be explained that the drawing of sectional diagram cannot be taken for granted, but must comply with the actual situation. The two fabrics of same weave may have different sectional diagrams. In Fig 1-2-6, the bending of the warps and wefts in a habotai (plain weave) is roughly the same, however, the warp bends more but the weft bends less in a poplin fabric (plain weave) as shown in Fig. 1-2-7. In compound weaves, some interlacing points only show the upper and lower relationship between warps and wefts without real interweaving. In the vertical sectional diagram of a double-layer weave shown in Fig. 1-2-8, the wefts are placed in two columns, indicating that there are two layers in warps. If it is drawn like the diagram in Fig. 1-2-6 and the wefts are arranged in single column, a correct sectional diagram will not be obtained. Therefore, the sectional diagram cannot be correctly drawn only based on a weave pattern.

三、飞数 /Step Number

同一系统纱线中，相邻两根纱线上相应的组织点之间间隔的纱线数，称为组织点飞数，用 S 表示。对于经纱，规定向上为正、向下为负；对于纬纱，规定向右为正、向

Step Number

In a full weave repeat unit, the number of threads spaced between the corresponding points on two adjacent threads in the same series is called the step, count or move. Step number S is actually a vector. For warp direction, S_j is positive if it is upward, or negative if downward. For

左为负。图1-2-9中，B 点相对于 A 点的经向飞数 S_j 为 3，A 点相对于 B 点的经向飞数 S_j 为 -3；C 点相对于 A 点的纬向飞数 S_w 为 2，A 点相对于 C 点的纬向飞数 S_w 为 -2。E 点相对于 D 点的经纬飞数为 1，而相对于 F 点则为 -1。

weft direction, S_w is positive if it is outward, or negative if towards the left. In Fig. 1-2-9, S_j of point A to point B is 3, in contrast, S_j of point B to point A is -3. S_w of point C to point A is 2, in contrast, S_w of point A to point C is -2. S_j and S_w of point E to point D are both 1 in two directions while S_j and S_w of point E to point F are both -1 in two directions.

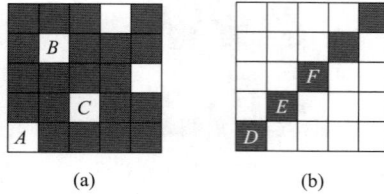

(a)　　　　　(b)

图1-2-9　飞数计算示意图
Calculating of step number

四、同面组织和异面组织 /Even Sided Weave and Unbalanced Weave

凡织物的组织，其正面和反面的经组织点数等于纬组织点数时，称同面组织，如图1-2-3、图1-2-4 和图1-2-5 中（a）~（d）所示的组织；反之，则为异面组织，如图1-2-9 和图1-2-10 所示的组织。经组织点数多于纬组织点数的织物组织称经面组织，如图1-2-9(a)和图1-2-10(b)所示；纬组织点数多于经组织点数的织物组织称纬面组织，如图1-2-9（b）和图1-2-10（a）所示的组织。将织物组织如图1-2-10（a）所示通过"底片翻转"、中心反向对称后得到其反面组织，如图1-2-10（b）所示。

A weave is called an even sided weave if the number of risers equals to the number of sinkers, or unbalanced weave if otherwise. The examples of even sided weave are shown in Figs. 1-2-3,1-2-4,1-2-5(a)(b)(c)(d), and unbalanced weaves are shown in Figs. 1-2-9 and 1-2-10. Figs. 1-2-9 (a) and 1-2-10(b) are also called warp dominated weaves/warp-faced weaves since they are more risers. Figs. 1-2-9 (b) and 1-2-10(a) are weft dominated weaves/weft-faced weaves due to more sinkers in the weave pattern. The weave shown at (b) in Fig. 1-2-10 is obtained by herringbone reversal of the weave at (a) in Fig. 1-2-10.

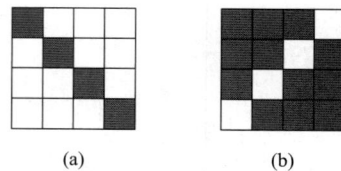

(a)　　　　　(b)

图1-2-10　异面组织
Unbalanced weave

Drawing of Regular Weave Patterns

五、规则组织图的绘制 /Drawing of Regular Weave Patterns

若要正确绘制组织图，必须正确计算 R_j 和 R_w，以及 $R_j \times R_w$ 个组织点的信息和位置。

Before drawing a weave pattern, it is necessary to calculate R_j, R_w, and the attributes (mark or bank) and positions of all the interlacing points ($R_j \times R_w$) .

（一）规则组织循环计算 /Calculation of Regular Weaves

1. 可用分子式表示的基础组织循环 /Weaves Represented by A Fraction

此类组织的 R_j 和 R_w 都有特定的计算式，将在后面章节叙述。

The sizes of repeat unit R_j and R_w of such regular weaves have their specific formulas, which will be described in the following sections respectively.

2. 间隔排列的合成组织 /By Rearranged Weaves at Intervals

有些组织是由两种或两种以上的组织在经向或者纬向按照一定的间隔排列而成，常见的排列顺序是 1：1、2：2、2：1 或者 1：2。图 1-2-11 所示为经纱按照 1：1 排列形成的组织。

Some weaves are composed of two or more weaves arranged at regular intervals warp-wise or weft-wise. The common arrangements are 1：1, 2：2, 2：1, and 1：2. Fig. 1-2-11 shows the weave combined with two weaves arranged at interval order of 1：1 warp-wise.

若两种基础组织的循环根数分别是 R_{j1} 和 R_{j2}，在经向的排列比为 $m：n$，则合成组织的经纱循环根数 R_j 计算式为：

If the warp repeat numbers of two base weaves are R_{j1} and R_{j2} respectively, and the ends from the two weaves are arranged at the ratio of $m：n$, then the formula for calculating the warp repeat number R_j of the new composed weave is given by the following formula:

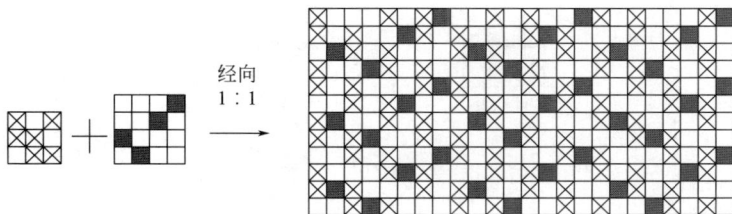

图1-2-11　两种组织经向间隔比例为1：1组合成新组织
Combining 2 weaves to form a new weave by interval arranging warps at ratio of 1：1

$$R_j = \text{lcm}\left(\frac{\text{lcm}(R_{j1}, m)}{m}, \frac{\text{lcm}(R_{j2}, n)}{n} \right) \times (m+n) \qquad （1-2-1）$$

这里 lcm 表示最小公倍数。

Here, lcm means the lowest common multiple.

合成组织的纬纱循环根数 R_w 为两种基础组织的纬纱循环 R_{w1} 和

The weft repeat number R_w of new composed weave is the least common factor (lcm) of the weft repeat numbers of two

R_{w2} 的最小公倍数，用下式表示为：

若两种基础组织在纬向间隔排列，合成后的组织循环根数可以类推。双层组织则是由两种基础组织在经、纬两个方向都是间隔排列得到，R_j 和 R_w 均采用式（1-2-1）计算。

base weave R_{w1} and R_{w2}, expressed by the following formula.

$$R_w=\mathrm{lcm}(R_{w1}, R_{w2}) \qquad （1-2-2）$$

If the two base weaves are alternately arranged in weft direction, the weft repeat number of the new weave can be analogized. Double-layer weave is obtained by arranging yarns alternately in both warp and weft directions, and the formula (1-2-1) is also applied to calculate R_j and R_w for double-layer weaves.

（二）组织点的信息计算 /Determining the Attributes of Interlacing Points

由于每个组织点中只有经组织点和纬组织点两种，若每一个组织点用一个字符来表示，那么，整个组织图就可以用 $R_j \times R_w$ 的矩阵来表示。通常经组织点用字符"1"表示，纬组织点用字符"0"表示。表示组织图至少需要三个信息：经纱循环根数 R_j、纬纱循环根数 R_w、包含 $R_j \times R_w$ 个"1"或"0"的组织点属性（经组织点或纬组织点）的矩阵 M。通过这种方法，组织图可以被数字化处理。图 1-2-12 所示为一个组织的数字化表示方法。

Since there are only two possibilities for each interlacing point: raiser or sinker, if each interlacing point is represented by a character or digital number, then the entire weave can be represented by a $R_j \times R_w$ matrix. As a practice, raisers are represented by the character "1" and sinkers by the character "0". At least three parameters are required to represent a weave: the warp repeat number R_j, the weft repeat number R_w, and a matrix M containing $R_j \times R_w$ symbols （"1" or "0"） for the attributes of interlacing points. In this way, the weave can be digitally processed as shown in Fig. 1-2-12.

```
1 0 1 1
0 0 1 0
1 1 1 0
1 0 0 0
```

图1-2-12　组织图的数字化表示图解
Digitalizing a weave

组织图信息的自动生成方法适合有一定规律、能用某种方法表示的组织。组织图自动生成的核心是：①能用表达式表征此类组织；②解释此类组织的表达方式；③计算组织图的经纬纱循环根数；④生成组织图中所有组织点的矩阵信息。

The method of automatic generation of weave is suitable for the designs that can be expressed in a certain way. The cores of automatic generation of a weave are ①representing such weaves with certain expressions, ② explaining the expressions of such weaves, ③ calculating the repeat numbers R_j and R_w of the weave, and ④ generating the matrix information of all interlacing points of the weave.

组织点的自动生成分两种：单个组织公式法自动生成和多个组织合成一个组织。能够自动生成的单个组织是能够用"分数式＋说明"表示的组织，包括三原组织及其变化组织、透孔组织、网目组织、凸条组织等联合组织；通过多个组织合成的有大循环单层组织包括绉组织、重组织和多层组织。

有些组织图设计软件对组织图数字化后，可以进一步处理，如进行底面翻转、旋转、复制等各种编辑操作。

The method of automatically generation can be divided into two kinds: automatic generation of single weave and combination of multiple base weaves into one new weave. The single weave that can be automatically generated is the weave that can be expressed by "fraction + statement", including three basic weaves and their derivatives, open gauze weaves, distorted effect weaves, bedford cord weaves, etc. Crepe weaves, backed weaves and multi-layer weaves are automatically generated by combining several base weaves.

After being digitalized, a weave can be further processed like reversing, rotating, copying by some software on weave design.

（三）组织图绘制 /Drawing Procedures

手工绘制组织图或者用计算机编程自动绘制组织图的步骤如下：

（1）计算经纱循环根数 R_j、纬纱循环根数 R_w、包含 $R_j \times R_w$ 个"1"或"0"的组织点属性（经组织点或纬组织点）的矩阵 M。

（2）确定每个组织点单格尺寸大小。

（3）根据 R_j 和 R_w，确定组织图在方格纸中的范围，并在外围方框画粗线。

（4）确定经组织点的形状和颜色。不同颜色和不同形状的经组织点（实心、点状、叉线、三角状、十字形等）表示组织点具有不同的作用。

（5）根据组织点矩阵中各个组织点顺序和经纬属性，在方格纸的相应位置绘制组织点。纬组织点一般不画，除非特别说明。

The steps for drawing a weave design manually or by using computer programming are described as below:

(1) Calculate the warp repeat number R_j, the weft repeat number R_w, and a $R_j \times R_w$ matrix M containing "1" or "0" for the attributes of interlacing points.

(2) Determine the size of square for a interlacing point.

(3) According to R_j and R_w, determine the scope of the weave on design paper, and draw 4 bar lines on the outer border.

(4) Determine symbols and colors of the marks for the raisers. Different marks (solid rectangle, dot, fork, triangle, cross, etc.) may represent different functions of the interlacing points.

(5) According to the order and attributes of each interlacing point in the matrix, draw marks at the corresponding square on the design paper. Squares for the sinkers are kept empty unless otherwise specified.

第三节　上机图 /Looming Plans

一、上机图的组成 /Components of Looming Plans

上机图是表示织物上机织造工艺条件的图解。生产、仿造或创新织物时均须绘制与编制上机图。根据上机图确定综框数量、穿综方式、穿筘方式、纹钉植入位置或者开口凸轮形状。上机图是由组织图、穿筘图、穿综图、纹板图四部分排列在一定的位置而组成。上机图的位置布置如图 1-3-1 所示。

The looming plans are a group of diagrams showing the weaving process conditions of the fabric. Looming plans are required for the production, imitation or innovation of fabrics. According to looming plans, the number of heald frames, order of drawing heald, reed threading-in (way of denting) and pegs implantation or cam profile are determined. Looming plans comprise of Weave Pattern, Denting Plan, Drafting Plan and Lifting Plan. Fig. 1-3-1 shows the arrangement of looming plans.

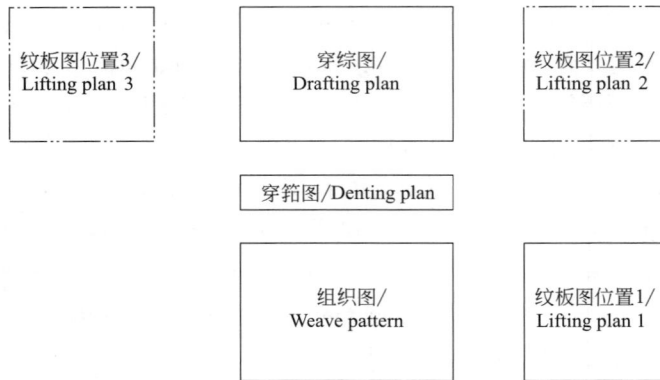

纹板图位置3/ Lifting plan 3	穿综图/ Drafting plan	纹板图位置2/ Lifting plan 2
	穿筘图/Denting plan	
	组织图/ Weave pattern	纹板图位置1/ Lifting plan 1

图1-3-1　上机图的布置
Arrangement of the looming plan

二、穿综图 /Draft

（一）穿综目的 /Purpose of Drawing-in

按照交织沉浮规律，将每次开口（沉浮）相同的经纱穿入同一综框内的综丝，可以一次全部提起，便于投纬。相对于手工生产而言，可大幅度提高织造效率，如

All the warps with same interlacing order are drawn in the heald wires of a heddle frame to raise or lower several threads at one time to form a shedding for picking. Comparing with hand working, the efficiency is improved dramatically by shedding, which is shown in Fig. 1-3-2.

图 1-3-2 所示。综丝被成组地放置在综框中，综框从机前开始顺序编号。综丝中间有综眼，经纱穿入综眼中（图 1-1-5）。每页综上仅挂有一列综丝的，称为单列式综框。为了减小综丝密度，每页综框上分挂几列综丝的，称为复列式综框。当综框上下运动时，综丝随之上下运动，带动经纱上下运动，形成梭口。

穿综顺序基本由经纬纱交错规律确定。对于织造图 1-3-2 所示的平纹织物，相邻经纱依次穿入第 1、2 页综框；在引纬时，轮流提升或者下降第 1、2 页综框，即可织造该织物。如果要织造图 1-3-3 所示的织物，第 1、2、3、4 根经纱依次穿入第 1、2、3、4 页综框。对于图 1-3-3（a）的组织 A，在织第 i 纬（i=1，2，3，4）时，仅仅提升第 i 页综框，其余综框均下沉即可。对于图 1-3-2（b）的组织 B，在织第 i 纬（i=1，2，3，4）时，同时提升第 i 页综框和第 $i-1$（若 $i-1<0$，则 $i=4$）页综框，其余综框下沉即可。

Heald wires are grouped and attached on the heald shaft, therefore, the heddles (actually, the threads) of same movement will be grouped together in a shaft to be raised or lowered simultaneously. Heddle shafts are ordered from the front of a loom. Each thread is controlled by being threaded in the mail eye of a heddle (see Fig. 1-1-5). Usually, there is only one row of heddle wires in a heald shaft. Sometimes, several rows of heddles are placed on a heald bar to decrease the density of heald wires to reduce the friction in weaving. When heald shafts move upward and downward, the heald wires move accordingly to raise or lower the threads to form the shed.

The order of drawing threads is based on the way of interlacing of warps and wefts. To produce the weave in Fig. 1-3-2, the 2 neighboring ends are threaded upon heald (or harness) No. 1 and No. 2 alternately. The healds No. 1 and No. 2 will be raised or lowered alternately, and the weave is made. However, if the weave in Fig. 1-3-3 is to be made, the ends No. 1, No. 2, No. 3 and No. 4 will be drawn upon the heald No. 1, No. 2, No. 3 and No. 4 respectively. When picking filling yarn No.i(i=1,2,3 or 4) in weave A, only the heald No.i is lifted. For the weave B in Fig. 1-3-3, when picking filling yarn No.i(i=1,2,3 or 4), the heald No.i and heald No.$i-1$(i=4 if $i-1$=0) will be raised simultaneously.

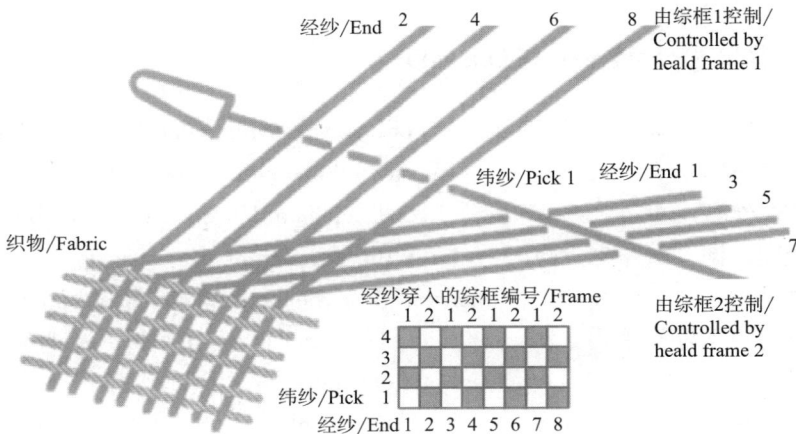

图 1-3-2　穿综目的
Purpose of drawing-in for plain weave

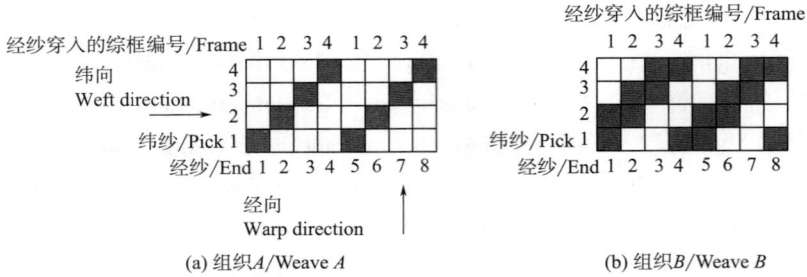

图1-3-3　其他组织的穿综思路
Drawing-in idea for other weaves

（二）穿综原则 /Principles of Drawing-in

穿综必须遵守以下原则：

（1）一根经纱仅可以穿入一页综内，不可以穿在2个综丝眼内（纱罗织物除外）。

（2）不同交织规律的经纱必须分穿在不同综框内。因为综框内所有综丝是整体上下运动的，穿在同一综框内各综丝眼中的经纱交织沉浮规律必须相同。

（3）交织沉浮规律相同的经纱一般穿入同一页综框中，也可根据需要，穿入不同综框中。

For drawing-in, the rules listed below must be followed:

(1) An end can only be threaded through a single heald wire and cannot be threaded through 2 heald wires (exceptions for lenos).

(2) The ends with different interlacings must not be threaded upon a harness shaft. As the heald is an entity, all ends drawn through the eyes of heald wire on a given heald frame must work alike. Therefore, the ends which work in different orders require separate healds.

(3) The ends with same interlacing can be threaded upon the different harness shafts if it is good for weaving.

（三）穿综优化 /Optimization of Drawing-in

为了织造方便，穿综时还需考虑以下因素，以便进一步优化。

（1）每列综丝数不可过大。为了减小综丝密度，同一页综框上可以挂多列综丝。

（2）穿综的顺序要简单、方便记忆。

（3）在清晰梭口的情况下，穿在前面综框中的纱线张力较小，穿在后面综框中的纱线张力较大。另

For easily weaving, other factors in drawing-in should be considered to optimize the drafting.

(1) Each heald frame shouldn't be overloaded. To reduce the density of the harness wires, more rows of harnessed can be attached on a heald frame.

(2) The order of the drawing-in should be simple and easy to remember.

(3) For a clear shed, the ends in front healds are relatively slack tensioned while the ends in back healds are relatively tight tensioned. It is easier to knot the broken

外，如果经纱在织造中断头，穿入前面综框容易操作。故下列情况的经纱，综框应尽量放置在前面：①穿入的纱线强力低；②在织造时承受了最大张力（交错最频繁）；③经纱密度大；④提综多的经纱。如果情形④与②矛盾，优先考虑情形②。

（4）尽可能减少综框数量，综片负荷尽可能均匀，每页综框上穿入的经纱根数尽可能相同。

ends if they are drawn upon the front healds. The healds should be placed nearest the front, which ① carry the weakest threads, ② carry the threads which are subjected to the most strain (most frequently interlaced), ③ are the most crowded with the threads, and ④ have more risers. If ④ is in conflict with ② , give priority to ② .

(4) It is better to minimize the number of healds. The healds should be loaded equally and carry an equal number of heald wires.

（四）穿综图的图形表示法 /Representing A Draft Plan

穿综图是表示组织图中各根经纱穿入各页综框的顺序的图解，说明了满足织造所需要的综框数量。穿综图位于组织图的上方，并与组织图保持一定距离。其左右两侧与组织图对齐。穿综图每一横行表示一页综框（或一列综丝），综框的顺序在图中是自下向上排列；每一纵行表示与组织图相对应的一根经纱。因为一根经纱只能穿在一页综上，故穿综图中每一列上至多只能有一个符号（■，☒或其他），其余都是空格。如需某根经纱穿入某页综框上，则在穿综图代表该经纱的纵行与代表该综框的横行相交的方格中填入符号即可。符号位置（X 行，Y 列）表示第 Y 列经纱被穿在第 X 页综综框上。图1-3-4说明该组织织造时，使用4页综框，经纱1、2、3、4、5、6、7、8分别穿在第1、2、3、4、2、1、4、3页综框中。

A draft plan indicates the number of healds used to produce a given design and the order in which the warp ends are threaded through the mail eyes of the heald wires. The draft plan is located above the weave pattern, and both the left side and the right side are aligned with the weave pattern. In a draft, each horizontal space indicates a heald frame or a row of healds while each vertical space indicates the corresponding end in the weave. The healds are ordered from bottom to top in a draft. Marks are inserted upon the small squares to indicate the healds upon which the respective threads are drawn. A mark on square (Row X, Column Y) means that the end No.Y is drawn upon heald No.X. In the example in Fig. 1-3-4, ends No. 1,2,3,4,5,6,7 and 8 are drawn upon heald No. 1,2, 3, 4 ,2 ,1,4 and 3 respectively.

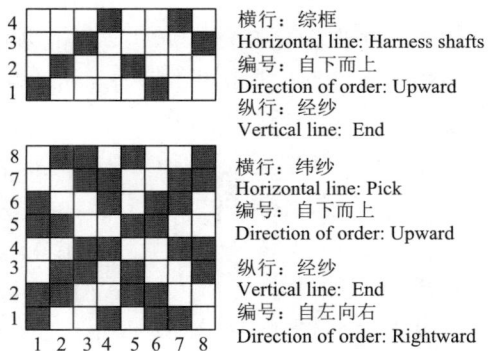

横行：综框
Horizontal line: Harness shafts
编号：自下而上
Direction of order: Upward
纵行：经纱
Vertical line: End

横行：纬纱
Horizontal line: Pick
编号：自下而上
Direction of order: Upward
纵行：经纱
Vertical line: End
编号：自左向右
Direction of order: Rightward

图1-3-4　穿综图
A draft plan

（五）穿综方法 /Systems of Drafting

穿综应根据织物的组织、原料、密度、操作来定。穿综方法包括照图穿法、顺穿法、飞穿法、山形穿法（对称穿法）、分区穿法、间断穿法等。山形穿法是一些特殊形状的照图穿法。

Drafting is determined by the interlacing of the weave pattern, yarn materials, fabric density and convenience of operation. Common drafting systems include drafting based on design, straight drafting, skip drafting, pointed drafting, and divided drafting, etc. Some special curved drafts such as pointed drafting are named by their shape.

1. 照图穿法 /Drafting Based on Design

该方法根据穿综原则，将组织中浮沉规律相同的经纱穿入同一页综框中。为简单起见，第 1 根纱线穿入第 1 页综框。后面的经纱如果沉浮规律与前面任何经纱都不同，就穿入下一页综框；如果与前面的经纱沉浮规律相同，就穿到与该根经纱相同的综框中。图 1-3-4 穿综图的两种绘制过程如图 1-3-5。照图穿法适用于经循环较大而其中含有经纱沉浮规律相同的组织（如绉组织、平纹小提花组织），但是穿综顺序较复杂。

According to the principle of drafting, it is natural for the warp threads in the weave that have the same interlacing or movement to be drawn upon the same heald. For simplicity, the 1st end is drawn upon the 1st harness, and the next end is drawn upon the next harness if it is different from any end that has been threaded, or drawn upon the same heald upon which the end shares the same movement. Fig. 1-3-5 shows two ways to design a draft plan for the weave pattern in Fig. 1-3-4. In the case that repeat of the weave is large and complex in warp direction, the manner can reduce the number of healds used, however, the drafting order may be complex.

2. 顺穿法 /Straight Drafting

在照图穿法中，如果组织图中每根经纱的沉浮规律都不一样，则穿综图中的符号将形成一条直线，称为顺穿法，如图 1-3-6 所示。也可认为顺穿法将一个组织循环中的各根经纱逐一、顺次地穿在每一页综框上。采用顺穿法时，所需的综框页数等于一个组织循环的经纱根数 R_j。此法简单、方便记忆，适合密度较小、经纱循环根数小的织物。

如果组织经向循环根数 R_j 大，使用的综框就多，会受到织机类型限制。如果 R_j 小，假设经纱密度恒定，则单位宽度上使用的综丝

If the interlacing of each end is different from any other ends in a weave, the marks of the draft by design will form a straight line. In this case the method is called straight drafting as shown in Fig. 1-3-6. Therefore, R_j, the warp repeat number equals to the number of the healds employed. In drawing a drafting plan, the successive ends of the repeat are drawn upon successive healds until the end of the repeat is reached. The system of drafting is easily remembered and usually applied to the fabric with low warp density and small number of the ends in the design repeat.

If the warpwise size of the repeat R_j is large, the number of harnesses increases, which limits the loom. If R_j is small (supposing the warp density is constant), the number of the heddle wires in the unit width increases and may result in difficulty in weaving. The denser the heddle wires,

1 2 3 4 5 6 7 8   1 2 3 4 5 6 7 8

(a)   1 2 3 4 5 6 7 8

(b)

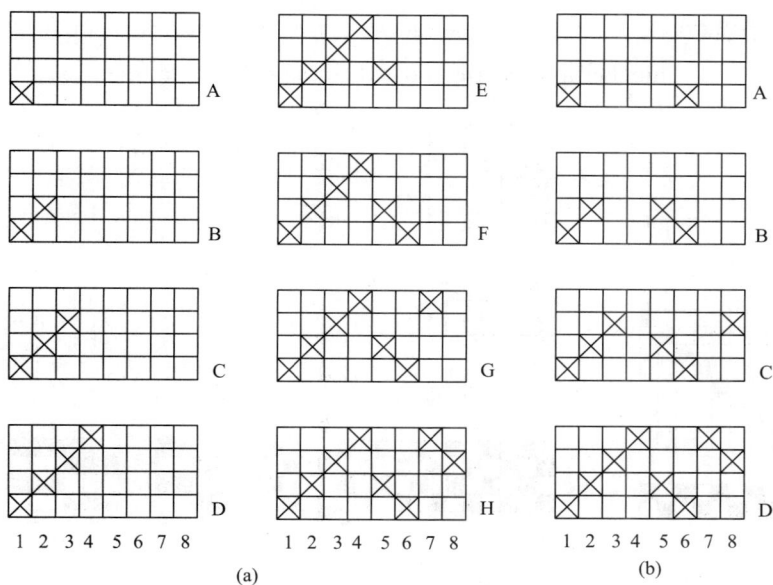

图1-3-5 照图穿法穿综图绘制步骤图解
Two ways of drawing a drafting plan by design

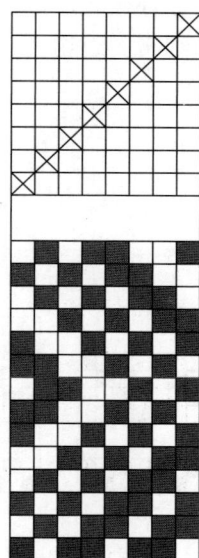

图1-3-6 顺穿法穿综图
Straight draft

就多，会导致织造困难。综丝密度越大，相邻综丝间距越小，综丝与综丝、综丝与纱线之间的摩擦就越大。

the smaller the split between the neighboring heddles, the greater the friction between the ends and the heddle wires, and the greater the friction between the heddle wires.

3. 飞穿法/Skip Drafting

综丝宽度为 0.3 ~ 0.65mm，对于各类织物来说，综丝的最大密度小于 12 根/cm。织造最简单的织物至少需要 2 页综框，而大多数织物的经纱密度大于 200 根/10cm，这意味着当遇到织物密度较大而经纱组织循环较小的情况时，无法简单使用单列式综框顺穿法织造。此时，有两种解决方案：①成倍增加单列综框数量，用顺穿法织造，如图 1-3-7 所示；②采用复列式或 4 列式综框（一页综框上有 2 ~ 4 列综丝）飞穿法，如图 1-3-8 的 C 和 D 所示，减少每页综上的综丝数，减少经纱与综丝的摩擦，使织造能顺利进行。飞穿法适合经密较大、

The width of a single heald wire is between 0.3mm and 0.65mm. For most fabrics, the maximum density for heald wires is 12 ends/ cm. As we know that there are at least 2 healds are required for weaving the simplest weave, and the warp density for majority woven fabrics is more than 200 ends/10cm. Therefore, the fabric with a high warp density and small weave repeat number in warp direction cannot be simply straight drafted on a loom with single-row heald frame. There are two solutions: ① multiplying the number of the healds used by using straight drafting as shown in Fig. 1-3-7; ② using skip drafting on a heald frame with 2 to 4 rows of heddle wires as shown in Fig. 1-3-8 (C and D). Several rows of heddle wires are combined and work as a single harness. The number of the heddle wires in each row is decreased and the friction and chafing are reduced for a smooth weaving. Skip drafting is applied to

经纱循环数较小的织物，如高密府绸、细布类织物。

the fabric with a high warp density and small weave repeat number in warp direction, such as poplin, percale, etc.

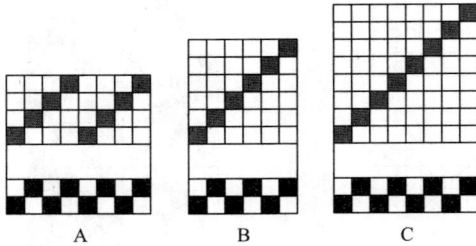

图1-3-7　成倍增加单列综框织造
Multiply the heald shaft to weave a denser fabric

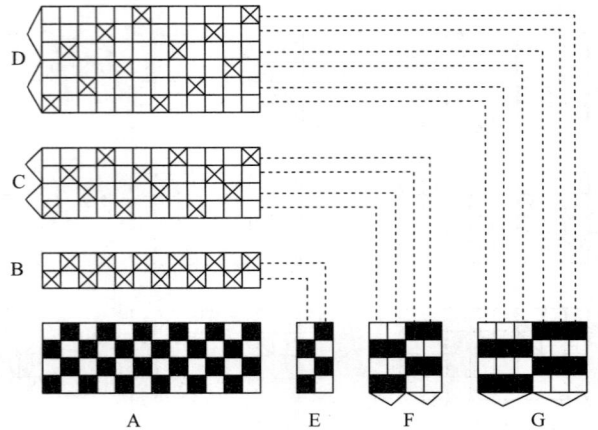

图1-3-8　飞穿法织造
Skip drafting

4. 山形穿法 /Pointed Drafting

山形穿法是特殊的照图穿法，适合中心对称的组织图。穿综图分成左右两部分，左边部分长度是综框个数，将经纱按顺穿次序从第一页综穿到最后一页综。然后在右边部分按相反的顺序穿，形成山形，如图1-3-9所示。因此，组织循环根数等于：综框数量 ×2-2。山形穿法降低了生产大组织循环织物的成本，适用于对称花纹织物，如山形斜纹、菱形斜纹等。

5. 分区穿法 /Divided Drafting

组织中纱线所起作用、原料或张力不同时，把所有综框分成前后若干区，各区中所包括的综框数可以相同，也可不同。经纱轮流地穿在前后不同区域的综框中，如图1-3-10所示。这种穿法适合重经组织、双层织物、起毛组织等复杂组织。

Pointed drafting is used for weaves which are symmetrical about the center. The draft area can be divided into left and right parts. The length of the left part is just the number of the healds used. In the left part, straight drafting is applied while in the right part, the order of drawing-in is reversed as shown in Fig. 1-3-9. The number of ends per repeat of design = 2 × number of healds-2. The system is frequently employed to produce waved or diamond effects. The main advantage is that it allows the production of quite large effects economically.

All heads shall be divided into several zones with different functions of threads or raw materials or tensions in the weave. The number of heads included in each zone may be the same or different. The ends are alternately threaded into the front or the back heald zone as shown in Fig. 1-3-10. Divided drafting is suitable for the compound weaves such as backed weaves, double-layer weaves and pile weaves.

6. 间断穿法 /Grouped Drafting

间断穿法是分区穿法的一种，只是穿综区域非上下交替并列，而是左右大块并列。即先把完全组织的某一部分经纱全部穿入一区内，然后把另一部分经纱全部穿入另一区内。间断穿法特别适用于条格、凸条等组织，如图 1–3–11 所示。

Grouped drafting is a special kind of divided drafting. The area of draft is divided in left and right blocks rather than front and back parts alternately. The neighboring ends are not alternately drawn through the front and back healds respectively, instead, all the ends of one block in the weave are drawn into the front healds, then, the remained ends in the other block will be drawn into the back healds. Grouped drafting as shown in Fig. 1–3–11 is suitable for stripe and check weaves, bedford cord weaves, etc.

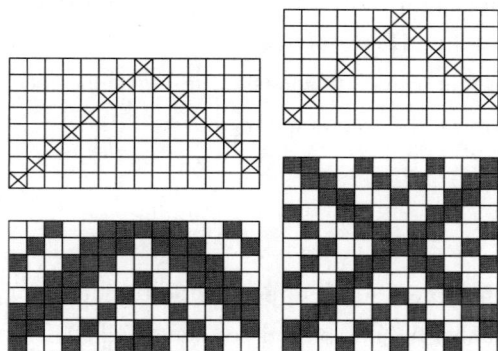

图1–3–9 山形穿法
Pointed drafting

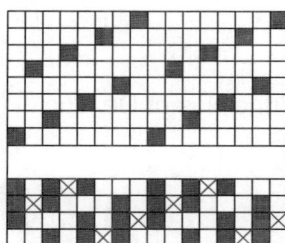

图 1–3–10 分区穿法
Divided drafting

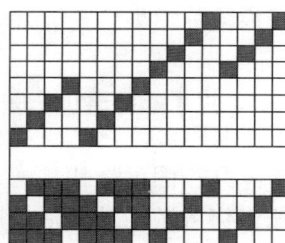

图1–3–11 间断穿法
Grouped drafting

（六）穿综图的文字表述法 /Representing A Draft by Numbering

穿综图还可以用文字表示，即直接写出组织循环中经纱所穿入的综框次序。图 1–3–5 所示的穿综次序为 1、2、3、4、2、1、4、3；图 1–3–11 所示的穿综次序为（1、2、3、4）×2，（5、6、7、8）×2。

The drafting plan can also be designated by numbering. The threads are successively drawn upon the healds in the order indicated by the numbers. For example, the drafts in Fig. 1–3–5 and Fig. 1–3–11 can be written as 1, 2, 3, 4, 2, 1, 4, 3 and (1, 2, 3, 4) × 2, (5, 6, 7, 8) × 2 respectively.

三、穿筘图 /Denting Plan

（一）穿筘的作用 /Roles of A Reed

Denting Plan

钢筘是横穿在织物幅宽方向上的梳子状钢片形成的框形部件，钢

In a loom, reed is a comb–like wired frame through which warp threads pass. Each partition is a "dent /split".

片间的缝隙称为筘齿，筘齿将一组组经纱分隔开。钢筘的作用是均匀排布经纱，控制经密和将刚刚引入梭口的纬纱推向织口（打纬）。织造中，经纱根据钢筘控制一定的经密均匀地分布在织物幅宽范围内。常见的穿筘方法是幅宽上每筘齿内均匀穿入 2 ~ 4 根经纱。

Wires separate the ends. Roles of the reed includes: ensure a desired warp density; arrange the ends uniformly; and put the picks to the fell (beating–up). During weaving, the warp ends are spaced out across the width of the warp sheet according to a desired density by the wires of the reed. The most frequent order for denting is 2, 3, or 4 ends per dent regularly across the width.

（二）钢筘规格 /Specification of A Reed

筘号是钢筘的规格，即 10cm 长度内的筘齿数目。英制筘号则是 2 英寸内的筘齿数目。筘号越大，筘齿间隙越小。

Reed number is a term traditionally specifying the number of dents per 10cm (Metric system) or per inch (Imperial system) on a reed. Obviously, the larger the reed number, the narrower the split of the reed.

（三）穿筘考虑因素 /Principles of Denting

穿筘图设计就是确定筘号与每筘齿穿入数，其原则是提高生产效率和织物外观。考虑因素包括：经纱密度（织物密度，10cm 幅宽方向中的经纱根数）、经纱线密度和织物组织。

选择穿入经纱根数少，筘号大的钢筘，经纱分布均匀，但筘片与经纱摩擦严重，断头增加。选择穿入经纱根数多，筘号小的钢筘，经纱分布不均匀，布面筘路明显，但钢筘价格便宜。经纱线密度大，每筘穿入数可以小。

每筘齿穿入经纱根数应等于织物组织经纱循环数 R_j 的约数或倍数。组织点的分布也会影响穿筘设计。例如，方平组织要将沉浮规律相同的相邻经纱穿在不同的筘齿中，以保证布面平整；而在一些复杂组织中，将具备相互重叠关系的

The design of denting is to determine the count of reed and the number of ends placed per dent (NEPD). The principles of designing a denting plan are to realize the purposes of the design, to improve production efficiency, and to enhance the fabric appearance. The factors to consider in designing includes: warp setting, warp linear density and weave.

If choosing a small NEPD with large reed number, the ends will be evenly distributed across the width of the fabric at a cost of serious rubbing with the reed wire and a high warp breakage. However, if more ends are threaded in a reed with small reed number, the reed mark is obvious due to the unevenly distribution of the ends. For a coarse fabric, small NEPD is recommended.

NEPD is recommended to be set as the factor or multiple of warp repeat number R_j. Weave should be considered in designing denting. For example, the adjacent ends which work in same order should be threaded into separated dent splits to ensure a clarity of the design. However, some adjacent ends are required to threaded into

一组相邻经纱穿入同一筘齿，帮助纱线更好地重叠。棉纤维经起毛组织的毛经则在紧邻筘齿，避免被夹持挂带，形成织造疵点。为了使布边坚牢，便于织造和后整理加工，边经穿入经纱根数一般比地经穿入经纱根数要多。

穿筘可用穿筘图和文字表示法两种方法表示。

the same dent split to help overlapping in backed weave or double-layer weave. The pile cotton warps of a terry fabric should be arranged near to the dent wire to avoid being trapped and forming defects. For a firm selvedge, NEPD is usually larger in selvedge part than that in body part.

The arrangements of denting can be described by denting plan and numbering.

（四）穿筘图的图形表示法 /Representing A Denting Plan

确定每筘齿穿入经纱数的图，称为穿筘图，在意匠纸上用两横行表示。在上机图中，穿筘图位于组织图与穿综图之间，左右两端与组织图（穿综图）对齐。

在穿筘图中两个横行表示相邻筘齿，以横向方格连续涂绘符号●、■、×的个数，表示一组纱线的根数，即筘齿穿入数，且这一组纱线穿在同一筘齿中。而穿在相邻筘齿中的纱线符号绘制在另一行中。空白符号相当于分隔符。图1-3-12所示为一个2纱/筘的穿筘图，第1、2经纱在同一筘齿中，第3、4经纱穿在另一筘齿中。

A denting plan is a diagram showing the number of warp threads per dent. The denting plan is also drawn on a design paper. There are only two horizontal spaces representing the adjacent dents. In looming plans, the denting plan is located between the weave design and the drafting plan.

In a horizontal space of the denting plan, the continuous number of squares of different marks or symbols such as ●, ■, and × represent that the number of the ends are grouped and the corresponding ends are threaded into same dent. Usually, the number of the marks for its neighbor dent are drawn in another horizontal space. Blank in a denting plan works as separator. Fig. 1-3-12 shows a denting plan with 2 ends/dent. The ends N.o 1 and No. 2 are placed in a dent while ends No. 3 and No. 3 are put into another dent.

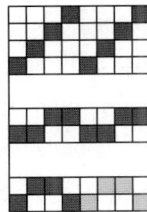

图1-3-12　2纱/筘的穿筘方式
2 ends/dent

（五）空筘、花式穿筘 /Empty Denting and Fancy Denting

根据织物要求，纱线可以均匀穿入钢筘，也可以不均匀穿入钢筘，部分钢筘内甚至不穿入任何经纱，称为空筘。空筘处在穿筘图的底部用符号"∧"或"e"表示，如图 1–3–13 所示。有些条格织物中的经纱密度不同，每筘齿中的经纱穿入数根据经密不同而变化，此时称花式穿筘。如图 1–3–14 所示，均表示在一个穿筘循环中，共有 6 个筘齿，第 1、2 筘齿各穿入 3 根经纱，第 3、4 筘齿各穿入 2 根经纱，第 5、6 筘齿各穿入 3 根经纱。

The ends can be evenly threaded in dents, or unevenly threaded in dents, or even none is threaded in some dents at all. An empty dent is usually indicated as "e" or "∧" at the bottom of the denting plan as shown in Fig. 1–3–13. For a crammed stripe, the warp density may vary from one stripe to another, and NEPD changes from one dent to another, which is called fancy denting. Four plans in Fig. 1–3–14 are actually the same denting: 3 ends are threaded into dents No. 1 and No. 2, 2 ends in dents No. 3 and No. 4, and 3 ends in dents No. 5 and No. 6 respectively.

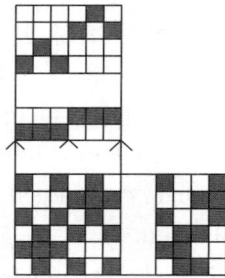

图1–3–13　空筘的穿筘方式
Empty split

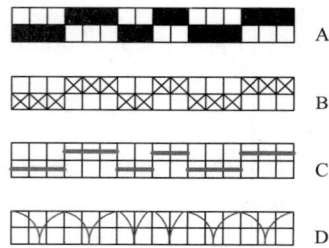

图1–3–14　花式穿筘
Fancy denting

（六）穿筘方法的文字表示 /Designation by Numbering

穿筘图还可以用文字表示，图 1–3–12 可表示为"2 入"，而图 1–3–13 的穿筘方法可表示为"3、0、3、0"。还可以同时表示穿综和穿筘，图 1–3–13 可表示为"[1 2 1 0 3 4 3 0]3 入"。

A denting plan can also be designated by numbering. For example, "2 ends/dent" is used for Fig. 1–3–12, and "3,0,3,0" for Fig. 1–3–13. Sometimes, the drafts and denting plan are designated simultaneously, e.g., "[1 2 1 0 3 4 3 0] 3ends/dent" for Fig. 1–3–13.

四、纹板图 /Lifting Plan

纹板图也称提综图，是织物织

Lifting Plan

Lifting plan (weaving plan, pegging plan) defines

制时控制综框运动规律的图解。改变纹板图，可得到不同组织的织物。纹板图可以放置在组织图右侧，也可以放置在穿综图的左右两侧，但不同位置，含义有所不同。纹板图仍然在方格纸上绘制，若在方格内记有符号，表示相应的综框在引纬时被提起。在多臂机或提花机开口提综时，可按纹板图进行纹板塑料纸的轧孔、在纹板上植纹钉或直接电子提综。踏盘开口织机综框的升降由凸轮的外形决定，一般不必绘制纹板图。

多臂开口织机可通过改变纹板图或穿综方法来织制不同组织的花纹织物，而踏盘开口织机一般通过改变穿综的方法来织制不同组织的织物。下面以多臂开口织机为例介绍纹板图的绘制。

The selection of heals to be raised or lowered on each successive insertion of the pick of weft. By changing the lifting plan, fabrics of different weaves can be woven. Lifting plan can be placed at the right side of the design or both sides of the drafts. However, a different position has a different meaning. Lifting plan is drawn on the point paper as well, and the mark at the square indicates the corresponding heald to be lifted. For a dobby loom, lifting plan determines the implantation of the pegs. If for a Jacquard loom, lifting plan determines which heddle to be raised directly by pattern cards. For a tappet loom, lifting plan is the base for designing the contour of the shedding tappet or shedding cam; or for placing the tappet cams. In fact, it is unnecessary to draw the lifting plan if a tappet loom is used.

On a dobby loom, the design can be modified by changing the lifting plan or drafts. But for a tappet loom, the design is usually achieved by varying the drafts. The drawing of a lifting plan is explained with a dobby loom.

（一）纹钉与多臂开口 /Pegs and Dobby Shedding

图 1-3-15 所示纹板上的孔洞是植入纹钉的地方，在复动式纹板编链上，每 2 纬间隙转动一下花筒。黑色圆点表示在纹板植入纹钉。该图所示为左手织机（右龙头或右花筒）用纹板控制织造某组织的多臂开口机构示意图。提刀 K 左右往复运动，在转动的花筒嵌入纹板。如纹板植有纹钉，则顶起平衡杆 F，牵动竖钩 H 落下钩住提刀，当提刀向右运动时，带动 J 形牵手提综开口；若没有纹钉，则竖钩提升，与提刀脱钩，综框在弹簧回弹力下落下。

纹板图上每一横行控制一根纬纱。假设穿综采用顺穿法，则第 1、

In Fig. 1-3-15, the holes in the lags are positioned such that they correspond with the location of the feelers. The pattern chain is turned intermittently by a wheel so that a new lag is presented every second pick. A filled circle represents a peg planted in the lag. A dobby shedding mechanism of a left-handed loom (with pattern cylinder on right side) is used to demonstrate how to weave a pattern by implanting pegs on the control lags. The knife K reciprocates. Pattern cylinder rotates counterclockwise. The control lags are embedded in the grooves of the pattern cylinder. If there are pegs on the control lag, the feeler F will be raised and let hook H falling to engage the knife K. When knife K moves towards right, the jack lever J drags the straps of the heald frame to lift to form a shedding. Otherwise, when there isn't any peg, the hook raises, and detaches the knife, the heald frame will be lowered due to

2、…、8 页综框分别控制 a、b、c、d、e、f、g 和 h 这 8 根经纱。若纹板图上有符号，说明此处要植入纹钉（纹板）或者纹板纸上冲孔，表示要提综。在图 1-3-15 中，织第一纬时，第 a 根经纱提起；织第二纬时，第 b、h 根纬纱提起；织第三纬时，第 a、c、g 根经纱提起……最终织造的组织如图 1-3-15（b）所示。

the resilience force of the springs.

Each row of a weaving plan controls the lifting of the ends over a filling yarn. Supposing the straight drafting is applied, heals No. 1, 2,…, 8 control the filling yarn No.a, b,…, h respectively. The mark on the lifting plan indicates that the pattern lag will be implanted with a peg. When the first filling yarn is inserted, end a is raised; for the 2nd filling yarn, ends b and h are raised; for the 3rd yarn, ends a, c and g are raised, and so on. The final weave formed is shown in Fig. 1-3-15(b).

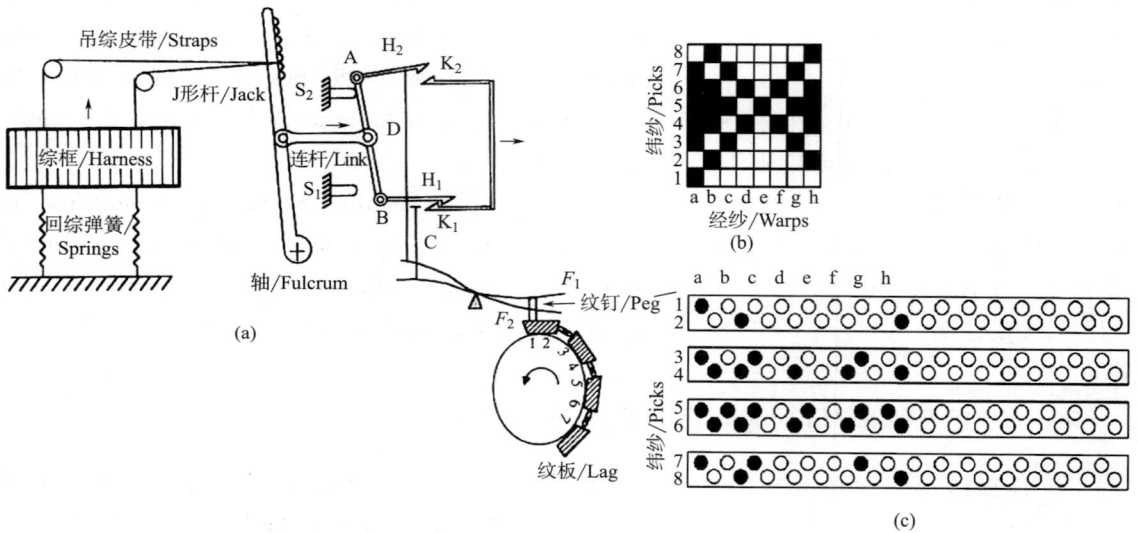

图1-3-15　多臂机纹板图提综原理
Schematic diagram of dobby shedding for a honeycomb weave

（二）纹板编链 /Lattice of Pattern Lags

目前，多臂机使用单动式与复动式两种纹板（图 1-3-16），纹板按照顺序编链后嵌入在花筒凹槽中（图 1-3-15）。由于花筒有 8 个凹槽，因此，纹板编链长度必须大于 8，才能循环织造。单动式纹板每织一纬转动一格凹槽，复动式纹板每织两纬转动一格凹槽。无论何种纹板，每一行均控制一根纬纱引入

Currently, two types of pattern lags, single-lift lag and double-lift lag, are employed in a dobby loom. These lags are shown in Fig. 1-3-16. The lags are chained and embedded in the grooves of the pattern cylinder. Since there are 8 grooves in a pattern cylinder, the length of the lattice of pattern lags must be larger than 8 to fulfill the weaving cycle. For a single-lift lag, pattern cylinder rotates one groove per weft while for a double-lift lag, 2 wefts per groove. On a pattern lag, the hole near the front of the loom

Lattice of Pattern Lags

时的开口。靠近机前的纹板孔控制第一列综框，靠近机后的纹板孔控制最后一列综框。新型多臂机使用纹板塑料纸替代纹板，冲孔则综框提起，如图 1-3-17 所示。

controls the first heald. For a new type dobby loom, the pattern lag is replaced by an electronic pattern plastic as shown in Fig. 1-3-17, and the punctuation means the lifting of the heald.

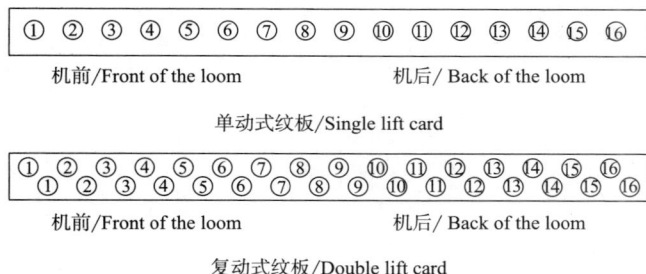

图 1-3-16　两种类型的空白纹板
Two types of pattern cards (single-lift, double-lift)

图 1-3-17　电子纹板纸
Electronic pattern plastic

　　由于左手车的花筒（在机器右方）逆时针转动，右手车的花筒（在机器左方）顺时针转动，两者纹板植入纹钉及编链的方向不同。图 1-3-18 是左手车复动式纹板图在右下方时，与纹板编链的图示。

　　纹板图也可放置于穿综图的左右两侧。右手车左龙头（花筒）的纹板图一般放置在穿综图左侧，而左手车右龙头（花筒）的纹板图一般放置在穿综图右侧。在穿综图两侧的纹板图与纹板植入纹钉的关系如图 1-3-19 所示。

On a left-hand loom, the pattern cylinder (at the right side) rotates counter clockwise, but on a right-hand loom, the pattern cylinder rotates clockwise. Therefore, the way of implanting the pegs and the direction of the lattice of pattern lags are different to the two types of looms. Fig. 1-3-18 shows the way of implanting the pegs and the direction of the lattice of pattern lags for a double-lift left-hand dobby loom.

Lifting plan can also be placed at left side of the drafting plan for right-hand loom or right side for left-hand loom. Fig 1-3-19 demonstrates how the pegs are implanted according to the pegging plan that is placed beside the draft.

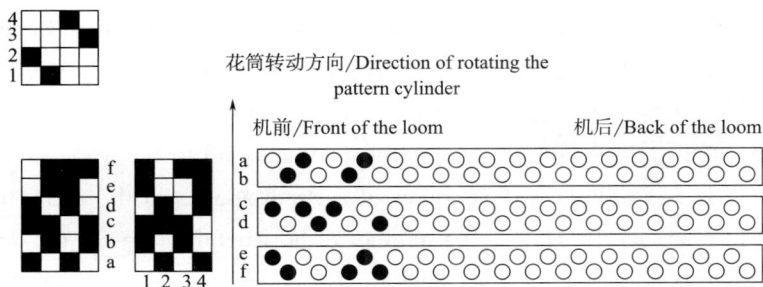

图1-3-18　复动式纹板图在右下方的编链顺序（左手车）
The pegs and the direction of lattice of pattern lag for a double-lift left-hand dobby loom

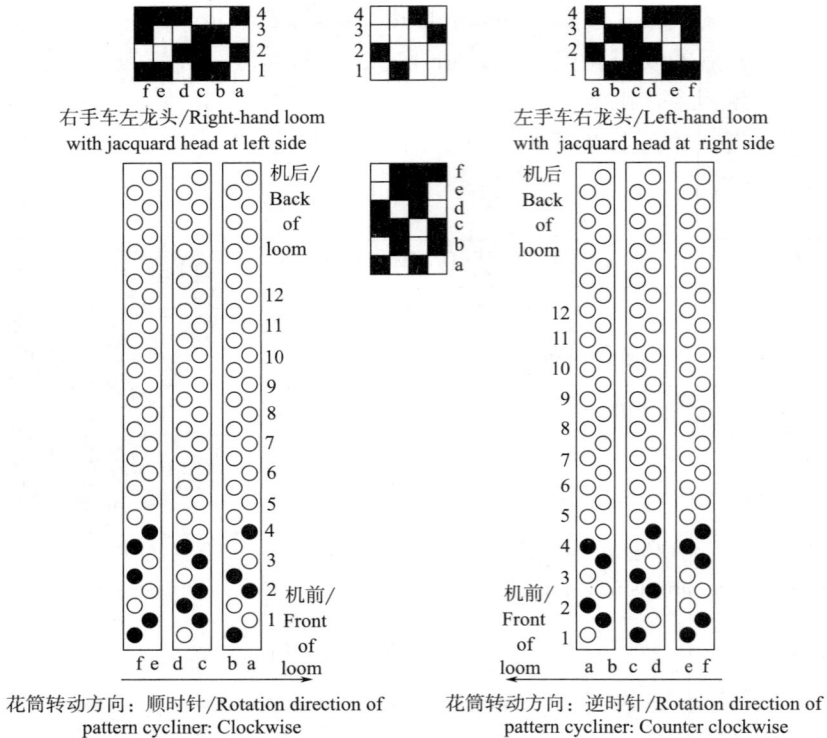

图1-3-19　纹板图在穿综图左右两侧时，复动式纹板分布在左右手车上的编链顺序示意图
Schematic diagram of implanting pegs into the double–lift pattern lag on dobby looms
(Lift: for left–handed looms; Right: for right–handed looms)

（三）纹板图绘制方法 /Drawing of the Lifting Plan

1. 纹板图在组织图右侧 /At the Right Side of Weave

如果纹板图处于组织图的右侧，其上下与组织图平齐，纹板图上每一横行代表与组织图相应的一根纬纱，即一块纹板，控制一次开口过程。纹板图的每一纵列代表一页综框，顺序自左向右，表示对应综片控制的某根经纱在整个组织循环内的沉浮规律。纹板图的纵列数等于穿综图的横行数（纹板页数）。纹板图的每一横行表示一次开口时综页的提升规律，或在某次开口的一瞬间所有经纱的沉浮规律。

If the lifting plan is on the right side of the weave pattern, its lower part and upper part level with the weave pattern. Each row on the lifting plan represents a filling yarn corresponding to the design, that is, a pattern lag controls one shedding process. Each column in the pattern weave represents a heald frame, from left to right, indicating the lifting and lowering of a warp controlled by the corresponding heald in the whole weaving cycle. The number of vertical spaces in the lifting plan is equal to the number of horizontal spaces in the draft (the number of healds). The marks on each column of the lifting plan indicate the lifting at one shedding, or the interlacing of all warp yarns at this shedding.

2. 纹板图在穿综图两侧 /Beside of Draft

纹板图位于穿综图的两侧时，此时纹板图与穿综图上下平齐，每一横行表示一页综框，其顺序自下而上；每一纵列表示一块纹板，或者一根纬纱，或者控制一次开口过程。对于左手车右龙头的多臂机，其顺序则是自左向右；对于右手车左龙头的多臂机，其顺序则是自右向左。

图 1-3-20 展示了纹板图在不同位置时，上机图各组成部分行与列之间的关系。

When the lifting plan locates beside the drafts, each row of the lifting plan represents a heald with the order upward, and each column represents a pattern lag or a weft thread. If for a left-hand loom, the order of the wefts is from left to right. If for a right-hand loom, the order of the wefts is from right to left.

Fig. 1-3-20 shows the meanings of the rows and columns of the lifting plan at different positions.

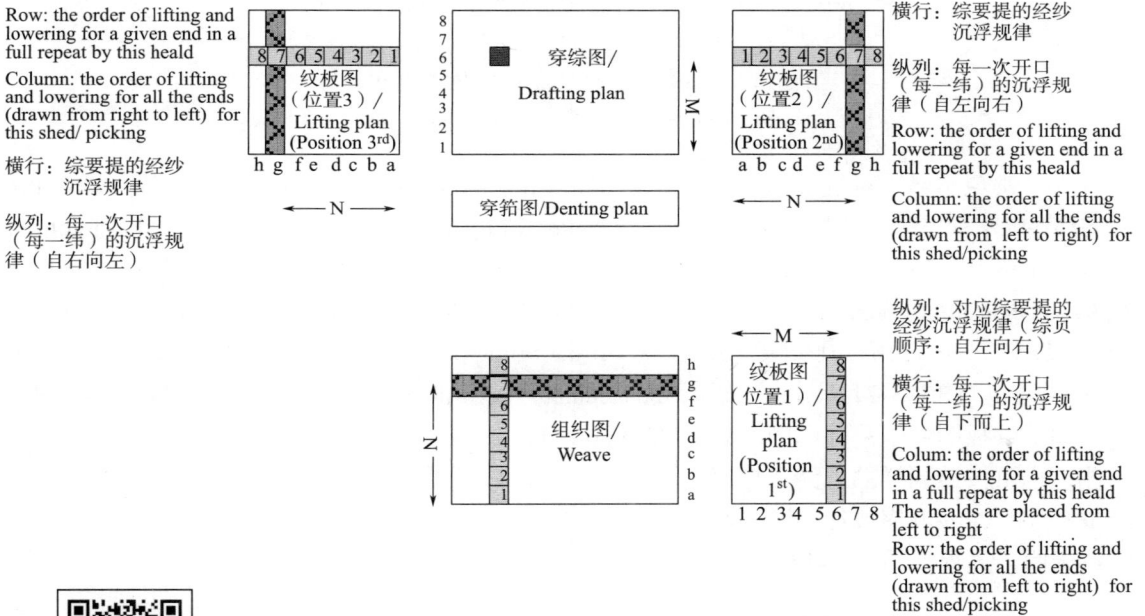

图1-3-20　不同位置的纹板图的行列意义示意图
The explanation of the lifting plan at different positions

Draw Lifting Plan Based on Design and Drafting Plan

（四）根据组织图、穿综图绘制纹板图 /Draw Lifting Plan Based on Design and Drafting Plan

纹板图的绘制方法：若某次开口需要引入某根纬纱时，需要提升某个综框上的综丝（即控制的经纱在此次交织时为经组织点时），在

In picking, if a heald is to be lifted which means that the end controlled by this heald is raised over the filling yarn, the square in the lifting plan where the vertical space for corresponding heald and horizontal space for the weft

纹板图相应的纵列与横行相交的方格中填入符号。图 1-3-21 所示为纹板图处于右下角时组织图与纹板图的关系。从该图中发现，此时如采用顺穿法穿综，纹板图与组织图完全一致。

cross is filled with mark. Fig. 1-3-21 shows the relationship between the design and the lifting plan when the latter is placed at the bottom right of the looming plans. If the straight drafting is applied, the lifting plan is just the same as the design/weave.

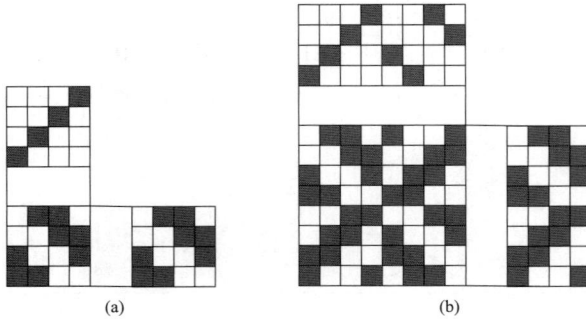

(a)   (b)

图1-3-21　已知组织图和穿综图绘制纹板图
Draw lifting plan based on design and drafts

为了避免出错，对纹板图、组织图和穿综图首先标上顺序，即一一对应，如图 1-3-20 所示。纹板图绘制方法可以按照综页顺序、经纱顺序或者纬纱顺序一一求解，其步骤分别如图 1-3-22 ～图 1-3-24 所示。

如果纹板图处于穿综图的左右两侧，也可按照同样的思路，按照综框顺序、经纱顺序或纬纱顺序绘制纹板图，过程如图 1-3-25 所示。

For a beginner, the lifting plans, designs and drafts are labelled with order as shown in Fig. 1-3-20. The lifting plan can be drawn according to the order of healds, ends or wefts as shown in Figs. 1-3-21 to 1-3-24 respectively.

If the lifting plan is set beside of the drafts, it can also be calculated according to the order of healds, ends or wefts. The procedure is explained in Fig. 1-3-25.

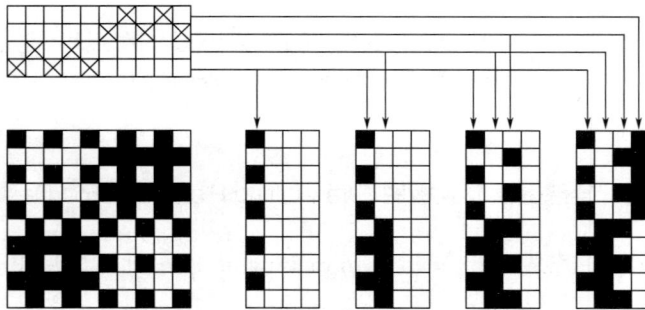

图1-3-22　按照综页顺序绘制纹板图
Draw lifting plan according to the order of healds

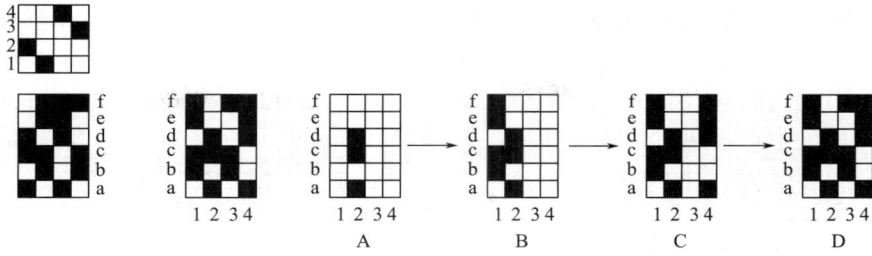

图 1-3-23 按照经纱顺序绘制纹板图的过程
Draw lifting plan according to the order of ends

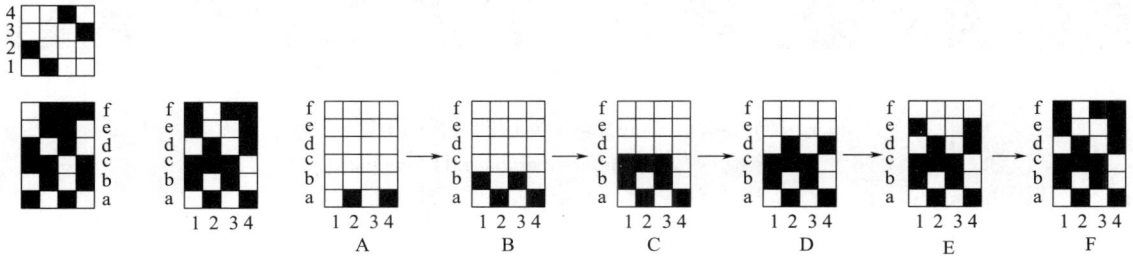

图 1-3-24 按照纬纱顺序（开口顺序）绘制纹板图的过程
Draw lifting plan according to the order of wefts

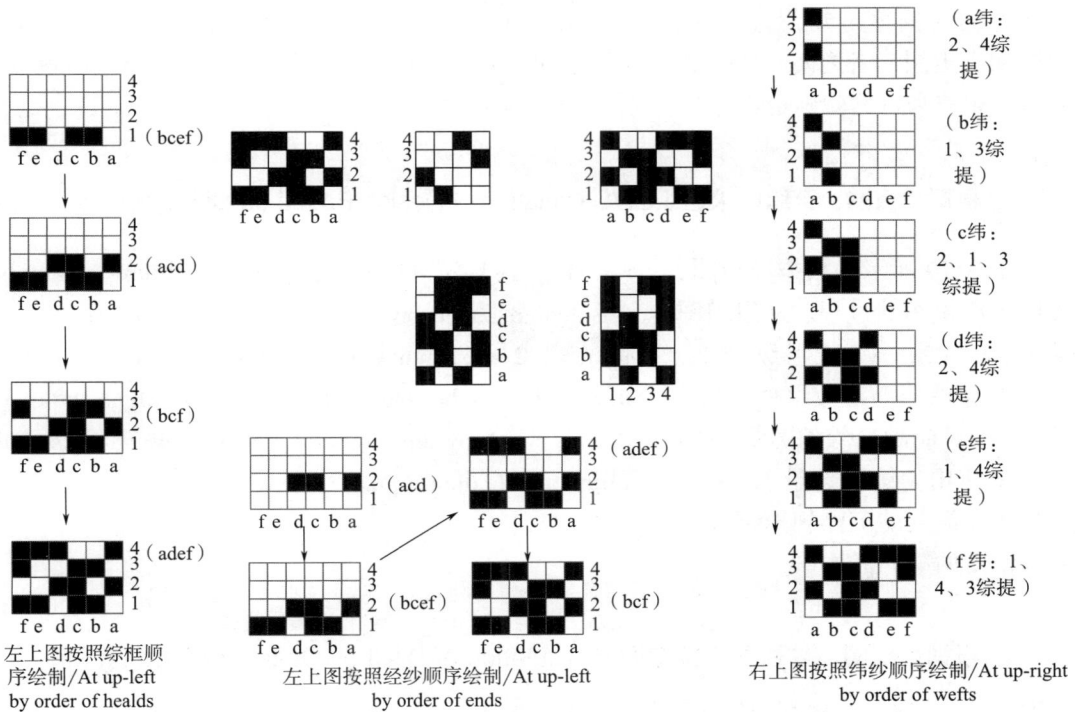

（a纬：2、4综提）
（b纬：1、3综提）
（c纬：2、1、3综提）
（d纬：2、4综提）
（e纬：1、4综提）
（f纬：1、4、3综提）

左上图按照综框顺序绘制/At up-left by order of healds

左上图按照经纱顺序绘制/At up-left by order of ends

右上图按照纬纱顺序绘制/At up-right by order of wefts

图1-3-25 纹板图处于穿综图的左右两侧不同顺序的绘制过程
The process of drawing lifting plan at different positions

在绘制如图 1-3-26 所示的纹板图过程中发现，即使组织图相同，若穿综图不同，纹板图肯定不同。

Fig. 1-3-26 shows that certain designs can be drafted in different ways, but a change in the order of drafting necessitates a corresponding change in the lifting plan.

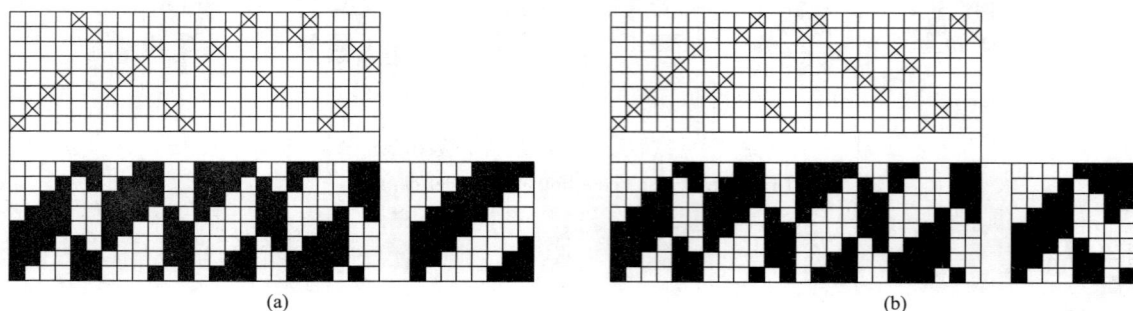

(a)   (b)

图1-3-26　不同穿综图对应的纹板图不同
The same design by different drafts and lifting plans

五、组织图、穿综图和纹板图之间的关系 /The Relationship Among Design, Draft and Lifting Plan

从上面的论述可知，组织图、穿综图和纹板图三者之间，任意确定两个，第三个自然就确定了。

From the above examples, the design, the draft and the lifting plan are very closely dependent on one another as already indicated.

（一）根据纹板图、穿综图求组织图 /Calculation of Design Based on Lifting plan and Draft

图 1-3-27 所示为根据纹板图、穿综图求组织图的过程。从组织图的第 i（$i=1$，2，$\cdots$，R_j）纵列上，找到其在穿综图上对应综框位置 k（即该经纱对应的穿综符号所在的横行，是使用的综页数量），将纹板图的第 k 纵列的沉浮规律符号复制到组织图的第 i 纵列上。

Fig. 1-3-27 demonstrates the procedure to calculate the design based on the lifting plan and draft. For end No.i($i=1, 2, \cdots, R_j$) in weave, find its heald No.k ($k=1, 2, \cdots$, n, n is the number of healds used) upon which the end is drawn, and copy the interlacing of the column No.k of the lifting plan to the weave.

（二）根据组织图、纹板图求穿综图 /Calculation of Draft by Design and Lifting Plan

图 1-3-28 所示为根据组织图、纹板图求穿综图的过程。对于判断组织图的第 i（$i=1$，2，3，$\cdots$，R_j）

Fig. 1-3-28 demonstrates the procedure to calculate the draft by design and lifting plan. For the end No.i ($i=1$, 2, 3, $\cdots$, R_j) in a weave, search the same interlacing from

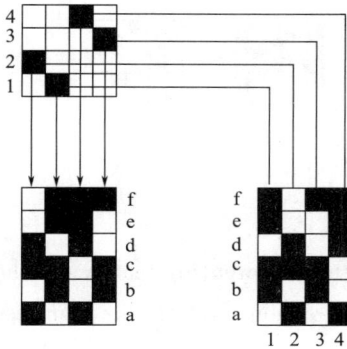

图1-3-27 根据纹板图、穿综图求组织图
Calculation of design based on lifting plan and drafts

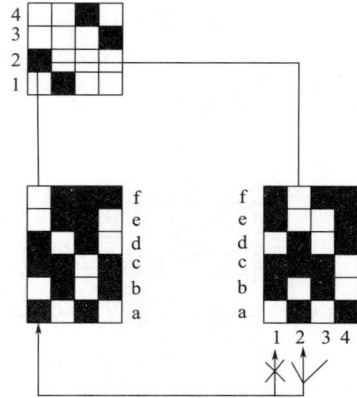

图1-3-28 根据组织图、纹板图求穿综图
Calculation of drafts based on design and lifting plan

纵列所穿综框来说，每次都是从纹板图的第 1 列开始，直至找到其在纹板图上对应的第 j 纵列，满足条件是这组织图的第 i 纵列与纹板图的第 j 纵列的符号完全相同。将穿综图的第 i 纵列与第 j 横行交叉处填充符号即可。根据该方法得到的穿综图可能不唯一，如图 1-3-29 所示。这时，需要考虑负载平衡，优化穿综图，如图 1-3-30 所示。

column No. 1 of the lifting plan each time and stop until column No.j meets the requirement. Fill in a mark at the square where column i and row j crossed in the draft. The procedure continues until the draft is finished. It should be noticed that the draft calculated by the method is not unique and should be optimized according to principle of even loading. Fig. 1-3-30 is an optimized draft based on Fig. 1-3-29.

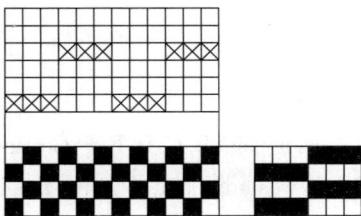

图1-3-29 未经优化的穿综图
Unoptimized drafting plan

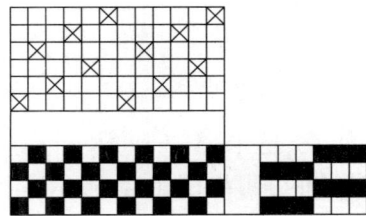

图1-3-30 经过优化的穿综图
Optimized drafting plan

习题 /Questions

1. 判断下列组织中（习题图 1-1），哪些是同面组织，哪些

1. Which weaves are belonged to the even-side weave, warp-face weave and weft-face weave respectively?

是经面组织，哪些是纬面组织？

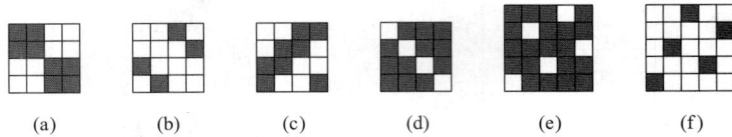

习题图1-1

2. 分别从经向与纬向来看，下列组织（习题图 1-2）的交织规律是什么（如果每根纱线都相同的话）？从经纬两个方向上看，每根纱线上组织点对应前一根纱线的飞数是多少？

2. What are the interweaving features of the following weaves from warp direction and weft direction respectively if they have the same interlacing at each thread? Find out the step number for each weave from both two directions.

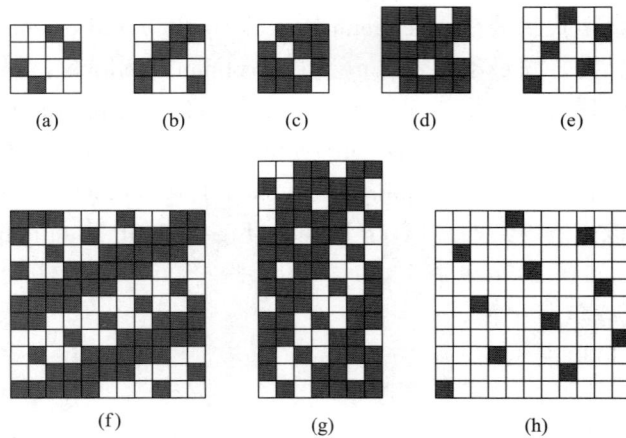

习题图1-2

3. 如习题图 1-3 所示，在 A 点左侧和右侧纱线上分别标出 S_j =−1、−2、−3、1、2、3 的组织点为 B、C、D、E、F、G 和 H、I、J、K、L、M 点。在 B 点的上侧和下侧纱线上分别标出 S_w =−1、−2、−3、1、2、3 的组织点为 b、c、d、e、f、g 和 h、i、j、k、l、m 点。

3. Mark the interweaving point B, C, D, E, F, G respectively with S_j =−1,−2,−3,1,2,3 to weaving point A at the neighboring thread at left side, and mark the interweaving point H, I, J, K, L and M respectively with S_j =−1,−2,−3,1,2,3 to weaving point A at its neighboring thread at right side. Similarly, mark the weaving point b, c, d, e, f, g respectively with S_w =−1,−2,−3,1,2,3 to weaving point A at the neighboring thread at upper side, and mark the weaving point h, i, j, k, l and m respectively with S_w =−1, −2,−3,1,2,3 to weaving point A at the neighboring thread at lower side.

4. 下面的 2 个组织（习题

4. Draw the new combined weave by placing the ends

图1-4），按照经纱排列比1∶1、纬纱排列比2∶2的规律，间隔排列得到新的组织图是什么？

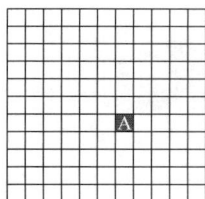

习题图1-3

5. 辨析题。

（1）习题图1-5（a）图采用了顺穿法。

（2）习题图1-5（b）图采用了飞穿法。

（3）习题图1-5（c）图采用了分区穿法。

（4）习题图1-5（d）图组织可以用踏盘织机织造。

and the picks at intervals from the 2 base weaves below. The ends are arranged at the ratio of 1 : 1 while the picks are arranged as 2 : 2.

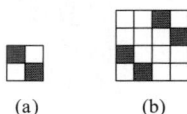

(a)　　(b)

习题图1-4

5. Judging and analyzing.

(1) Straight drafting is used in exercise figure 1-5(a).

(2) Skip drafting is used in exercise figure 1-5(b).

(3) Divided drafting is used in exercise figure 1-5(c).

(4) The weave as shown in exercise figure 1-5(d) can be woven on a tappet loom.

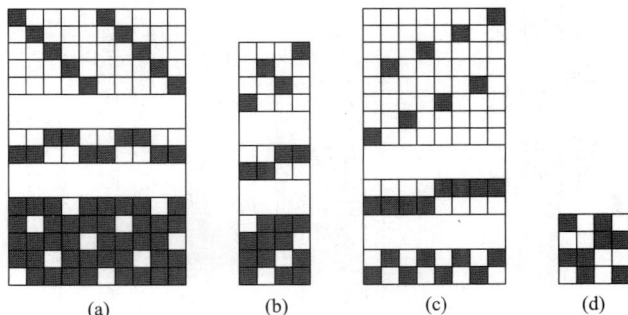

(a)　(b)　(c)　(d)

习题图1-5

6. 根据习题图1-6的组织图和穿综图，分别画出在右下角、左上角和右上角三种不同位置时的纹板图。

7. 习题图1-7填入缺失的组织图、穿综图或者纹板图。

8. 已知在上机图在右下方的纹板图分别如习题图1-8（a）和（b）所示。在左手车、右手车上的纹板

6. Draw the lifting plans at different locations according to the designs and drafts.

7. Fill in the missing design, draft or lifting plan.

8. The lifting plans at the bottom-right of the looming plans are shown at (a) and (b) respectively. Trying to implant the pegs corresponding to the pegging plan in the lags at left-handed loom and right-handed loom respectively.

对应的位置植入纹钉。

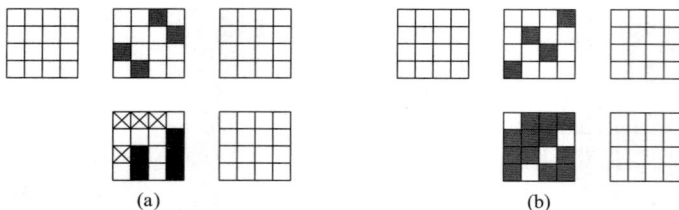

习题图1-6

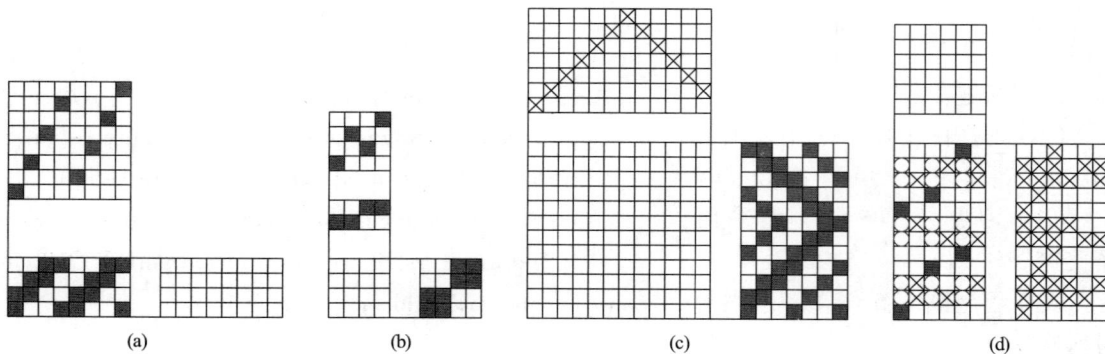

习题图1-7

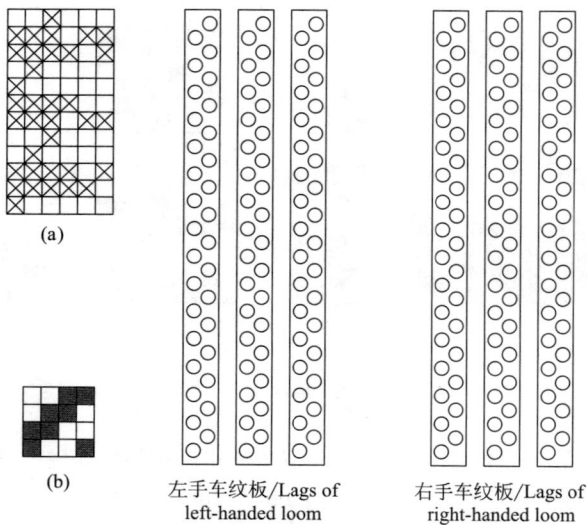

左手车纹板/Lags of
left-handed loom

右手车纹板/Lags of
right-handed loom

习题图1-8

第二章　机织物几何结构 /Geometric Structure of Woven Fabrics

第一节　机织物几何结构参数 /Parameters of Woven Geometric Structure

一、机织物几何结构的概念 /Concept of Geometric Structure

机织物几何结构就是经纱和纬纱在空间中的几何形态，可用纱线中心线的形态和纱线的截面形状来描述。它影响织物外观、表面形态、力学性能（拉伸、抗冲击、弯曲、悬垂、手感等）、透气、透湿、渗透、保温、散热、阻燃、电磁等物理性能及生产加工的难易程度。

若能精确建立机织物几何结构模型，根据织物设计参数与加工条件，对每根纱线在空间的形态进行精确计算和描述，则可利用计算机真实模拟织物的光泽图案等视觉效果，准确预测织物各项性能，可用于纺织品物理性能预测、外观图案效果设计等，能显著缩短设计和生产周期，为民用和产业用织物性能设计提供基础数据。

在织物设计中，织物几何结构是概算织物经纬纱密度、经纬纱线密度、经纬纱缩率等参数的依据。

The geometric structure of woven fabric is the extending and bending form of warp and weft yarns in the three-dimensional space, which can be described in 2 aspects: cross-section of yarn and central curve of yarn. The geometry of the woven fabric has a great influence on the appearance, surface morphology, mechanical properties (including tensile strength, ballistic impact resistance, flexibility, drapability, handle), air ventilation, vapor permeability, penetrating performance, thermal insulation property, flame resistance, electromagnetic performance and the processing convenience of fabric.

If an accurate woven geometry model is set up, the contour and crimp of each yarn in the structure will be estimated and described based on the fabric specifications and looming conditions, which can be used to imitate the appearance and pattern of the fabric, estimate the properties of the fabric, shorten the design and production cycle significantly, and provide the basic data for apparel and industrial fabric design.

In fabric design, woven geometry is also the basis for estimating linear density of the threads, fabric setts, contraction rate and other parameters.

二、织物中纱线的几何形态 /Contour and Crimp of Yarns

1. 前提假设 /Hypothesis

关于织物几何结构的研究，一般假设为纱线是柔软可自由弯曲、不可压缩与伸长的圆柱体。

The study on the woven structures is usually based on the following hypothesis: yarn is a limp, non–compressible, non–extended cylinder.

2. 纱线的直径系数 K_d/Calculation of the Diameter of A Circular Yarn

既然纱线是圆柱体，其截面则是圆形，则有如下公式（所有的物理量都按照公制标准单位计算）：

Since the yarn is supposed to be a cylinder, the cross–section of the yarn is round. Then, the following formula holds:

$$G = \frac{\text{Tt} \cdot L}{1000 \times 1000} = \pi \cdot \left(\frac{d}{2} \right)^2 \cdot L \cdot \delta \cdot 1000 \qquad (2-1-1)$$

$$d = \frac{1}{1000} \sqrt{\frac{4\text{Tt}}{1000\pi \cdot \delta}} \, (\text{m}) = 0.035678 \cdot \frac{\sqrt{\text{Tt}}}{\sqrt{\delta}} \, (\text{mm}) = \frac{0.035678}{\sqrt{\delta}} \sqrt{\text{Tt}} \qquad (2-1-2)$$

式中：G 是纱线重量（g）；L 是纱线长度（m）；Tt 是纱线的细度（tex）；d 是纱线直径；δ 是体积密度。部分纱线的体积密度见表 2–1–1。

Where, L is the length of yarn (m), Tt is the linear density of the yarn (tex), G is the weight of a given length of yarn (g), d is yarn diameter (mm), and δ is cubic density of the yarn (g/cm^3). Table 2–1–1 lists the cubic density of the yarn of some materials.

表 2-1-1　部分纱线体积密度
The cubic density of the yarn of some materials

纱线种类/Types of the Yarn	体积密度δ/ Cubic Density of the Yarn（g/cm^3）
棉纱/Cotton yarn	0.80 ~ 0.90
精梳毛纱/Worsted yarn	0.75 ~ 0.81
粗梳毛纱/Woolen yarn	0.65 ~ 0.72
亚麻纱/Linen yarn	0.90 ~ 1.00
绢纺纱/Schappe silk yarn	0.73 ~ 0.78
黏胶纤维纱/Rayon yarn	0.80 ~ 0.90
涤/棉纱/Polyester/cotton yarn（65/35）	0.80 ~ 0.95

若将式（2-1-2）简写为：

Formula（2-1-2）can be written as :

$$d = K_d \cdot \sqrt{\text{Tt}} \qquad （2-1-3）$$

K_d 称为纱线的直径系数，可以根据表 2-1-1 的纱线体积密度换算。

K_d is called coefficient of diameter and can be calculated according to Table 2–1–1.

3. 织物中纱线的截面形态/Cross-Section of Yarn in Fabric

Peirce 为了方便研究棉织物几何结构，认为纱线在织物中不伸长，截面呈现圆形（图2-1-1），后来又修订为椭圆形（图2-1-2）。但 Peirce 的椭圆形纱线截面理论在解决密度较大的织物结构的计算时并不适用，于是，Kemp 提出了跑道形纱线截面（图2-1-3），Hearle 和 Shananan 提出凸透镜形纱线截面，分别如图2-1-4和图2-1-5所示，还有研究者为了造型方便，提出了碗形纱线截面模型（图2-1-5）。

Peirce first prompted the theory that the cross-section (Fig. 2-1-1) of a yarn is round then extended to an ellipse (Fig. 2-1-2). However, the theory of elliptical cross-section is not suitable to calculate the geometry of the woven fabrics in high sett. Later Kemp proposed a racetrack cross-section (Fig 2-1-3), Hearle and Shananan proposed a lens model(Fig. 2-1-4, Fig. 2-1-5). For simplicity, bowl-type cross-section(Fig. 2-1-5) was also proposed.

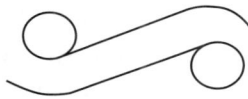

图2-1-1　圆形纱线截面
Circular cross-section

图 2-1-2　椭圆形纱线截面
Elliptic cross-section

图2-1-3　跑道形纱线截面
Race-track cross-section

图2-1-4　凸透镜形纱线截面
Lens cross-section

图2-1-5　碗形纱线截面
Cross-section of bowl type

提出各种纱线截面模型主要是为了在特定情形下计算方便。最广为接受的是椭圆形纱线截面形态，该方法简单、方便，并用压扁系数 η 反映纱线在织物中的压扁状态，其计算如下，η 在织物中的范围为 0.5 ~ 0.95，一般取值0.85。

Various models of the contour of the yarn cross-section were introduced just for the particular occasion or for the convenience of calculation. The most widely accepted is Peirce's elliptic cross-section model for its simplicity and convenience. Flatness coefficient η is defined in the following formula and is used to better describe the contour of the yarn. Generally, η ranges between 0.5 and 0.95 in fabric, and η is usually set as 0.85.

$$压扁系数\eta = \frac{纱线在织物切面图上垂直于布面方向的直径}{纱线计算的圆形直径\ d}$$

（2-1-4）

$$Flatness\ coefficient\ \eta = \frac{Minor\ radis\ of\ the\ ellipse\ d}{The\ theoretic\ diameter\ of\ the\ circular\ yarn\ d}$$

4. 织物内经纬纱的纵向屈曲状态/Yarn Path in Fabric

Perice 等以纯粹的几何图形关系角度考虑织物中经纬纱的弯曲

Perice and his followers considered the bending form of the yarns in the fabric from the perspective of pure

形态，认为纱线的中心线是弧形结合直线的状态，如图2-1-1~图2-1-5所示。若织物紧密，则直线部分消失，纱线完全呈弧线状态。在该理论体系中，同一类型组织点（经或纬）具有同样的高度，经纬纱之间紧密接触。图2-1-6所示为紧密织物的几何结构示意图。

Peirce体系的机织物几何结构计算简单，推导严密，避免了采用其他方法导致的纱线之间可能因为计算不精确而出现的穿刺现象，是首选的机织物结构建模方法。但该模型没有考虑纱线材料、上机张力、纱线伸长等因素，对织物结构的成因和外观形成原理的解释并不理想。

geometric graph relations, and proposed that the center curve of the yarn was the state of arc combined with straight line, as shown from Figs. 2-1-1 to 2-1-5. If the fabric was tight, the straight line part disappeared and the yarn was completely curved. In this theoretical system, the same weaving points had the same height and warp yarn contact weft yarn closely, as shown in Fig. 2-1-6.

Peirce system was the preferred structural modeling method for woven fabrics because of its simple calculation and rigorous derivation, avoiding the puncture phenomenon among yarns that might be caused by inaccurate calculation caused by other methods. However, this model was inaccurate due to the lack of considering yarn materials, weaving tension, yarn elongation and other factors, so it was not ideal to explain the cause of fabric structure and appearance formation.

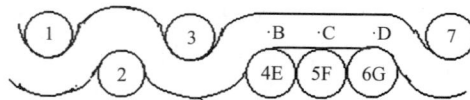

图2-1-6　Peirce的紧密织物几何结构
Peirce's geometry model for jammed woven fabrics

Olofesson Grosberg

Olofesson、Grosberg、Hearle等试图用最小能量原理，从力学角度计算织物结构，用多项式来表示纱线中心线形状，但仅适用于非紧密织物。从织物本身弯曲和悬垂等力学性能研究出发，川端季雄认为平纹织物中纱线的中心轴线呈曲折直线，如图2-1-7所示。还有研究者认为平纹织物中纱线的中心轴线呈正弦曲线，也有研究者提出弹簧—质子模型，但这些理论并不能解决织物本身几何结构的问题。

Olofesson, Grosberg, Hearle, et al. tried to use the principle of minimum energy to calculate the fabric structure from the mechanical point of view, and used polynomials to represent the yarn center curve, but this method only applied to non-dense fabrics. Starting from the research on the mechanical properties of the fabric itself, such as bending and draping, Yoshio Kawabata believed that the yarn center line of plain weave fabric was a zigzag straight line, as shown in Fig. 2-1-7. Some people think that the central curve of plain grain was sinusoidal curve, or put forward spring—particle model, but these theories failed to solve the problem of fabric geometry.

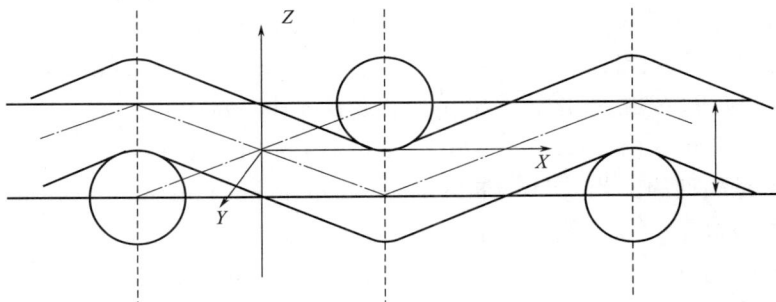

图2-1-7　曲折中心轴线的几何结构
Fabric geometry of zig-zag model

Crimp Height and Geometry Phase

三、屈曲波高与几何结构相 /Crimp Height and Geometry Phase

1. 屈曲波高 /Crimp Height

织物中，纱线的中心轴线呈屈曲状态，其波峰与波谷之间的垂直距离称屈曲波高，用 h 表示，分经纱屈曲波高 h_j 和纬纱屈曲波高 h_w，如图 2-1-8 和图 2-1-9 所示。

（1）纬纱完全伸直时（图 2-1-10），$h_w=0$。此时，$h_j=d_j+d_w$，$h_j+h_w=d_j+d_w$。

（2）经纱完全伸直时（图 2-1-11），$h_j=0$。此时，$h_w=d_j+d_w$，故 $h_j+h_w=d_j+d_w$ 成立。

The crimp of yarn is the waviness or distortion of a yarn that is due to interlacing in the fabric. The distance between the peak and the trough of a certain yarn is called crimp height h. Warp crimp height h_j and weft crimp height h_w are shown in Figs. 2-1-8 and 2-1-9 respectively.

(1) If weft is totally straighten as shown in Fig. 2-1-10, $h_w=0$, then, $h_j=d_j+d_w$, $h_j+h_w=d_j+d_w$.

(2)If warp is totally straighten as shown in Fig. 2-1-11, $h_j=0$, then, $h_w=d_j+d_w$, $h_j+h_w=d_j+d_w$.

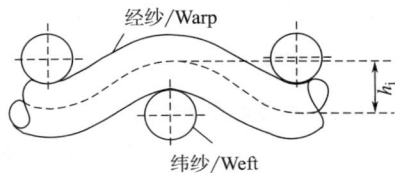

图2-1-8　经纱屈曲波高
Crimp height in warp direction

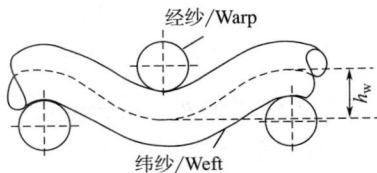

图2-1-9　纬纱屈曲波高
Crimp height in weft direction

图2-1-10　纬纱完全伸直状态
The status of unbent weft yarn

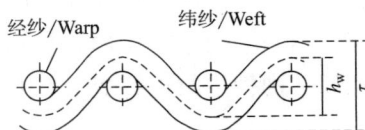

图2-1-11　经纱完全伸直状态
The status of unbent warp yarn

（3）若经纬纱处于其他中间状态，则经纬纱屈曲波高等于常数。根据假设，经纱与纬纱紧密接触，纱线之间没有压缩变形。由于经纱和纬纱上下运动是联动的，若一个方向的纱线屈曲波高增大 Δ，则必然导致另一个方向的纱线屈曲波高减小 Δ（图 2-1-12）。若变化后的屈曲波高为 h_j'、h_w'，则有：

(3) If the warp and weft is at mediate phase, the sum of the two crimp heights is a constant. According to the assumption that the warp and the weft contact closely, while the crimp height increases Δ in one direction, the crimp height in another direction will decrease Δ which is demonstrated in Fig. 2-1-12. After the movement, the crimp heights change to h_j' and h_w' respectively, then:

$$h_j'-h_j=\Delta=h_w-h_w'$$
$$h_j'+h_w'=h_w+h_j=d_j+d_w=L \qquad (2-1-5)$$

式（2-1-5）说明，无论何种织物结构，经纬纱屈曲波高之和为一常数，等于经纬纱直径之和。

Formula (2-1-5) indicates that the sum of the warp crimp height and weft crimp height equals to the sum of the warp diameter and the weft diameter.

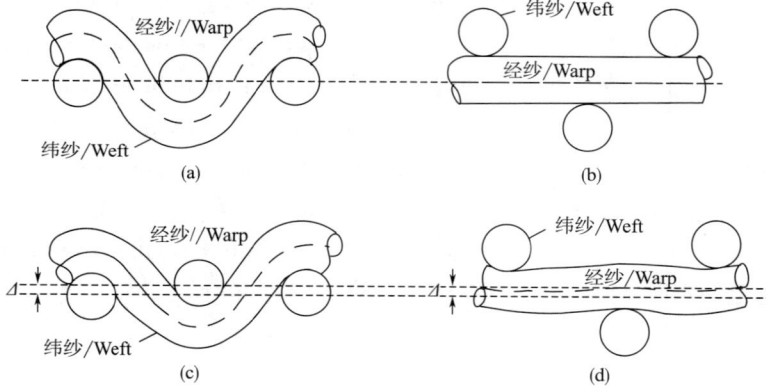

图2-1-12　屈曲波高的变化
Changing of the crimp height

2. 织物的几何结构相 /Geometry Phase of Woven Structure

织物中，经纬纱的波形是连续变化的。为了描述经纬纱处于某种位置，将其连续编号，把具有某个状态的结构称为结构相（阶序）。以两种极端状态的序号 1 和 9 作为起始相位和终止相位。图 2-1-13 所示为织物几何结构相位的连续变化过程。

The waveform of warp and weft yarns changes continuously. In order to describe the position of warp and weft yarns, different waveforms are successively numbered, and the structure with a certain state is called structural phase (order). The structure phase starts from phase No. 1 and terminates at phase No. 9. Fig. 2-1-13 shows how the fabric geometry changes.

在第 1 相位时：

At phase No. 1:

$$d_j=d_w=2r$$

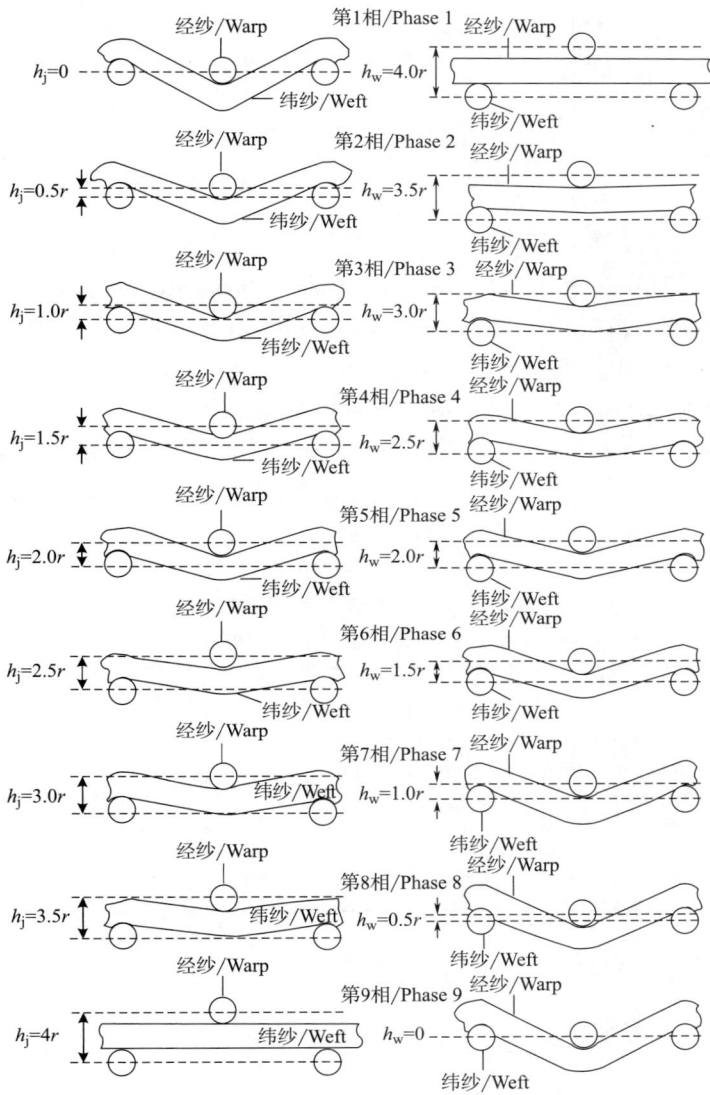

图2-1-13　织物几何结构相位的连续变化过程
Continuous changing of the geometry of the woven structure

$$h_{\mathrm{j}}=0, \quad h_{\mathrm{w}}=d_{\mathrm{j}}+d_{\mathrm{w}}$$

在第 9 相位时：　　　　　　　　　　　At phase No. 9:

$$h_{\mathrm{w}}=0, \quad h_{\mathrm{j}}=d_{\mathrm{j}}+d_{\mathrm{w}} ;$$

在中间各相位时：　　　　　　　　　　At middle phase:

$$N=9-8\frac{h_{\mathrm{w}}}{d_{\mathrm{j}}+d_{\mathrm{w}}} \tag{2-1-6}$$

45

$$h_j = \frac{N-1}{8}(d_j + d_w), \quad N = \frac{8h_j}{d_j + d_w} + 1 = \frac{8h_j}{h_j + h_w} + 1 = \frac{9 \cdot \frac{h_j}{h_w} + 1}{\frac{h_j}{h_w}} \tag{2-1-7}$$

$$h_w = \frac{9-N}{8}(d_j + d_w), \quad N = 9 - \frac{8h_w}{h_j + h_w} \tag{2-1-8}$$

织物结构在各相位时的特征参数见表 2-1-2。由表可知，经面支持时大于 5 相，纬面支持时小于 5 相，在同支持面时为 5 相。当经纬纱直径不等且处于同面支持时，称为 0 结构相，如图 2-1-14 所示。

Table 2-1-2 shows the geometric parameters at various phases. Obviously, the phase order exceeds No. 5 when warps support the fabric. On the contrary, the phase order is less than No. 5 if wefts support the fabric. When the fabric is supported by both warps and wefts as shown in Fig. 2-1-14, the fabric is at the geometry phase No. 5 when warps and wefts are equal in thickness , or phase No. 0 if unequal.

表 2-1-2　织物结构相的特征参数
Parameters at different geometry phases

结构相/ Phase	1	2	3	4	5	6	7	8	9	0
h_j	0	1/8 L	1/4 L	3/8 L	1/2 L	5/8 L	3/4 L	7/8L	L	d_w
h_w	L	7/8 L	3/4 L	5/8 L	1/2 L	3/8 L	1/4 L	1/8 L	0	d_j
ϕ	0	1/7	1/3	3/5	1	5/3	3	7	∞	d_w/d_j

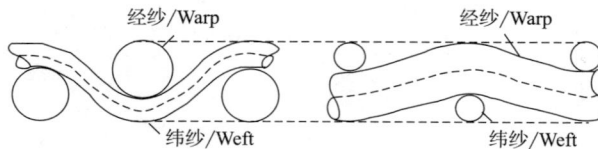

图 2-1-14　经纬纱直径不等的同支持面织物（0 结构相）示意图
Schematic diagram of geometric structure of the phase No. 0

在 0 结构相时：

故：

或者：

At phase 0:

$$d_j + h_j = h_w + d_w = d_j + d_w$$

Therefore:

$$H_j = d_w, \quad h_w = d_j$$

Or:

$$\frac{h_j}{h_w} = \frac{d_w}{d_j} \tag{2-1-9}$$

四、影响织物屈曲波高的因素 /Factors Affecting Crimp Height

影响织物屈曲波高的主要因素有纤维材料、纱线结构、上机张力、织物密度、织物组织等。

1. 纤维材料 /Fiber Materials

纤维弹性模量高，在同等张力下，伸长量低，弯曲程度小；反之，纤维弹性模量低，弯曲程度大。其他情况相同时，经纬纱中，采用模量高的纤维材料的纱线弯曲程度小。

2. 纱线结构 / Yarn Structure

纱线粗，抗弯弯矩大，不容易弯曲，屈曲波高小。

3. 上机张力 /Looming Tension

织造中，承受初始张力大的纱线不容易弯曲，承受初始张力小的纱线则容易弯曲。

4. 织物密度 /Fabric Density

经向密度大，单位长度内纬纱的交错次数多。在幅宽也就是纬纱长度大致一定的情况下，纬纱弯曲小，在织物中相对平直；因为经纬屈曲波高为定值，故经纱弯曲程度大，几何结构相高，经纱织造缩率也就越大。同理，纬向密度大，几何结构相低。

5. 织物组织 /Weave Pattern

织物组织通过影响纱线的可密率来影响织物几何结构。平纹织物交织频繁，限制双向织物密度提高，纱线弯曲程度相差不大，难以达到过高或者过低的结构相；而单位长度内缎纹织物交错次数少，纱线可密性强，可以在较大的纱线密

The main factors affecting the crimp height of woven fabrics include fiber materials, yarn structure, looming tension, fabric density and weave pattern.

The fiber of a higher elastic modulus extends less and bends less under the same tension. On the contrary, the fiber of a lower elastic modulus is likely to extend and bend more. Therefore, the threads made of higher elastic modulus fibers will have a lower crimp height.

The thick yarn has a large bending moment. It is not easy to bend, and tends to be straight in the fabric.

The threads undertaking heavier tension tend to be straight while the threads undertaking lighter tension tend to bend around the threads of another series.

In a high warp sett fabric, there are more intersections for weft thread per unit space. At the circumstance that the width or the length of the weft thread is almost the constant, weft thread can only bend little and relatively straight in the fabric. Since the sum of the warp crimp height and the weft crimp height is constant, a small amount of the weft crimp height means a large amount of the warp crimp height and high order of the geometric phase and higher contraction rate. Similarly, the fabric with high weft-sett will stay at a lower geometric phase.

By affecting the packing ability, the weave pattern is closely related to the geometric structure of the woven fabrics. There are more intersections in a unit space for plain weave, both the warp density and the weft density are restricted, and a little difference appear in the warp crimp height and the weft crimp height thereby a middle geometric phase is achieved. However, there are lesser intersections

度下织造，容易达到较高或较低的织物结构相位。

in a unit space of fabric for satin weave, more threads can be placed into fabrics per unit space, resulting in a higher or lower geometric phase.

五、织物其他主要参数 /Other Parameters of Fabric

根据表征织物规格的参数，可计算织物紧度、织物缩率、织物厚度和孔隙率等参数，进一步反映织物的几何结构特征。

The specifications for woven fabric can be utilized to calculate parameters like covering factor, rate of contraction, fabric thickness, porosity to reflect the geometric structure of the fabric.

1. 织物紧度 /Covering Factor

织物紧度 E 是指纱线投影面积占织物面积的百分比，本质上是纱线占织物空间的覆盖率。根据图 2-1-15 所示的经纬纱在整个组织循环中的占据比例，计算织物的经向紧度 E_j、纬向紧度 E_w 和总紧度 E_z。计算式如下：

Covering factor E is the ratio of the yarn area to the fabric area. It is the measure of the percentage area covered by one or more threads. Covering factor can be related to weft (E_w), warp (E_w) or woven fabric (E_z). Based on Fig. 2-1-15, covering factor is calculated by the following formula:

$$E_j = \frac{d_j}{a} \times 100\% = d_j \cdot P_j \qquad (2-1-10)$$

$$E_w = \frac{d_w}{b} \times 100\% = d_w \cdot P_j \qquad (2-1-11)$$

$$E_z = \frac{d_j \cdot b + d_w(a-d_j)}{ab} \times 100\% = E_j + E_w - \frac{E_j \cdot E_w}{100} \qquad (2-1-12)$$

式中，a、b 分别表示一根经纱、纬纱占据的空间长度；P 表示织物密度，即单位长度内的纱线根数；下标 j 表示经向（经纱），下标 w 表示纬向（纬纱）。若 E_j 或 E_w

In the above formulas, a and b denote the warp spacing and weft spacing respectively, which are represented in Fig. 2-1-15, P denotes the number of the yarns per unit space. j and w denote in warp direction and weft direction respectively. If either E_j or E_w exceeds 100%, which means

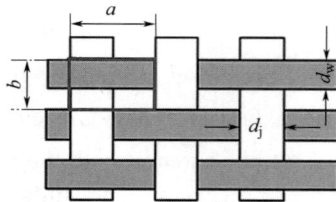

图2-1-15　机织物紧度计算示意图
Schematic diagram for calculating the covering factor

有一个大于 100%，则说明在该方向纱线系统有挤压现象发生。

that yarn compressing happens in this direction.

2. 织物缩率（织缩率）/Rate of Contraction

交织导致纱线弯曲，从而使织物产生收缩。α_j 和 α_w 分别表示经、纬纱缩率，其计算式为：

Interlacing makes the yarns bent and results in the contraction in weaving. The rate of contraction is calculated by the following formulas respectively:

$$\alpha_j = \frac{L_{j0} - L_{j1}}{L_{j0}} \times 100\% \qquad (2-1-13)$$

$$\alpha_w = \frac{L_{w0} - L_{w1}}{L_{w0}} \times 100\% \qquad (2-1-14)$$

式中：L_{j1} 和 L_{w1} 分别表示织物中经、纬向的长度；L_{j0} 和 L_{w0} 分别表示将织物中经、纬纱抽出伸直的长度。

Where L_{j1} and L_{w1} are the length of the fabric in warp and weft direction respectively. L_{j0} and L_{w0} are the length of the extended warps and the wefts that are removed from the fabric respectively.

3. 织物厚度 /Fabric Thickness

图 2-1-16 所示为织物的各种厚度。厚度有以下几种：①表观厚度 T_a，包括织物表面的绒毛、毛羽等所占空间；②织物加压厚度 T_p，是一定压力下，织物正反面距离，是用仪器测量得到的厚度；③织物空间厚度 T_t，是织物几何结构计算时的理论厚度；④织物实体厚度 T_b，即经纬纱直径之和，其中 $T_b = d_j + d_w$。根据图 2-1-16，有以下关系成立：

There are 4 ways to represent the thickness of a woven fabric as shown in Fig. 2-1-16. ① Apparent thickness T_a, the thickness including tufts and hairiness of the fabric. ② Pressed thickness T_p, the thickness measured at a given pressure by apparatus. ③ Theoretic thickness T_t, which can be calculated by the geometric structure of the fabric. ④ Body thickness T_b, $T_b = d_j + d_w$.

According to Fig. 2-1-16:

$$T_a \geqslant T_t \geqslant T_p \geqslant T_b \qquad (2-1-15)$$

一般将织物厚度定义为在一定的压力下，织物正反面的距离，即 T_p。T_t 表示理论厚度，是织物几何结构中的厚度，简写为 T。

Generally, T_p, the distance of the face side and the back side measured at a given pressure is defined as the theoretic thickness of the fabric, and written as T_t, or abbreviated as T.

Theoretic thickness T_t can be obtained by the following formulas.

$$T_t = \max(d_j + h_j,\ h_w + d_w) \qquad (2-1-16)$$

在低相时：

At lower phase:

$$T_t=h_w+d_w$$

在高相时：

At higher phase:

$$T_t=d_j+h_j$$

在其他相位时：

At other phase:

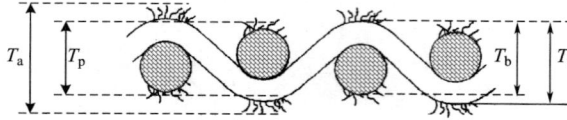

图 2-1-16　织物各种厚度示意图
Various thickness of the woven fabric

$$T_{1\text{相}}=T_{\text{phase 1}}=h_w+d_w=d_j+d_w+d_w=d_j+2d_w$$

$$T_{9\text{相}}=T_{\text{phase 9}}=h_j+d_j=d_j+d_w+d_j=2d_j+d_w$$

$$T_{0\text{相}}=T_{\text{phase 0}}=d_j+d_w$$

$$T_{5\text{相}}=T_{\text{phase 5}}=d_j+d_w=2d$$

若 $d_j=d_w=d$，则织物的厚度在 $2d \sim 3d$。表 2-1-3 是几种常见织物的厚度范围。

If $d_j=d_w=d$, the thickness of the woven fabric is between 2d and 3d. Table 2-1-3 lists the thickness of common woven fabrics.

表 2-1-3　几种常见织物厚度
The thickness of common woven fabrics

单位：mm

类型/ Type	棉织物/ Cotton fabrics	精纺毛织物/ Worsted fabrics	粗纺毛织物/ Woolen fabrics	丝织物/ Silk fabrics
薄型/Thin	<0.25	<0.4	<1.10	<0.14
中厚型/Medium	0.25 ~ 0.40	0.40 ~ 0.60	1.10 ~ 1.60	0.14 ~ 0.28
厚型/Thick	>0.40	>0.60	>1.60	>0.28

4. 孔隙率 /Porosity

（1）孔隙形状。纺织品是纤维的多孔集合体，孔隙的形态、数量、大小对纺织品的力学性能、透气、保温、传湿、渗透、吸附、防护、光电磁传播、化学性质有极大影响。机织物中的孔隙分为 3 个层次：①存在于织物中纱线间的孔

(1) Shape of Pores. Textile is assembly of porous fibers. The shape, quantity and size of pores have great influence on the mechanical properties, air permeability, heat preservation, moisture transfer, permeability, adsorption, protection, photoelectromagnetic transmission and chemical properties of textiles. The pores in woven fabrics can be divided into three levels: ①the pores among threads in

隙；②存在于纱线中纤维间的孔隙；③存在于纤维内空腔与各级原纤间的孔隙。这三个层次的孔隙可能是贯通孔隙，也可能是非贯通孔隙。这些孔隙排列有序，方向性强。孔隙的尺寸分布很广，纤维内部的孔隙在 1nm ~ 1μm，纤维间的孔隙在 1μm 以上，而纱线间的孔隙在 40μm 以上。

织物中纱线间的孔隙多为贯通孔隙，基本沿织物平面法线取向，孔隙的截面有一定规律。孔隙的大小、结构、形状与织物组织、经纬纱排列密度、纱线结构及后整理工艺有关。孔隙大小与织物紧度、浮长呈现明显的负相关，但与厚度关系不明显。

以平纹为例，假设纱线截面为圆形，在平衡结构下，将其高度分成9层，则每层孔隙的水平孔隙形状如图 2-1-17 所示。可以看出，平纹织物的纱线间孔隙是一对外口大，内口小，中间小口连接的喇叭形状。其他组织织物的孔隙也有类似的特征。

（2）孔隙的表征。织物中纱线间的孔隙可以用孔隙率、当量直径和孔隙数量表征。

孔隙率基本内涵就是孔洞体积占据织物总体积的百分比。因此，织物内纱线间孔隙率 ε 计算式为：

式中：δ_y 为纱线的体积密度；δ_f 为织物的体积密度。

纱线间孔隙的当量直径 d_e 计算式为：

woven fabrics; ②the pores that exist among the fibers in a thread; ③the pores that exist in the fiber cavity and the microfibrils. These orderly and directional pores in these three layers may be through or non-through. There is a wide distribution of the size of pores. The size of pores at the inner of fibers ranges from 1nm to1μm, larger than 1μm among the fibers, and more than 40μm between the threads.

The pores between the threads in the fabric are mostly through pores, which are basically oriented along the normal plane of the fabric. The cross-section of the pores has certain rules. Pore size, structure and shape are related to weave, fabric density, yarn structure and finishing technology. The pore size is negatively correlated with fabric tightness and float length, which is not correlated with fabric thickness.

Taking plain weave as an example, assuming that the cross-section of yarn is round and its height is divided into 9 layers under balanced structure. The schematic diagram of horizontal pore shape of pores in each layer is shown in Fig. 2-1-17. It can be seen that the pore between the threads of plain weave fabric is a horn shape with large external mouth, small internal mouth and small middle mouth connected. The pores of other woven fabrics have similar characteristics.

(2) Characterization of pores. The pores between the threads can be characterized by porosity, equivalent diameter and number of pores.

Porosity is basically the percentage of the volume of the pores in the total volume of the fabric. Therefore, the porosity ε between the threads is calculated by the following formula:

$$\varepsilon = \frac{\delta_y - \delta_f}{\delta_y}$$

Here, δ_y is the cubic density of threads, and δ_f is the cubic density of the fabric.

The equivalent diameter d_e of the pores between threads is calculated by the following formula:

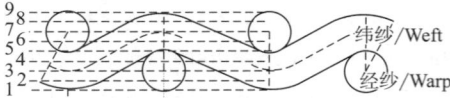

The pores between the threads in the fabric

(a) 第9层/Layer No.9   (b) 第8层/Layer No.8   (c) 第7层/Layer No.7

(d) 第6层/Layer No.6   (e) 第5层/Layer.No 5   (f) 第4层/Layer No.4

(g) 第3层/Layer No.3   (h) 第2层/Layer No.2   (i) 第1层/Layer No.1

图2-1-17　平纹织物各层孔隙形状示意图

Schematic diagram of horizontal pore shape of pores in each layer of a plain fabric

$$d_e = \left[\frac{10^4 \, T_c}{P_j P_w} - \frac{\pi}{4} \, (h_j d_j^2 + h_w d_w^2) \right] \frac{1}{\pi \, (h_j d_j + h_w d_w)}$$

织物中纱线间的孔隙数量 N_{py} 的计算式为：

The number N_{py} of the pores between the threads is calculated by the following formula:

$$N_{py} = P_j P_w \times 10^{-4}$$

织物组织对织物几何结构的影响

第二节　织物组织对织物几何结构的影响 /Influence of Woven Structure on Woven Geometric Structure

上一节介绍了织物组织对织物屈曲波高的影响，仅仅涉及织物的

The effect of the weave on the crimp of woven structure has been discussed in the previous section but

厚度。本节将继续讨论织物组织对织物几何结构的影响。

only the fabric thickness is involved. The influence will be elaborated further in this section.

一、交错与浮长 /Intersections and Float Length

由经组织点变成纬组织点，或者由纬组织点变成经组织点，称为一次交错。交织一次，交错两次。如图 1-2-3 中的平纹组织，在一个经向或者纬向循环内，经纬纱均交织 1 次，交错 2 次。一次交错从织物正面到反面，另一次交错从织物反面到正面。

浮长指连续浮在另一系统纱线上的纱线长度。图 1-2-3 中，经纬纱上所有组织点的浮长均为 1。

织物组织的平均浮长 F 是指组织循环纱线数 R 与一根纱线在组织循环内交错次数 T 的比值。在组织循环内，某根经纱与纬纱的交错次数用 T_j 表示，某根纬纱与经纱的交错次数用 T_w 表示。因此平均浮长为 $F_j = R_w/T_j$，$F_w = R_j/T_w$。

织物组织不同，体现在交织次数和交织位置的变化、浮长长度及其起始位置分布的差异，对织物的几何结构、外观与性能的影响很大。

Where a weave changes from warp-up point to weft-up point, and vice versa, the warp and weft threads correspondingly change from one side of the cloth to the other, or "intersect" each other. There are two intersections or one interlacing in a complete repeat of a weave shown in Fig. 1-2-3 for each thread, one in passing from the face to the back, and another in passing from the back to the face.

Length of float means the number of the threads crossed over on the threads of the other series. The lengths of the floats in Fig. 1-2-3 are all 1.

Average length of float, denoted as F, in a weave means the ratio of the repeat number R of the weave to the intersections T. Hence, $F_j = R_w/T_j$, $F_w = R_j/T_w$, where the j and w designate the direction in warp and weft respectively.

The difference in weave is actually the difference in intersections and positions of interlacing, the difference in the length of the floats, and the difference in the two endpoints of the floats. The variation in weave will result in changing of the structure, appearance and properties of the fabric.

二、交织对纱线形态和织物性能的影响 /Effects of Interlacing on the Yarn Contour and the Properties of Fabrics

经纬纱线的交织，对纱线的几何形态、织物性能产生以下影响。

（1）交织使纱线弯曲，产生织造缩率；同时，纱线伸长，织物张力增加。在某系统（经、纬）纱线在长度一定时，交织越频繁，该系

The effects of interlacing on the yarn contour and the properties of fabrics are reflected in the following respects.

(1) Interlacing bends the threads and causes the contraction of the fabric as well as extends the threads and increases the tension. In a given length of thread of a series (warp or weft), the more frequent the interlace, the less the

列纱线的弯曲程度越小。如平纹府绸织物的经密大，在纬纱方向上，单位长度内纬纱与经纱交织的次数多，故纬纱弯曲程度小。在经纱方向则刚好相反，经纱弯曲程度大。

（2）纱线弯曲，使同系统（经、纬）的相邻纱线分开，不容易靠近，织物的可密性降低。即使覆盖紧度不大，织物也显得紧密，如平纹组织织物。

（3）纱线弯曲，使不在同一浮长下的纱线之间的移动受到阻碍，织物柔曲性降低，织物硬挺度增加；同时，撕裂强度下降。单位长度内交错越少，平均浮长越长，更多组织点移动相对自由，织物越松软。

（4）纱线弯曲，对另外一个系统纱线的支撑作用加强。拉伸时，摩擦阻力增加，增加了拉伸强力。如果经纬纱的弯曲程度大致相同，织物耐磨性能提高。

（5）飞数体现了交织位置的改变。飞数为1时，相邻纱线弯曲处距离较近，可以相互支撑，故平纹、斜纹组织如卡其织物较硬挺。飞数大于1时，相邻纱线弯曲处距离较远，纱线之间支撑力量减弱，织物柔软度增加，如缎纹组织。

（6）交织使同一系统相邻纱线之间插入了另一个系统的纱线，织物更加紧密。在同等密度和纱线直径条件下，一个组织循环内，交错次数越多，平均浮长越小，织物越紧密；交错次数越少，平均浮长越大，织物越稀松。

bending degree of the thread of the series. A good example is poplin fabric with a high warp density, where the wefts interweave more times per unit length but with a small degree of bending. The opposite is true in the warp direction and warp threads bend more.

(2) Thread bending makes adjacent threads of the same series (warp or weft) separate from each other, making it difficult to get close to each other and reducing the aggregating of the threads. Even if the covering factor is not high, the fabric will appear firm and tight, such as plain fabric.

(3) Thread bending hinders the movement between threads not under the same float, and decreases the fabric flexibility, with the increasing of fabric stiffness. At the same time, the tear strength decreases. The less intersections in a weave, the longer average lengths of floats, the more weave points can move relatively freely, therefore, fabric touches more pliable and softer.

(4) Thread bending also increases support for threads of another series. When stretched, friction resistance among the threads increases, which helps to increase tensile strength. If the warp and weft threads are bent roughly the same level, the wearing resistance of the fabric is improved.

(5) The step number represents the change in interweaving position. When the step number is 1, the adjacent threads are both bent near each other and will support each other, so plain weave and twill weave such as khaki fabric are stiffer. With a step number is greater than 1, the thread bends farther apart, the support between the threads decreases, and the fabric becomes softer, such as satin and sateen.

(6) Each intersection causes the neighboring threads of same series to be separated and filled of a thread of other series, which makes the fabric tighter. Therefore, other things being equal, the more frequently the intersections occur, the shorter the average float, and the firmer the fabric. On the contrary, the less the intersections occur, the longer the average float, and the slacker the fabric might be.

三、浮长对织物结构和外观的影响 /Effects of Floats on the Structure and Appearance of Fabrics

Effects of Floats on the Sturcture and Appearance of Fabrics

1. 浮长作用 /Functions of Floats

（1）浮长在织物表面形成凹凸，织物松弛变厚。浮长越长，织物交织次数越少。在纱线定长的情况下（送经量大于卷取量，或者定幅宽下纬向收缩），一定浮长范围内（2～6左右），浮长越长，交织弯曲越少，因纱线有一定的刚度，只能在空间垂直方向上增加高度，使纱线凹凸程度增加；组织点浮长短，凹凸程度低。图2-2-1中，截面 A 交织规律是1/1，浮长均为1，纬纱屈曲波高为 h_1；截面 B 的交织规律为1/3，经纱 J_3 与纬纱 W_1 的交织点变为纬组织点，纬纱浮长增加。原本（a）图中下沉的 W_{31} 在（b）图中反向后，肯定在 W_{21} 的上方高度继续增加到 W_{31}^1 的位置，故纬纱屈曲波 h_2 增加，即 $h_2>h_1$，织物变厚。需要说明的是，若浮长太长，纱线弯曲刚度不足，高度不再增加，松弛浮在织物表面。交织次数减少，使织物的坚牢度下降。

(1) Floats protrude on the surface or drop under the back of the fabric making the fabric slack and thickened. The longer the float, the less the intersections. In weaving, the mount of let-off is greater than the amount of taking-up and the weft yarn will contract off-loom due to interlacing. Supposing the length of the yarn is constant and within a certain range of float length (about 2 to 6), the longer the length of float, the less interweaving and the less of thread bending. Because of the certain stiffness, thread can only extend and raise itself in the vertical direction of space which increases the height of crimp and the thickness of fabric. On the contrary, the crimp height of the thread of a short float is lower. In Fig. 2-2-1, the warps and the weft interlace at 1/1, so all the float lengths are 1 at sectional diagram A, and the crimp height of weft threads is h_1. At sectional diagram B, the warps and the weft interlace at 1/3, the interweaving point at J_3 and W_1 changes from a riser to a sinker, so the length of the weft float increases. W_{31} at A reverses at B, so its position, say W_{31}^1, must be higher than W_{21}. Therefore, the crimp height h_2 in weft direction increases, viz., $h_2>h_1$, and the fabric becomes thicker. However, if the float is too long, the stiffness of the thread is insufficient to support the bending, the crimp height no

J: 经纱/Warp　　　　0/NULL: 相当于平纹时位置/Position similar to plain weave
W: 纬纱/Weft　　　　1: 浮长线作用后位置/Position changed due to floats
　　　　　　W_{21}^0
经纱顺序/　　　　纬纱顺序/
Order of warp　　　Order of weft

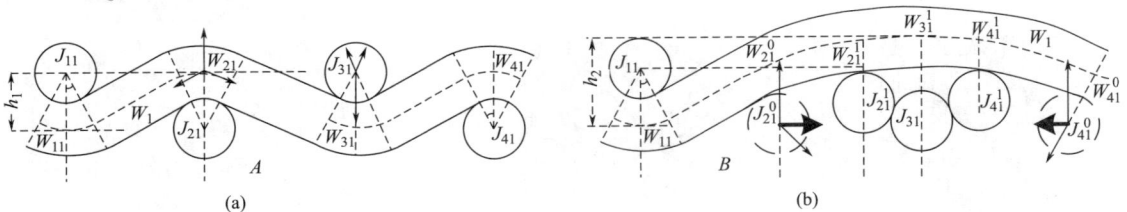

(a)　　　　　　　　　(b)

图2-2-1　纱线长度一定时，在不同交织次数下的屈曲高度对比图

Contrast diagrams of the crimp height at different intersections

（2）浮长线的凸起产生光照反射强度的差异，在织物表面形成纹理和明暗阴影效果。各组织点所在浮长不同，在织物表面高度不同，使织物表面凸凹程度不同。根据 Lambert 余弦定律，物体表面的明亮度直接取决于光线向量和表面法线两个向量夹角的余弦值。故在光照条件下，织物较高部分接受光照充分，亮度高；而照射到较低部分的光线容易被遮挡，亮度低，因此形成明暗效应或者立体花纹。同浮长内尽管各点的浮长长度相同，但离浮长中心的距离不等，高度也略有差异。浮长边部低，中间高。纱线在空间位置的高低不同是织物具有立体感的主要因素。如果高处或者低处有明显的分布规律，比如成斜向分布，则形成斜纹路（图2-2-2）。平布（平纹组织，经纬浮长均为1，图2-2-3）织物纱线处有弯曲，但各组织点表面最高点高度大致相同，故表面从明到暗变化均匀，光泽柔和。缎纹（图2-2-4）各组织点表面大部分高度大致相同，纱线基本平直，少量的低凹容易被周围高处遮挡成阴影，在漫反射下，基本不可见，故缎纹表面各处光亮度高。

longer increases, and the long floats slack and collapse on the surface of the fabric. When the number of interlacing is reduced, the firmness of the fabric is also decreased.

(2) The raised float leads to a difference in the intensity of light reflection, creating a texture and shading effect on the fabric surface. The length of floats varies, and the height of the thread on the surface is different, which makes an uneven fabric surface. According to Lambert's law of cosine reflection, that the radiant intensity observed from a "Lambertian" surface is directly proportional to the cosine of the angle between the observer's line of sight and the surface normal. Under lighting conditions, the higher portion of the fabric receives sufficient illumination and has high degree of brightness. However, the light illuminating the lower portion is easily blocked, thereby a lower degree of brightness is formed. The difference in brightness results in the formation of three-dimensional pattern. To each weaving point of a float, although the length of float is the same, its distance to the center of the float varies, and the height is also slightly different. The portion at the center of the float is higher. The difference of threads position in space is the main factor to give the fabric three-dimensional sense. If there is a clear pattern of distribution at high or low places, such as an oblique distribution, a diagonal path is formed (Fig. 2-2-2). In a fabric of plain weave as shown in Fig. 2-2-3 where the length of each float is 1, all the threads bend equally and lowly. It can be imaged that the highest point on the surface of each weaving point is roughly the same, so the surface changes evenly from light

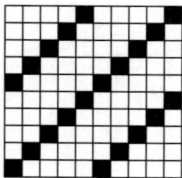

图2-2-2　斜纹组织
A twill weave

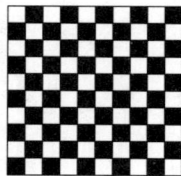

图2-2-3　平纹组织
A plain weave

图2-2-4　缎纹组织及截面图
A sateen weave and its cross-section diagrams

（3）使同浮长覆盖的纱线聚拢和不同浮长覆盖的纱线分离。经浮长线拉拢纬纱，纬浮长线拉拢经纱。纱线交织后弯曲，一般要伸长，在应力作用下有收缩的趋势，并支撑另一系统纱线。如图 2-2-1 所示，若另一系统纱线（如 J_{21}^0、J_{41}^0）不在浮长中间，则支撑力不平衡，使之推向中间新位置（如 J_{21}^1、J_{41}^1），使被浮长线覆盖并拉拢的另一系统的若干根纱线距离减小甚至紧密接触，长浮长线的中间组织点（如 J_{31}）受两侧的挤压力大致均衡，织物平面内前后或左右位置基本保持不动。由于高度可能不同，使它们距离更近（浮长线的拉拢作用）。长浮长线内每个组织点所占的空间要比整个组织循环内每个组织点平均占的空间要小。同时，由于交织作用，使某个浮长线与相邻浮长线所覆盖的另一系统的几根纱线分隔开（即分离作用）。图 2-2-1 显示了浮长的拉拢与分离作用。尽管不同的浮长之间距离增加，短浮长有向相邻的长浮长方向移动的趋势，以保持受力平衡。

（4）浮长线的聚集作用使织物的可密性增加，织物可以在厚重的同时保持柔曲性，且悬垂性变好。由于浮长内纱线没有弯曲交织，单位长度内可以容纳的纱线根数增加，即可密性增加（图 2-2-5）。由于伸直的纱线应力较小，在同一浮长下拉拢的另一系统纱线之间有可以自由活动的余量，织物挠曲性提高。平纹织物平均浮长最小，交织最频繁，可密性最差；缎纹织物平均浮长最大，交织最少，可密性

to dark everywhere, and the luster is soft. In a sateen weave (Fig. 2-2-4) where most of the floats are 4, the surface of each long float has roughly the same height, and the threads are basically straight. At the place of a small number of single weaving points, the lower thread segments are easily shaded by the surrounding high places. Under diffuse reflection, they are basically invisible. Therefore, such fabrics are smooth, thick and bright.

(3) Long float gathers the threads covered and keeps away the threads of different float. The weft threads will be pushed by the long warp floats while the warp threads will be pushed by the long weft floats. The threads in a fabric have a tendency to contract and support the threads of another series due to the stress in weaving, so that the distances among several threads of the other series that are covered and drawn by the floats are reduced to even close contact. As you may see in Fig. 2-2-1, none of the point J_{21}^0 or J_{41}^0 is at the center of the weft float, and receive unbalance forces from weft threads, thereby is pushed to new position J_{21}^1 or J_{41}^1. The middle weaving point J_{31} of the float is roughly balanced by the pushing pressure from two directions, therefore, its position is basically unchanged at the center of the float within the plane of the fabric. The other weaving points covered by the same float are drawn close together toward the center of the float. The distances can be further reduced if these weaving points are located at different heights. Obviously, each weaving point in a float occupies less space than the average weaving point in a weave repeat unit. Meanwhile, due to interlacing, there is a relatively larger distance between the two adjacent weaving points of different float, which means the threads of different long floats will be separated. The gathering and separating effects are illustrated in Fig. 2-2-1. Although the distance between the neighboring threads of different floats increases, the threads in the shorter floats have the tendency to move towards the longer floats to achieve a force balance.

(4) The gathering effect of the float increases the packing ability for the threads, weight per square unit space,

The gathering effect of the float

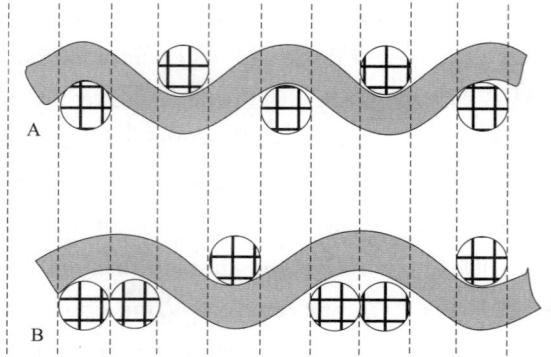

图2-2-5 交织次数与可密性关系示意图
The relationship of intersections and packing ability

最好，织物覆盖紧度可能性最大，柔曲性好，手感柔软。另外，可密性大，若密度低，则织物稀松，容易纰裂。

（5）浮长线高度差使同一系统内的相邻纱线重叠。重组织和多层组织中，下层纱线上任意组织点的浮长长度都比其相邻的上层组织点的浮长小，即下层纱线上任意点的垂直高度值比相邻上层纱线上对应点的垂直高度值要小。如果该系统的纱线排列密度大，在周边挤压下，可能导致重叠。

flexibility and drape performance. As the threads within the floating length are not bent and interlaced, more threads can be contained under the float and the number of yarns per unit space can be increased (Fig. 2-2-5). Due to the smaller stress for a straight yarn, the natural elasticity of the material will produce a reasonable amount of "give" in warp or weft directions, which provides the fabric with flexibility. Plain weave fabric has the lowest average float length, the most frequent interlacing and the worst ability of gathering yarns. Satin weave fabric has the longest average float length, the weakest interlacing, the best ability of packing threads, the potential largest covering factor. The satins are pliable, flexible and touched soft. Very close packing of threads is possible and quite heavy constructions can be achieved by the satin or sateen weave of long floats with properly set. Fabrics with insufficiently thread density, however, exhibit poor seam strength in made up articles due to seam slippage arising on account of the excessive freedom of the threads.

(5) The height difference of floats causes adjacent threads in the same series to overlap. The floating length of any weave point on the lower threads is lower than that of the adjacent upper threads in the backed weaves and multi-layer weaves, that is, the vertical height of any weave point on the lower thread is lower than that of the corresponding point on the adjacent upper thread. If the threads of this series are arranged in high density, overlapping may result

under the pressure.

2. 浮长线分布与组织层数分析 /Distribution of the Floats and the Analysis on the Fabric Layer

浮长线差异导致相邻纱线重叠，可以应用在织物组织层数的判断上。为了方便判断，对浮长线长度引入正负值，浮在上层为正值，沉在下层为负值。对于经纱层数的判断，使用经浮长值；对于纬纱层数的判断，使用纬浮长值。经组织点的经浮长为正，纬组织点的经浮长为负；经组织点的纬浮长为负，纬组织点的纬浮长为正。图2-2-6（b）所示组织图的经向浮长分布为图2-2-6（c），其纬向浮长分布为图2-2-6（d）。

判断的方法如下：

（1）与相邻组织点的浮长值有差异，意味着纱线局部比相邻纱线高度低或高。如果高度差达到一个纱线直径的大小，理论上可能产生重叠。如果同一根纱线上的组织点的重叠高度相互交替过于频繁，相邻纱线间高度差异小，无法超过一根纱线的高度；同时，交替引起纱线交错而产生较大的移动阻力，使其无法滑移至邻纱上而不能发生重

The overlapping of the floats can be applied for judging the layer of a weave that might have the possibility to overlap. For the convenience of judgment, positive and negative values are introduced for the length of the floats, which is positive when floating on the surface and negative when covered on the other side of the fabric. To judge the warp layers, the value of warp floats is used while the weft float value is used to judge the weft layers. The value of warp floats is positive for warp-up points or negative for warp-down points. Similarly, the value of weft floats is negative for warp-up points or positive for warp-down points. The distribution of the warp floats shown in Fig. 2-2-6(b) is shown at Fig. 2-2-6(c) and Fig. 2-2-6(d) shows the distribution of weft floats.

The overlapping of floats can be judged according to the following statements.

(1) The difference in value of neighboring floats means the difference in height of the threads at the local portion, which will cause overlapping theoretically if the height difference exceeds one diameter of the thread. However, too frequent alternate of the height of the neighboring threads will cause the decrease of the height difference to less than one diameter of the thread while a large resistance force due to the interlacing will stop overlapping of the neighboring

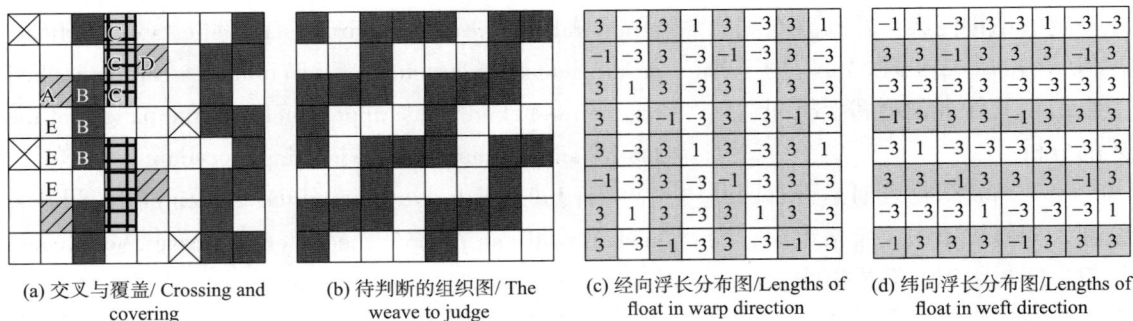

(a) 交叉与覆盖/ Crossing and covering

(b) 待判断的组织图/ The weave to judge

(c) 经向浮长分布图/Lengths of float in warp direction

3	-3	3	1	3	-3	3	1
-1	-3	3	-3	-1	-3	3	-3
3	1	3	-3	3	1	3	-3
3	-3	-1	-3	3	-3	-1	-3
3	-3	3	1	3	-3	3	1
-1	-3	3	-3	-1	-3	3	-3
3	1	3	-3	3	1	3	-3
3	-3	-1	-3	3	-3	-1	-3

(d) 纬向浮长分布图/Lengths of float in weft direction

-1	1	-3	-3	-3	1	-3	-3
3	3	-1	3	3	3	-1	3
-3	-3	-3	3	-3	-3	-3	3
-1	3	3	3	-1	3	3	3
-3	1	-3	-3	-3	3	-3	-3
3	3	-1	3	3	3	-1	3
-3	-3	-3	1	-3	-3	-3	1
-1	3	3	3	-1	3	3	3

图2-2-6 根据组织图计算浮长线并判断组织层数

Judging the layers of the weave by the lengths and the distribution of the floats

叠。因此，一个平纹组织上，所有经纱和纬纱都无法重叠。经过大量组织分析，若浮长上所有组织点的浮长长度都比相邻的纱线低，且连续值在 3 个以上，则高度重叠可能性大，才可能导致该组织分层。

（2）只要两个相邻浮长线的同类型组织点产生交叉，即浮长起始端点互不包含，就不会重叠；即只有被长浮长包含在内，才有可能局部重叠分层。以图 2-2-6（a）中经浮长为例，浮长 A 被 B 包含，浮长 C 包含 D，但是浮长 B 和 C 之间交叉。因此，经纱在 B 处有可能重叠在 A 之上，经纱在 C 处有可能在 D 之上，但是经纱在 B 和 C 处不可能局部重叠。注意，这里只有同类型浮长才比较交叉关系，不同类型的经浮长如 B（正向）和 E（负向）之间即使交叉，也有可能重叠。斜纹和缎纹织物的经向或纬向浮长都是交叉关系，经纱或者纬纱也无法重叠。

3. 浮长原理对织物结构的分析应用 /Application of the Floats Theory on the Analysis of Woven Structure

正是由于浮长线的凹凸、聚集、分离、重叠等作用，即使是原色或者单色织物，也能因为纱线的有规律分布和反射，在织物表面形成特定的结构图案。浮长线对织物几何结构的形成机理，可应用于各类组织织物外观预测和上机织造工艺的分析。

（1）判断织物表面是否有凹凸纹理，如斜纹组织、条格组织、绉组织、蜂巢组织、浮松组织、海绵组织（缎纹变化）等织物。

（2）判断织物可密性，如平纹组织、斜纹组织、缎纹组织、变化

threads. Therefore, the warp threads and weft threads can't overlap in a plain weave. In practice, the overlapping and multiple layers may occur if the difference of height of successive 3 weaving points of the neighboring yarns appears.

(2) If two neighboring floats cross each other, no overlapping may happen. Only one float is completely covered by its neighboring float, then the two yarns can overlap locally and make two layers. Take the warp floats shown in Fig. 2-2-5(a) as example, float A is covered by float B, and float C contains float D, but float B crosses float C. Therefore, float B might overlap float A for warp threads, float C might overlap float D, but float B can't overlap float C since they cross. It should be noted that the crossing or containing must be compared between the same type of floats (both warp-up or warp-down). The different type of the floats (one warp-up and another warp-down) such as float B and E may overlap even if they cross each other. For twills or satins, all the neighboring floats cross, therefore, no overlapping occurs at all.

Judging the overlapping and number

Because of the raising, aggregation, separation, overlapping and other functions of the floats, even the fabrics of grey color or monochrome, due to the regular distribution and reflection of the yarn, can form a specific structural pattern on the surface. The effects of the floats on the geometric structure of woven fabrics can be applied for the prediction of the appearance and the pattern of the fabrics and for analyzing the looming procedure.

(1) Judging the texture and the pattern of the fabrics, such as twills, stripes and checks, crepe weaves, honeycomb weaves, huckaback weaves, spong weaves.

(2) Judging packing ability for the plain weaves, twill weaves, satin or sateen weaves, hopsack weaves, etc.

(3) Judging the gathering and separating effects of

平纹组织等织物。

（3）判断织物外观的集聚性与分隔性，如凸条组织、透孔组织、方平组织、网目组织等织物。

（4）应用于织物组织重叠判断与层数判定，如凸条组织、透孔组织、重组织、各类双层组织、表里换层组织、三层及多层组织。重纬组织中相邻纬纱间交替出现长纬浮长和短纬浮长，经纱密度小，使相邻纬纱长浮长和短浮长之间垂直高度差较大；又因纬密远远大于经密，每纬占的空间小，纬纱在挤压下容易重叠。而重经组织中，相邻经纱上一根经纱经浮长长，另一根经纱经浮长短，且经密远远大于纬密，显然，经纱之间产生重叠。经浮长大的经纱在引纬纱时通常处于梭口的上层，更容易重叠在经浮长小的经纱之上。多层织物组织特点既是重经组织，也是重纬组织，且经密和纬密都大，经纱首先重叠分层，相当于每层经密降低，满足重纬组织的条件，于是纬纱也重叠，形成多层织物。由经向浮长分布图 2-2-6（c）和纬向浮长分布图 2-2-6（d）可知，奇数位置的经纱浮长高度总是高于偶数位置的经纱浮长，故奇数根经纱比偶数位置的经纱层数高；同理，奇数根纬纱也比偶数位置的纬纱层数高。故待判断的组织图 2-2-6（b）是双层组织。

该判断方法也适用于经起花、纬起花组织、表里换层（重）组织、凸条组织中芯线、管状组织中特线、双幅组织织物中的特线和缝线的判断。部分单层组织也可能形

the weaves like Bedford cord weaves, open gauze weaves, hopsack weaves, distorted effect weaves.

(4) Judging the overlapping and number of layers of the weaves such as Bedford cord weaves, open gauze weaves, backed weaves, various double-layers, treble-layer weaves. In the fabric of a backed weft weave, the long weft floats and the short weft floats alternately appear on the neighboring weft threads, thereby form a big difference in height form between the neighboring weft threads due to lower warp density. At the same time, the weft sett is much higher than the warp sett, therefore, there is only narrow space for each weft thread and pressure occurs between the neighboring weft threads. Finally, overlapping happens. In the fabric of a warp-backed weave, the long warp floats and short floats alternately appear on the neighboring warp threads, and the warp sett is much higher than the weft sett. Naturally, the warp threads will overlap based on the same principle. Another condition favorable for overlapping is that the warp of longer float is usually at the top line of the shed, which makes it easier to overlap the warp of the shorter float that is usually at the bottom line of the shed. For a double-layer fabric, the weave is actually both warp-backed and weft-backed, the setts are high in both warp and weft directions. First, the warp threads tend to overlap in shedding, and the warp density in single layer decreases, which is benefit for the overlapping in weft threads. Therefore, two layers form. For the weave shown in Fig. 2-2-6(b), the diagrams of the value of the warp floats and the weft floats are shown in Fig. 2-2-6(c) and (d) respectively. From the above figures, the values of the warp floats in odd ends are always exceed the ones in even ends, therefore, the odd ends are higher than the even ends. Similarly, the odd weft threads are always higher than the even ones. Obviously, the weave shown in Fig. 2-2-6(b) is a double-layer weave.

The judgment is also applied to the extra figured weave, interchanging double-layer weave, the wadded yarn in a Bedford cord weave, the coarse wadded yarn in a tubular weave and the stitching thread in a double-width

成局部重叠效应，更易于理解其外观形成原理，如凸条组织、透孔组织、网目组织。

（5）应用于织物局部区域松紧度的判断，如蜂巢组织、凸条组织、条格组织、浮松组织、海绵组织等织物。

weave. Even in some single-layer weave, local overlapping may happen which is help to understand the formation of the woven structure such as the Bedford cord weaves, open gauze weaves and distorted effect weaves.

(5) Judgment of the firmness in local part of the weave, such as the honeycomb weaves, Bedford cord weaves, tripe and check weaves, huckaback weaves, and sponge weaves.

四、紧密织物与紧密度 /Tight Fabrics and Tightness

1. 紧密织物 /Tight Fabric

紧密织物是指经纬纱不交错时，相邻两根纱线之间无间隙的织物。经向紧密织物中，在经纬纱不交错时，相邻两根经纱线之间无间隙；在经纬纱交错时，相邻两根经纱之间的距离是纬纱直径。而纬向紧密织物中，在经纬纱不交错时，相邻两根纬纱之间无间隙；在经纬纱交错时，相邻两根纬纱之间的距离是经纱直径。由于纱线的交织弯曲，一般不能同时得到双向均为紧密结构的织物。紧密结构织物的交织示意图如 2-2-7 所示。

Tight fabric refers to a fabric with no gap between adjacent threads when the warp and weft threads are not interlaced. For a warp-way tight fabric, there is no gap between two adjacent warp threads covered by a long weft float. Where the warp and weft threads interlace, the distance between two adjacent warp threads is the diameter of the weft. For a weft-way tight fabric, there is no gap between two adjacent weft threads that are covered by a long warp float. However, the distance will be the warp diameter where the warp and weft threads interlace. Due to the interweaving and bending of the threads, it is generally impossible to obtain a fabric with tight structure in two directions at the same time. The geometric structure of a tight fabric is shown in Fig. 2-2-7.

2. 不同组织的紧密织物紧度计算 /Calculation of the Covering Tightness of Tight Fabrics

紧密织物的紧度就是理论最大紧度，为了简单起见，其计算仍然按照 Peirce 几何结构理论计算。L_j

The covering tightness of the tight fabric is actually the maximum theoretic weavability limit. For simplicity, the calculation of covering tightness is based on Perice's

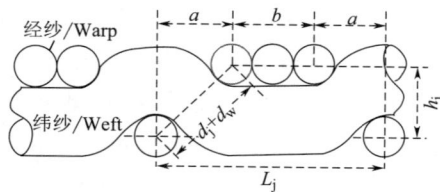

图2-2-7　紧密结构织物的交织示意图
Geometric structure of a tight fabric

是一个循环所占的纱线空间尺寸；a 是有交错的同系统相邻纱线之间的距离；b 是浮长内的同系统相邻纱线之间的距离。根据图 2-2-7，有以下计算式成立：

geometric structure. L_j is the space of a weave repeat unit, a is the distance of the neighboring threads covered by the long float of anther series, and b is the distance of the neighboring threads where there is an interlacing. According to Fig. 2-2-7, the following formula holds:

$$L_j=(R_j-t_w)d_j+t_w a \qquad (2-2-1)$$

$$a=\sqrt{(d_j+d_w)^2-h_j^2} \qquad (2-2-2)$$

故紧度计算式为：

Thus, E_j and E_w are calculated by the following formulas:

$$E_j=\frac{R_j d_j}{L_j}\times100\%=\frac{R_j d_j}{t_w\sqrt{(d_j+d_w)^2-h_j^2}+(R_j-t_w)d_j}\times100\% \qquad (2-2-3)$$

$$E_w=\frac{R_w d_w}{L_w}\times100\%=\frac{R_w d_w}{t_j\sqrt{(d_j+d_w)^2-h_w^2}+(R_w-t_j)d_w}\times100\% \qquad (2-2-4)$$

式中：t_w 是一个经纱组织循环根数（R_j）内，纬纱与经纱的交错次数；t_j 是一个纬纱组织循环根数（R_w）内，经纱与纬纱的交错次数。

故，当经纬纱等直径，同支持面紧密织物的最大紧度为：

Here, t_w is the number of intersections in a weave repeat unit along weft direction; t_j is the number of intersections in a weave repeat unit along warp direction.

Therefore, the maximum covering tightness of a balanced weave of similar warps and wefts is calculated by the following formulas:

$$d_j=d_w=d,\ h_j=h_w=d$$

$$E_j=\frac{R_j d_j}{L_j}\times100\%=\frac{R_j d}{t_w\sqrt{4d^2-d^2}+(R_j-t_w)d}\times100\%=\frac{R_j}{t_w\sqrt{3}+(R_j-t_w)} \qquad (2-2-5)$$

$$E_w=\frac{R_w d_w}{L_w}\times100\%=\frac{R_w d}{t_w\sqrt{4d^2-d^2}+(R_w-t_j)d}\times100\%=\frac{R_w}{t_w\sqrt{3}+(R_w-t_j)} \qquad (2-2-6)$$

若经纬纱直径不等，任意结构相的紧密织物的最大紧度为：

If the warps and wefts are of unequal sized, the maximum covering tightness of a tight fabric is calculated as:

$$h_j=\frac{N-1}{8}(d_j+d_w) \qquad (2-2-7)$$

$$h_w=\frac{9-N}{8}(d_j+d_w) \qquad (2-2-8)$$

$$E_j=\frac{R_j d_j}{L_j}\times100\%=\frac{R_j d_j}{t_w\sqrt{(d_j+d_w)^2-\left[\frac{N-1}{8}(d_j+d_w)\right]^2}+(R_j-t_w)d_j}\times100\%$$

$$E_j = \frac{R_j}{\frac{1}{8}\left(1 + \frac{d_w}{d_j}\right)t_w\sqrt{(9-N)(7+N)} + (R_j - t_w)} \times 100\% \qquad (2-2-9)$$

同理 Similarly： Similarly：

$$E_w = \frac{R_w}{\frac{1}{8}\left(1 + \frac{d_j}{d_w}\right)t_j\sqrt{(17-N)(N-1)} + (R_w - t_j)} \times 100\% \qquad (2-2-10)$$

最大紧度虽然是理论织造上的极限紧度，但因为纱线的压扁变形、同系统纱线可能不在同一平面等原因，实际织造的最大紧度可以超过理论计算值。

Although the covering tightness of the tight fabric is the theoretic weavability limit, the actual covering tightness in a fabric may exceed the calculation due to flatting of the threads or that the threads covered in a long float are not at a level.

3. 紧密率 /Tightness

织物紧度用来对比同组织、不同线密度、不同织物密度的织物的紧密程度，但是不适用于比较不同组织的紧密程度，不同组织用紧密率来比较。紧密率是织物的实际紧度与相同组织、相同结构相的紧密织物结构的紧度的比值。紧密率就是织物的实际紧度与理论最大紧度的比值，用下式表示：

The covering tightness is used to compare the compactness of the fabrics of same weave while tightness is used to compare the firmness or compactness of the fabrics with different weaves. Tightness is defined as the ratio of the actual covering tightness and the maximum theoretic covering tightness of a weave, and it is calculated by the following formula:

$$K = \frac{E_{\text{实}}}{E_{\text{紧}}} \times 100\% \left(K = \frac{E_{\text{Actual}}}{E_{\text{Theoretic}}} \times 100\% \right) \qquad (2-2-11)$$

4. 不同织物的紧密率比较 /Comparison of Tightness Between Different Weaves

例：$\frac{2}{2}$ 斜纹织物和平纹织物，经纬纱细度相同，均为第5结构相织物。斜纹织物的实际紧度为60%，平纹织物的实际紧度为50%。试比较这两种织物的紧密程度。

解：（1）因为经纬纱细度相同，同支持面的紧密织物的紧度为：

Example: There are two fabrics of $\frac{2}{2}$ twill weave and plain weave respectively. Both are made of same yarns at 5th geometric phase. The actual covering tightness is 60% for twill fabric and 50% for plain fabric. Try to compare the tightness of the two fabrics.

(1) Calculating the covering tightness of the even-supported tight fabric for twill and plain respectively:

$$E_j = \frac{R_j}{t_w\sqrt{3} + (R_j - t_w)} \times 100\%$$

$$E_{\text{平纹紧 j}}=E_{\text{平纹紧 w}}=\frac{R}{R+(\sqrt{3}-1)T}\times100\%=\frac{2}{2+0.732\times2}\times100\%=57.8\%$$

$$E_{\text{斜纹紧 j}}=E_{\text{斜纹紧 w}}=\frac{R}{R+(\sqrt{3}-1)T}\times100\%=\frac{4}{4+0.732\times2}\times100\%=73.3\%$$

（2）紧密率计算：

(2) Calculating tightness:

$$K_{\text{平纹}}=\frac{E_{\text{实}}}{E_{\text{紧}}}\times100\%=\frac{50}{57.8}\times100\%=86.5\%$$

$$K_{\text{斜纹}}=\frac{E_{\text{实}}}{E_{\text{紧}}}\times100\%=\frac{60}{73.3}\times100\%=81.9\%$$

由此，可以看出，平纹织物紧密程度较高。

Thus, the conclusion is made that the plain weave is tighter.

五、紧密织物的结构相 /Geometric Phase of Tight Fabrics

把平纹组织、三页斜纹组织、四页斜纹组织、五枚缎纹组织的组织循环根数与交错次数带入式（2-2-9）和式（2-2-10）中，得到从第1结构相到第9结构相下，不同组织的紧密织物的紧度数值，分别绘制紧密织物结构相与双向紧度的对应关系（图2-2-8）、紧密织物结构相与经向紧度的对应关系（图2-2-9），紧密织物结构相与纬向紧度的对应关系（图2-2-10）。

从图中可以发现，平纹织物在等支持面的第5相，到达紧密结构所需的紧度最小；但是在特高或者特低的结构相，要达到紧密结构，所需的紧度最大。在中间相位附近，相位的增加（或减少），紧密织物的经向（纬向）紧度增加值较少；但在极端相位附近，紧度变化值很大才能导致相位变化，该现象称为相位滞后性。极端相位结构会增加原料的消耗和增大织造困难，甚至使织物手感僵硬。

According to formula (2-2-9) and formula (2-2-9), E_j and E_w of the tight fabrics from geometric phase 1 to 9 of the plain weave, 3-shaft twill weave, 4-shaft twill weave, 5-shaft satin weave are calculated. The calculated data are transformed into Fig. 2-2-8 showing the relationship between the geometric phase and the dual-direction covering tightness in both direction, Fig. 2-2-9 showing the relationship between the geometric phase and E_j, and Fig. 2-2-10 showing the relationship between the geometric phase and E_w.

From the three figures above, it can be found that plain fabric at geometric phase No. 5 has the minimum tightness required to reach tight structure. But in the very high and very low geometric phases, the maximum covering tightness is required to achieve a tight structure. Near the intermediate phase, the increase of the warp (weft) covering tightness of the compact fabric is less with the increase (or decrease) of the phase. But in the vicinity of extreme phase, a big change of covering factor is required to lead to the change of phase, which is called phase hysteresis. Extreme geometric phase increases raw material consumption and weaving difficulty, and even makes the fabric feel stiff.

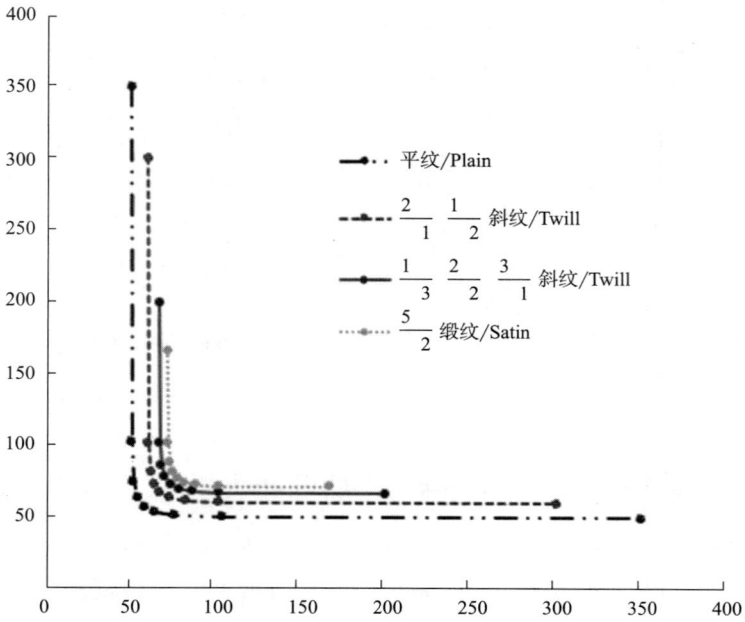

图2-2-8　紧密织物结构相与双向紧度的对应关系
The relationship between the geometric phase and the dual−direction covering tightness

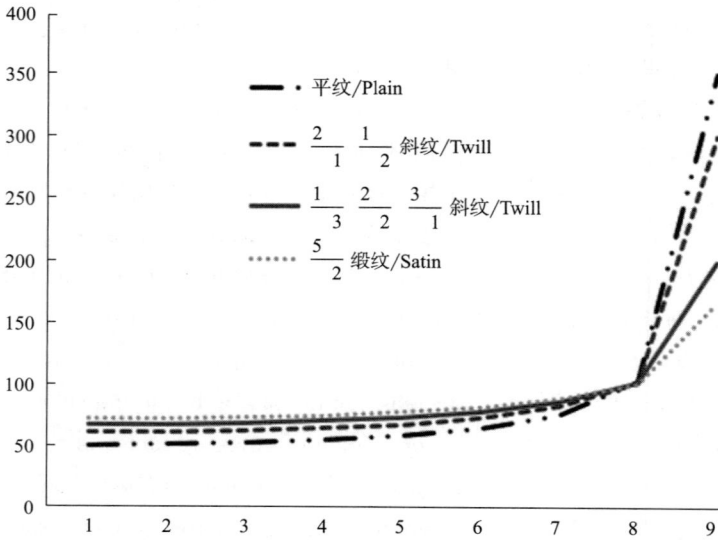

图2-2-9　紧密织物结构相与经向紧度的对应关系
The relationship between the geometric phase and E_j

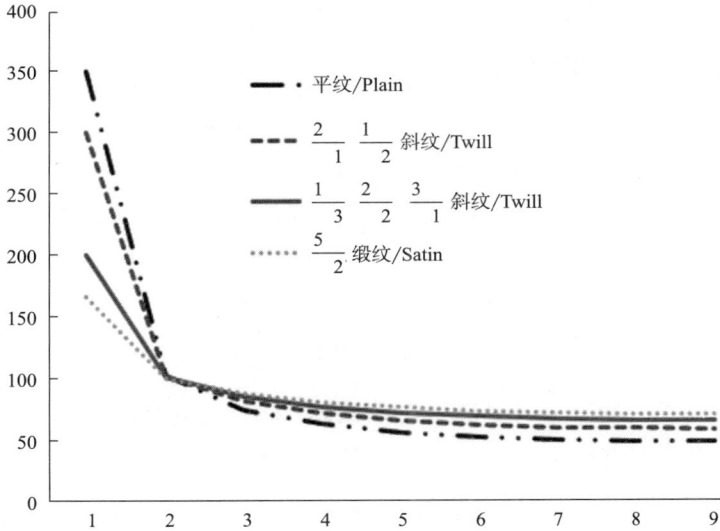

图2-2-10　紧密织物结构相与纬向紧度的对应关系
The relationship between the geometric phase and E_w

六、紧密织物的织缩率 /Contraction Rate of Tight Fabrics

为了研究方便，常研究 $d_j=d_w=d$ 时紧密结构织物各结构相的经、纬缩率。根据平纹、三页斜纹、四页斜纹、五枚缎纹、八枚缎纹织物在不同结构相下紧密织物经向缩率 a_j 和纬向缩率 a_w 实际数据，绘制图 2-2-11，发现有以下特点。

（1）平纹组织适合制织中结构相织物。平均浮长短的组织，由于在高或低结构相时，总有一个纱线系统的织缩率很大，因此，不适合在高或低结构相织造，适合制织第 5 结构相附近的织物。

（2）缎纹组织适合制织高、低结构相织物。缎纹组织在高或低结构相时，纬纱的织缩率并不很大，所以可以制织高或低结构相的织物。

（3）同一结构相不同组织织物的织缩率变化特征。在同一结构相

For simplicity of the study, the following experiments for tight fabric structure are obtained on the basis of $d_j=d_w=d$. Various tight fabrics are made under plain weave, 3-shaft twill weave, 4-shaft twill weave, 5-shaft satin weave and 8-shaft sateen weave at different geometrical phases, and the contraction rates in both warp direction and weft direction are measured. The relationship between the contract rates a_j, a_w and the different structures are drawn in Fig. 2-2-11. Based on the diagram, the following conclusions are made.

(1) Plain weave is suitable for the structure of middle phase. Due to its short average float length, there is always one series of threads that contracts much in either high or low structure phase. Therefore, the weave is not suitable for weaving in high or low structural phase, so it is suitable for weaving the fabrics near geometric phase No. 5.

(2) Satin/sateen weave is suitable for the fabrics with high or low geometric phases. The contraction rate of the satin weave is not very large in the case of high or low

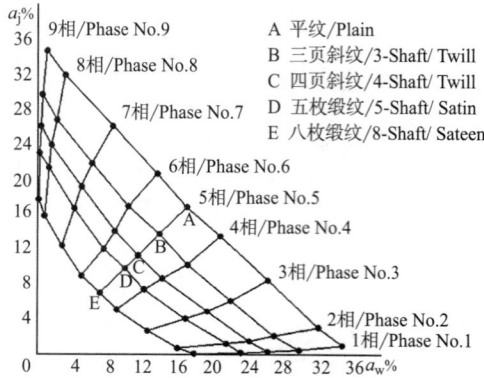

图 2-2-11　不同组织不同结构相下紧密织物的缩率

The contraction rate of the tight fabrics of various weaves at different geometric phases

中，织缩率随织物组织平均浮长的增加而减小。

（4）织缩率的变化。各种组织的织缩率是在有限的范围内连续变化的，平均浮长短的组织，其织缩率的变化范围大；平均浮长长的组织，其织缩率的变化范围小。

（5）在纵条纹织物设计中的应用。两种或两种以上的组织并列制织纵条纹织物时，不同的组织必须选择不同的结构相，使其织缩率相等，织物表面才会平整，才能消除由于织缩率的差异而引起的起绉、纵条交界不清和歪斜、松经停车以及经纱张力不匀造成的织造开口不清和跳纱织疵等现象。

geometric phases, so fabrics of high or low geometric phases can be woven.

(3) For the same geometric phase, the contraction rate decreases with the increase of the average float length of the fabrics.

(4) The contraction rates of various weaves change continuously within a certain range. For the weaves of shorter average float length, the contraction rates change in a wider range; however, for the weaves of longer average float length, the rates variate in a narrower range.

(5) In fabrics of stripes by arranging two or more weaves in parallel, different geometrical phases are designed for these weaves to ensure the equal contraction rates in warp direction, so that the surface is flat and level. The weaving faults such as wrinkle surface, unclear stripe border, slant threads, stoppage and unclear shed by loose warps, and scum threads due to the differences of weaving contraction rates can be relieved.

几何结构对织物外观性能的影响

Influence of Woven Geometric Structure on the Properties of Fabrics

第三节　几何结构对织物外观性能的影响 /Influence of Woven Geometric Structure on the Properties of Fabrics

织物几何结构会影响织物全部

The geometry of fabric affects all its physical properties,

的物理性能，包括外观、力学、电磁、传热导湿、透气、过滤、声音吸收性能等。

including appearance, mechanics, electromagnetism, heat and moisture conduction, air permeability, filtration, and sound absorption, etc.

一、外观效应 /Appearance

织物外观效应实际就是对光线的反射效应。颜色和织物组织对织物外观的影响最大，在前面章节已经略有讲述，后面章节还要一一详述。即使织物组织、纱线原料和线密度相同，不同的织物密度也会导致几何结构不同，从而对织物外观产生很大影响。

细平布与府绸织物均为平纹组织，采用棉纱细度大致相同，细平布的经密、纬密大致相同，府绸的经密约是纬密的 2 倍。府绸经纱屈曲波大，受到挤压较多，因此，织物有均匀的横向纹路和较明显的颗粒效应。

哔叽与华达呢均采用 $\frac{2}{2}$ 右斜纹组织，经纬纱粗细相同。哔叽的经向密度比纬向密度略大，经纬向紧度小；华达呢织物中，经密∶纬密约为 3∶2 ~ 2∶1。故哔叽的斜纹线倾角为 45° ~ 52°，纹路清晰、宽且平，质地松软。华达呢的斜纹线倾角为 63° ~ 70°，手感厚实、较挺括，斜纹线凸起，峰谷明显，耐磨但不折裂。

The appearance of the fabric is actually the effect of light reflection. Color and weave have the greatest influence on the appearance of the fabric, which has been described in the previous sections and will be detailed in the following chapters. Even if the weave, yarn material and linear density are the same, different fabric density will lead to different geometric structure, which will have a great impact on the appearance of the fabric.

Broadcloth and poplin fabric are both made of plain weave, using cotton yarn of similar density. The broadcloth is of a balanced structure however, the warp density of poplin fabric is about twice the weft density. The ends in poplin fabric bend more and get more extrusion, therefore, the fabric has uniform transverse ribs and obvious particle effect.

Serge and cabardine fabrics are both made of $\frac{2}{2}$ right twill weave with the same warp and weft density. The serge fabrtic has a slightly higher warp density than weft and a lower tightness in warp and weft direction. In a gabardine fabric, the ratio of the warp density and weft density is about 3∶2 to 2∶1. So the angle of the twill lines in serge fabrics ranges between 45° ~ 52°, and the diagonal lines are clear, wide and flat. The fabric touches soft. However, the twill angle of gabardines ranges from 63° to 70°. The fabric is crisper, wear-resistant but not rigid, thick and solid. The diagonal lines of the gaberdines are prominent, and the furrows contrast strongly with the ridges.

二、力学性能 /Mechanical Properties

1. 拉伸强力与强度 /Tensile Strength and Tenacity

织物经向或纬向断裂强力和断裂伸长的大小，主要由经纬纱的强

The tensile strength and breaking elongation of the fabrics in warp and weft directions are mainly determined

力和经纬纱的密度决定。若织物的某一方向紧度大，织物的紧密程度增加，该方向纱线的屈曲程度大，将使织物中该方向的纱线强力利用系数和断裂伸长有所增加。

织物密度变化，织物的强力也随之变化。若纬向密度增加，则织物纬向断裂强力增加；但经纱在单位长度内的交织次数增加，导致反复拉伸与摩擦次数增加，因此经纱疲劳加剧，故经向强力与强度均下降。若经向密度增加，则织物经向断裂强力增加，而纬纱由于与经纱交错次数增多，经纬纱之间的摩擦阻力增加，故纬向强力也有增加的趋势。

在其他条件相同的情况下，织物在一定长度内纱线的交错次数越多，浮线长度越短，则强力与伸长越大。故平纹组织织物的强度与伸长大于斜纹组织织物，而斜纹组织织物的强度与伸长又大于缎纹组织织物。

若经纬向单位长度内交织次数相差明显，则对织物不同方向的强力利用系数影响十分显著，如 $\frac{1}{3}$ 纬重平组织织物的纬向强度利用系数只有 60% 左右。

2. 撕裂强力 /Tearing Strength

织物经（纬）纱撕裂强力的大小主要取决于经（纬）纱的强力、断裂伸长和撕裂时受力区域的大小。在织物不过分紧密的条件下，凡是能够增加织物强力和断裂

by the strengths of threads and fabric setts. If the covering tightness of a certain direction of the fabric is high, the tightness of the fabric will increase, and the crimp height of the threads in this direction will increase, thereby the strength utilization coefficient and breaking elongation of the fabric in this direction are enhanced as well.

The tensile strength of the fabric changes as its density changes. If the weft density increases, the tensile strength increases in weft direction. However, more intersections of warp threads per unit length lead to the increase of repeated tensile and friction times, so the warp fatigue is aggravated, thereby the tensile strength and the tenacity in warp direction are both decreased. If the warp density increases, the tensile strength in warp direction increases, and the tensile strength in weft direction also tends to increase due to the increasing number of intersections with warp threads and the increasing friction resistance between warp and weft threads.

Under the same other conditions, the more times the threads are interlaced within a certain length of the fabric, the shorter the length of floats, and the greater the tensile strength and elongation. Therefore, the tensile strength and elongation of plain weave fabric are greater than that of twill fabrics, and that of twill weave fabrics are greater than that of satin weave.

If the intersections per unit length in warp and weft direction are significantly different, the tensile strength utilization coefficient of fabric in different directions is very significant. For example, the tensile strength utilization coefficient in weft direction of $\frac{1}{3}$ weft rib weave fabric is only about 60%.

The tearing strength of warp (weft) threads depends on the tensile strength of warp (weft) threads, the breaking elongation of warp (weft) threads, and the size of the loading area during tearing. Under the condition that the fabric is not too tight, all the factors that can increase the fabric tensile strength and

伸长的因素，都能提高织物的撕裂强力。高结构相的织物，经纱的缩率大。当撕裂经纱时，屈曲的经纱在断裂以前长度伸展，将扩大织物承受外力的区域，从而提高经向撕裂强力。紧密织物中织缩小，或织物中纱线断裂伸长小，当织物内纱线撕裂时，因受力区域狭小而产生应力集中，导致该系统纱线的撕裂强力下降。在织物撕裂时，纱线移动程度越大，承受力的纱线根数越多，撕裂强力越大。故府绸的经向撕裂强力远远高于纬向。

组织不同，纱线交错次数不同，纱线能够作相对移动的程度也不同。当其他条件相同时，平纹组织织物的撕裂强力小，而浮长线长、相对松散的方平组织织物的撕裂强力最大，密度较大的斜纹组织织物与缎纹组织织物的撕裂强力介于两者之间。

3. 耐磨性能 /Resistance to Wearing

织物表面受到的磨损，取决于哪种纱线构成织物的支持面。由经纱构成支持面的织物，经纱浮点的波峰被削平，经纱首先会受到严重磨损，但纬纱基本没有损伤。故磨损破洞处经纱断裂，形状一般为横向裂口。

经面织物若经纱密度大，则织物在摩擦面上的接触点多，磨损程度小；而纬面织物若纬纱密度大，磨损程度也会降低。织物密度大，使单位长度内交织次数增加，纤维不容易在磨损过程中抽拔出来，增加了织物的耐磨性能。但过大的织物密度会导致织物刚硬，纤维相对移动困难，容易造成应力集中而损

breaking elongation will increase the tearing strength. Fabrics with high geometrical phase have high rate of contraction in warp direction. When the warp threads are torn, the length of the warp is extended before breaking, enlarging the area of the fabric to which the force is applied, thereby increasing the longitudinal tearing strength. In a tight fabric which has a low contraction rate, or the fabric with a low breaking elongation, if the threads in the fabric are torn, the stress is concentrated due to the narrow stress area, which leads to the decrease of the tearing strength of the threads in this direction. When the threads are torn, the more moving the threads are, the more threads will bear the force and the higher tearing strength the fabric has. Therefore, the tearing strength of poplin fabric in warp direction is much higher than the weft direction.

The degree of relative movements of the threads in the fabrics of different weaves is different due to different intersections. When other conditions are similar, the tearing strength of plain weave fabric is the smallest while the tearing strength of hopsack weave fabric with long float thread and relatively loose is the largest, and the tearing strength of twill or satin fabric with higher sett is between them.

The wear on the surface of the fabric depends on which thread series forms the support surface of the fabric. For a fabric of warp supporting surface, the wave crest of the warp floating point is cut flat, the warp is first severely worn out, but the weft is basically undamaged. Therefore, the warp threads are broken and transverse crack appears on the surface of the fabric.

If the warp-sided fabric with high warp density, the fabric has more contact points on the friction surface and is unlikely wear out. Similarly, the weft-sided fabric with high weft density is less likely wear out. The high density of the fabric increases intersections per unit length, and the fiber is not easy to be pulled out in the process of wearing, which increases the wear resistance. However, the excessive fabric density leads to the rigidity of the fabric, the fiber is relatively difficult to move, and it is easy to cause stress

坏。故在经纬密度小的疏松织物中，平纹组织织物交织点多，耐磨性最好。但在紧密结构织物中，斜纹组织织物内的纤维附着相当牢固，由于交织少，纱线柔顺性好，故耐磨性能要优于平纹组织织物；缎纹组织织物的交错少，耐磨性最好。在斜纹组织内，$\dfrac{3}{1}$ 斜纹的浮长大于 $\dfrac{2}{2}$ 斜纹的浮长，支持面大，耐磨性能更好。中等紧密程度的绉组织织物耐磨性较好。总之，过松与过紧的结构均不利于织物的耐磨性能。表 2-1-4 所示为织物结构与织物耐用性能的关系。

concentration and damage. Therefore, in the open fabric with low warp and weft density, plain weave fabric has many intersections and has the best wear resistance. But for the tight fabrics, the fibers in twill fabric are fairly fixed and the threads are pliable due to less intersections, thereby the twill fabrics are superior to the plain fabrics as regard to the wearing resistance. Satin fabrics have the least intersections and more pliable and show the best wear resistance. $\dfrac{3}{1}$ twill has the longer float than $\dfrac{2}{2}$ twill, so it has large supporting surface and better wear resistance. Crepe weave fabrics with medium covering tightness wear better. In short, too loose and too tight structure are not conducive to the wear resistance of the fabrics. Table 2-1-4 shows the relationship between fabric construction and fabric durability.

表 2-1-4　织物结构与织物耐用性能的关系
The relationship between fabric construction and fabric durability

织物结构紧密程度/Fabric tightness	拉伸强力/Tensile strength	撕裂强力/Tearing strength	耐磨性能/Wearing resistance
较松/Loose	低/Low	高/High	低/Low
适中/Medium	中/Medium	中/Medium	高/High
较紧/Tight	高/High	低/Low	中/Medium

4. 起毛起球与钩丝性能 /Pilling and Snagging

织物紧度大，纤维之间的摩擦阻力大，与外界摩擦时不易产生毛绒；即使产生毛绒也不容易滑移到织物表面，减轻了起毛起球现象。

交织次数多，织物结构紧密，对纤维束缚较紧，降低了纤维头端滑移到织物表面的机会，不容易起毛起球。反之，浮长长，交织少，织物结构松散，纤维头端在摩擦后容易滑移到织物表面而起毛起球。故平纹组织织物抗起毛起球效果好，缎纹组织、蜂巢组织织物抗起毛起球效果差。

In a tight fabric, there are high friction resistance among the fibers, and it is unlikely to produce hairiness when chafing with other objects. And it is not easy to slip the hairiness to the fabric surface, therefore, the phenomenon of pilling is reduced.

More intersections cause a firmer structure and bind the fibers tightly, which decreases the possibility of the end of the fibers to slip on the surface of the fabric to form pills. On the contrary, if the float is long, the intersection is less, the fabric structure is loose, and the end of the fiber has the tendency to slip to the fabric surface after friction. Therefore, plain fabric has a good anti-pilling effect, while satin and honeycomb fabrics have a poor anti-pilling effect.

浮长线短，不容易被钩丝。浮长线长，容易被钩丝而影响外观。

It is not easy for the short float to get snag. However, the fabrics with long float thread are easily affected by snagging and appearance is damaged.

5. 弯曲性能与手感 /Bending Properties and Handle

（1）弯曲刚度和手感。织物的弯曲刚度与厚度成正比。织物中纱线的屈曲率越大，织物的弯曲刚度越小。织物组织中纱线浮长增加，经纬交错次数减小，弯曲刚度降低，手感柔软。同一种组织，在经向（纬向）紧度一定的情况下，随着纬向（经向）织物密度与紧度增大，单位长度内纱线排列根数增加，纬向（经向）弯曲刚度线性增加。

（2）抗皱性能。纱线弯曲后，在外力消除后，不可回复的塑性变形就是织物的折皱。浮线长，交错次数少，纱线间滑动余地大，抗皱性能好。三种基本组织对折皱回复角的影响顺序为：缎纹组织＞斜纹组织＞平纹组织，即缎纹组织织物抗皱性能最好，平纹组织织物抗皱性能最差。当经向紧度一定时，增加纬向紧度，折皱回复角减小。

(1) Bending Stiffness and Handle. The bending stiffness of a fabric is proportional to its thickness. The higher the thread deflects in the fabric, the lower the bending stiffness of the fabric. If the average length of floats in the weave increases, the intersections decrease, the bending stiffness decreases, and the fabric feels softer. For the fabrics of same weave, the warp (weft) bending stiffness increases linearly with the increase of warp (weft) density if the number of weft (warp) threads per unit length is given.

(2) Wrinkle Resistance. If the thread is bent, the irrecoverable plastic deformation is the wrinkle of the fabric after the external force is eliminated. If there are less intersections in a long float and large slippage between threads, the fabric will have a good crease resistance. When the wrinkle recovery angles of the three basic weaves are tested, the result shows that satin > twill > plain, which means satin fabrics have the best wrinkle resistance, and plain fabrics have the worst wrinkle resistance. When the covering factor of the fabric in warp direction is given, the crease recovery angle decreases with the increasing of the covering factor in weft direction.

三、与孔隙率、厚度相关的特性 /Properties Related to Porosity and Thickness

Properties Related to Porosity and Thickness

织物中的孔隙是气体流通、气态水分扩散和液体水分传输的通道，影响织物的保暖性能、透气性能、透湿性能、抗紫外性能、过滤性能、噪声吸附和电磁性能。孔隙多、孔隙大，利于水分、空气、各种颗粒和电磁波流通。

The pores in the fabric are the channels of gas circulation, gaseous water diffusion and liquid water transmission, which affects the fabric's heat preservation, air permeability, moisture permeability and energy absorption, UV resistance, filtration performance, noise adsorption and electromagnetic properties. Porous fabric with large pores is conducive to the circulation of water, air, various particles and electromagnetic waves.

1. 热学性质 / Thermal Properties

较大的气孔、直通气孔有利于

Larger pores and straight pores are beneficial to

增加织物的透气性；较小的气孔、曲折状的气孔，利于织物储存静止空气，提高保暖性；减少织物中静止空气的含量，则降低织物的保暖性能。多层织物、起绒织物中有较复杂、曲折的气孔形态，保暖性好；而透孔织物中，有较大气孔，透气性好。

热量在织物内部的纤维、纱线界面间发生一次或多次能耗湍流，到达织物另一侧的能量减小，故纺织品可以实现热防护性能。重量相同时，双层织物的热防护性能优于单层织物，多层织物优于双层织物。

increase the permeability of the fabric; small pores and zigzagging pores are conducive to the storage of static air fabric, thereby improve the warmth of the fabric. If the content of static air in the fabric is reduced, the thermal insulation performance of the fabric will be reduced. There are complex and zigzagging pores in multi-layer fabrics and fleecy fabrics, which have good warmth preservation. Open gauze fabrics show good air permeability due to larger pores formed.

Energy loss due to turbulence occurs one or more times between fiber and yarn interface inside the fabric, and the energy decreases when reaching the other side of the fabric. Thus, textiles can be employed to achieve thermal protection. At the same weight, the thermal protection performance of double-layer fabric is better than that of single-layer fabric, and multi-layer fabric is better than double-layer fabric.

2. 导湿性能 /Guide Wet Performance

织物厚度越大，湿阻越大，两者呈线性关系。织物紧度增加，透湿性能下降。但紧度程度在较低水平时，各类织物湿阻差异不大。孔隙率低于60%时，吸湿性好的纤维织物湿阻较小，但疏水性纤维织物湿阻增幅较大。

There is a linear relationship between fabric thickness and wet resistance. With the increasing of the covering tightness of the fabric, the moisture permeability decreases. However, when the covering tightness is at a lower level, there is little difference in moisture resistance of all kinds of fabrics. When the porosity is lower than 60%, the moisture resistance of fabric with good hygroscopicity is small, but the moisture resistance of fabric with hydrophobicity increases greatly.

3. 拒水防污性能 /Water Repellency and Anti-Fouling Performance

外界水滴的直径为 20 ~ 30μm，人体湿汽直径范围为 0.0003 ~ 0.0004μm。若织物中的微孔缩小到 3 ~ 4μm，可保护织物不被雨水浸透，同时可以使人体的湿汽排出。细特低捻高密重平棉织物微孔尺寸约 10μm，吸湿后纤维膨胀，微孔大小仅 4μm 左右，可以实现拒水防污性能。

The minimum diameter of external water droplets ranges from 20 to 30μm, and the wet vapor diameter of human body is 0.0003 ~ 0.0004μm. If the micropores in the fabric are reduced to 3 ~ 4μm, the fabric can be protected from being soaked by rain and sweat and steam can be discharged from the human body. The size of the microporous in a high-sett cotton rib fabric of fine low-twist yarn is about 10μm. The size of the pores decreases to 4μm when wet due to fiber expansion, therefore, the fabric can achieve water repellency and anti-fouling properties.

4. 介电性能 /Dielectric Property

纺织纤维的介电常数比空气大，故孔隙率越高的织物介电常数越小。

The dielectric constant of textile fiber is larger than that of air, so the fabric with higher porosity has a smaller dielectric constant.

5. 抗紫外性能 /Ultra Violet Resistance (UV Resistance)

织物紧度越高，则孔隙率越小，紫外透过率越低；织物越厚，吸收紫外线越多，紫外透过率越小。

The higher the covering tightness of the fabric, the smaller the porosity and the lower the ultra violet transmittance. The thicker the fabric, the more UV light it absorbs and the lower the transmittance.

6. 声音吸收性能 /Sound Absorption Performance

普通纤维材料的声音透射系数大，不能单独作为吸声材料，主要用来吸收中音区和高音区的噪声。纤维材料的吸声率受厚度影响最大。织物密度、纱线粗细对声音的吸收能力有一定的影响，但不大。故表面凸凹、地毯、帷幕等厚重起绒织物的声音吸收效果好等。

The average fiber material has a relatively large sound transmission coefficient and can not be used as sound absorbing or sound insulation material alone. The fiber material is mainly used to absorb the noise of the treble area and alto area. The sound absorption of fiber material is most affected by thickness of the fabric. Fabric density and yarn thickness have certain influence on sound absorption ability, but not much. The rough surface fabric, bottom weight piled fabric such as carpet, curtain have good sound absorption performance.

7. 过滤性能 /Filtration Performance

机织物孔隙率较低，一般只有30%～40%，孔眼直通，对流体阻力小，适合于压力较大的液体过滤和表面过滤。平纹组织织物交织紧密，交织点位置固定，用于较细颗粒过滤，但堵塞较快；缎纹组织织物堵塞少，但捕集能力较差。起绒织物滤尘效果好，可以通过起圈高度调节孔隙大小。起圈高度高，孔隙大，过滤阻力小，但过滤效率降低。

The porosity of woven fabric is generally as low as 30% to 40%. The through holes have small resistance to fluid, so the woven fabrics are suitable for liquid filtration at large pressure drop and surface filtration. Plain fabrics interweave firmly, so the positions of interweaving points are fixed. The fabric is used for fine particle filtration, but blocking faster. There is less chance for the satin clogging, but capture ability is not good. The filtration performance of the piled fabric is good, and the sizes of the pores can be controlled by changing the height of loops. The higher the loops, the larger the pores, and the smaller the filtration resistance, however the lower the filtration efficiency.

习题 /Questions

1. 若经纱的直径为 0.14mm，

1. The diameters of the warp thread and the weft thread

纬纱的直径为 0.16mm, 则在第 6 结构相时, 织物的理论厚度为多少?

2. 某 $\frac{3}{1}$ 斜纹织物和平纹织物, 经纬纱细度相同, 均为第 5 结构相织物。斜纹织物的实际紧度为 60%, 平纹织物的实际紧度为 50%。试比较这两种织物的紧密程度。

are 0.14mm and 0.16mm respectively. What is the theoretic thickness of the woven fabric at geometrical structure phase No. 6?

2. Two woven fabrics, viz., the $\frac{3}{1}$ twill and plain, have the same thicknesses of warp and weft threads. Both the fabrics are at geometrical structure at No. 5. The actual covering tightness of the twill fabric and plain fabric are 60% and 50% respectively. Try to compare the tightness of the two fabrics.

第三章　原组织 /Elementary Weaves

第一节　原组织概述 /A Survey of Elementary Weaves

一、定义 /Definition

原组织也称基本组织，满足下列条件：①每根经纱或纬纱上只有一个经（纬）组织点，其他均为纬（经）组织点。这两个条件也意味着 $R_j=R_w$，即原组织的组织循环经纱数 = 组织循环纬纱数。②组织点飞数 S 是常数。

Elementary weave, or basic weave, is a kind of structure meeting the following requirements: ① There is only one mark or one blank on each horizontal space or vertical space in a weave design, and the remains are all blanks or marks. The two conditions indicate that $R_j=R_w$. ② The step number S is a constant.

二、原组织的特征 /Characteristics of Elementary Weaves

（1）在一个组织循环中，$R=R_j=R_w$；否则，组织点飞数 S 不可能是常数。

（2）由于在一个完整组织循环中，经纬纱仅交织一次。所以，组织循环越大，交织点间距离越大，织物手感越松软。

(1) In a basic weave repeat, $R = R_j = R_w$. Otherwise, the step number won't be a constant.

(2) Since there is just one interlacing in a full weave repeat unit, the larger the size of the weave repeat, the longer the distance between the corresponding interlacing points, and the slacker the fabric handle.

三、原组织的类型 /Types of Elementary Weaves

原组织有三种类型，也称三原组织，即平纹组织、斜纹组织和缎纹组织。

There are three types of elementary weaves, viz., plain weaves, twill weaves and satin/sateen weaves.

第二节　平纹组织 /Plain Weaves

平纹组织也称为平组织、平布组织，可以用"平整、平凡、平常、平价"表示其特征。平整指织物表面没有较大起伏；平凡指通常情况下，该织物没有明显的特点，平淡无奇，朴实无华；平常指该织物最常见；平价指该织物与其他织物相比，织造成本低。

The plain weave is also called the tabby weave or calico weave. The fabric is usually flat, lack of distinguish features, most widely used in daily life and low in cost compared with the other fabrics.

一、平纹组织表示方法 /Representation of Plain Weaves

在平纹组织图中，每根纬纱都上下交替浮于一根经纱之上而沉于另一根经纱之下，每根经纱都上下交替浮于一根纬纱之上而沉于另一根纬纱之下，如图 3-2-1 所示。平纹组织中，经纱的一半跨过纬纱，另一半沉于纬纱之下，其后又沉于另一根纬纱之下而跨过其相邻下一纬上面，其截面如图 3-2-2 所示。

In a plain weave, each yarn of weft passes alternately over and under a yarn of warp and each yarn of the warp passes alternately over and under a yarn of the weft as shown in Fig. 3-2-1. In the section of interlacing of a plain weave shown in Fig. 3-2-2, half the ends pass over one pick and the other half pass under, then the action is reversed on the next pick.

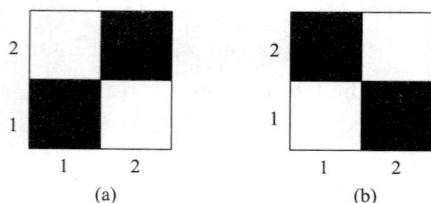

图 3-2-1　单起/双起平纹组织
2 forms of plain weaves

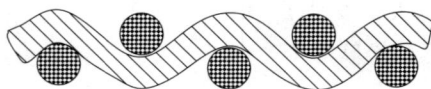

图 3-2-2　平纹组织交织截面图
Section of interlacing of a plain weave

平纹组织记作 $\frac{1}{1}$，读作"一上一下"。其中，分子是连续的经浮点（方格纸上画有符号或标记的点）长度，分母则是连续的纬浮点

Plain weaves can be represented as $\frac{1}{1}$, and read as "one up one down". The numerator is the number of successive marks in the design while the denominator is the number of successive blanks.

（方格纸上没画符号或空白的点）长度。

　　组织图上第一经与第一纬的交织点位于方格纸上组织图的左下角，称为起始点。平纹组织只有两种绘制方法。若平纹组织的起始点为经组织点，称为单起平纹［图3-2-1（a）］；否则，称为双起平纹［图3-2-1（b）］。当平纹组织在提花织物中与其他组织结合时，起点位置会影响花型轮廓的清晰度。

Usually, the starting point is the square where the first warp and the first filling cross in the lower left corner of the design. Plain weave can only be drawn in 2 forms. In Fig. 3-2-1(a), the starting point is "warp over weft" while the starting point of the plain weave is "weft over warp" in Fig. 3-2-1(b). When a plain weave is combined with other weaves in dobby spots or Jacquard fabrics, the starting point of the plain weave will affect the clarity of the pattern.

二、平纹组织的特征 /Characteristics of Plain Weaves

　　平纹组织具有如下特点：①在所有组织中，交织方式最简单；② $R_j=R_w=2$, $S_j=S_w=\pm1$；③交织最频繁，平均浮长为1，组织结构稳定；④组织正反面相同。

Plain weave has the following characteristics: ① plain weave provides the simplest interlacing among all the weaves; ② $R_j=R_w=2$, $S_j=S_w=\pm1$; ③ the weave has the most frequently interlacing with an average float length of 1, and is balanced and stable in construction; ④ the face side and the back side look the same, or, the fabric is reversible.

三、平纹织物上机 /Looming for Plain Weaves

　　对于低经密的织物，如巴厘纱、帆布可采用顺穿法穿综，使用2页综即可织造［图3-2-3（a）］；2页4列式飞穿法用于中等经密的

Straight drafting [2-shaft as shown in Fig. 3-2-3(a)] applies to the fabrics with low warp density, such as voile and canvas. For the fabrics with medium warp density, e.g., calico, skip draft in Fig. 3-2-3(b) applies. However, for

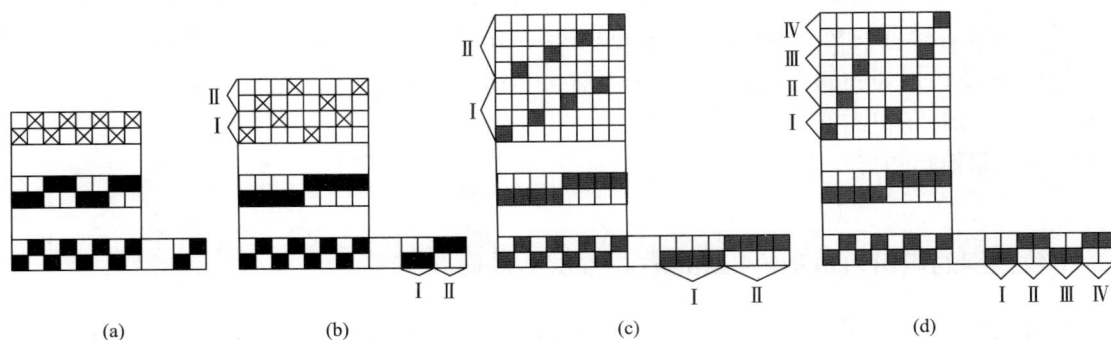

图 3-2-3　不同经向密度下的平纹织物上机图

The looming plans for the plain weaves at different warp densities

织物，如印花平布［图3-2-3(b)］；2页8列式或4页复列式飞穿法适用于高经密的府绸织物［图3-2-3(c)(d)］。穿筘方式一般为2纱/筘或4纱/筘，也有采用3纱/筘的方式（图3-2-3）。

fabrics of high warp density, such as poplin and percale, skip drafting with more rows of healds as shown in Fig. 3-2-3(c) and (d) is suitable. Usually, 2 ends or 4 ends are dented per split of reed. However, 3-end per dent is occasionally used.

四、平纹织物结构与外观性能特点 /Appearance and Properties of Plain Weaves

（1）2页综即可织造平纹织物。生产简单，织物价格低廉。

（2）织物正反面相同，表面平整，无明显特征，除非纱线颜色对比强烈、纱线直径或者张力差异大。纱线交替交织，如果织物结构平衡，即经纬纱粗细及经纬向密度接近时，经纬两个系统的纱线弯曲程度大致相等。

（3）交织最频繁，同等长度纱线的条件下，织造的织物最长。由于从另外一个纱线系统中获得更多的支撑，平纹织物具有很好的耐用性（取决于纱支和密度），手感质地坚硬、挺括。

（4）织物的紧密程度主要取决于经纬纱的交织情况，紧密的大小主要取决于两个系统的纱线交织的频繁程度。对于平纹织物，所有纱线都得到了最大程度的交织，每根纱线都能够得到相邻纱线的支撑。因此，在其他条件相同的情况下，平纹织物比其他织物更牢固。

(1) The plain fabrics can be 2-harness made; therefore, it is simple and easily produced which makes the fabrics inexpensive.

(2) The fabrics are reversible, flat unless using the threads in contrasting colors, thicknesses or tensions. The threads interlace in alternate order, and if the warp and weft threads are balanced—that is, are similar in thickness and number per unit space, the two series of threads bend about equally.

(3) The fabrics are of most frequently interlacing, maximum yardage. The fabrics get more support from another system of yarn, thus, they have good relative durability. The fabrics are hard and stiff.

(4) In large, the degree of firmness of texture in woven fabrics is determined by the manner of interweaving warp and weft, and will be greater or less according as the two series of threads interlace more frequently or less frequently, respectively. In a plain weave, all threads interlace to the uttermost extent, and each thread is supported by its neighbors, so, it is stronger than the other fabrics under the same conditions.

五、平纹组织的应用 /The Application of Plain Weaves

平纹组织采用最简单的交织方式，但它却比其他任何组织常用。从稀松轻薄到高厚高密等各类

Plain weave produces the simplest form of interlacing, yet it is used to a greater extent than any other weaves. Plain woven fabrics range in weight and compactness from thin

织物，都可以用平纹组织采用任何纤维、任何纱线织造而成。属于平纹组织的棉织物包括白棉布、印花布、细平布、马德拉斯布、平纹布、高密棉布、府绸、穆尔纱、巴里纱、色织布和帆布等；粗纺和精纺毛织物中的钢花呢、派力司、凡立丁、法兰绒等；丝织物中电力纺、雪纺绸、塔夫绸、乔其纱、绉绸和双绉等；麻织物（包括亚麻和苎麻）中的细麻布、上等细麻布。

lightweights to compact heavyweights. The weave can be constructed from any type of yarn that composed of any type of fiber. In cotton fabrics, the plain weave is used in calico, chintz, broadcloth, madras, muslin, percale, poplin, mull, voile, gingham and canvas etc.. As regard to woolen and worsted fabrics, homespun, palace, valetin and flannel etc. are constructed in plain weave. For silk fabrics, there are habutai, chiffon, taffeta, georgette, crepe de chine etc .while cambric, lawn and batiste etc .are made from flax or ramie.

六、平纹织物的特殊效应 /Special Effects of Plain Fabrics

平纹组织结构虽然简单，但变化方式繁多，可形成特殊效应。变化形式包括：① 纱线通过不同颜色、原料、粗细、捻度、捻向等，或多种方法组合；②纱线的每筘穿入数或纬密发生变化，如间隙式卷纬；③经纱可以用两种或两种以上不同张力的经轴，使经纱交替松紧；④借助异型筘对纱线的上下提升作用在织物上形成 Z 字线；⑤织造完成后的染色、印花和后整理。

Although the plain weave has simple structure, it has many kinds of variations to form special effects. Diverse methods of ornamenting and of varying the structure are employed: ① threads are different in color, material, thickness, or twist are combined; ② the number of threads per split of the reed, or of picks in a given space is varied in succeeding portions of a cloth; ③ the ends are brought from two or more warp beams which are differently tensioned, or are passed in sections over bars by which they are alternately slacked or tightened; ④ by means of a specially shaped reed which rises or falls the thread are caused to form zig-zag lines in the cloth; ⑤ after weaving, the fabrics are further processed by dyeing, printing and finishing.

1. 隐条隐格效应 /Faintly Visible Stripe Effects/Shadow Stripe

利用不同纱线捻向对光线反射不同的原理，经纬纱采用不同捻向的纱线，按一定规律相间排列，在平纹织物表面会出现若隐若现的条纹，形成隐条隐格织物，如图 3-2-4 所示。

Yarns of different twist directions reflect the light differently. By alternatively using the threads of different twist direction warp-wise, weft-wise or in both directions, the stripes or checks are faintly visible due to the reflection of the light as shown in Fig. 3-2-4.

2. 凸条效应 /Rib and Cord Effects

采用不同的经纬密、纱线细度和上机张力都会导致部分纱线弯曲程度更高，形成凸条效应。如果经

Rib and cord effects are produced by different fabric setting, thickness of yarns and weaving tensions, which result part of yarns bend more than others. If P_j (the number

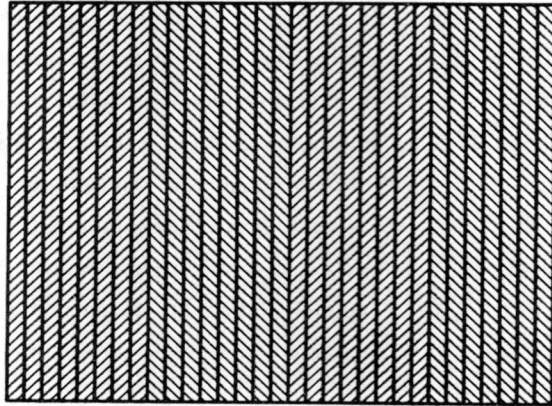

图3-2-4 隐条效应示意图
Faintly visible stripe effects

Rib effects can be
formed by

密远大于纬密，纬纱趋于平直，经纱绕纬纱弯曲，形成横凸条效应；如果纬密远大于经密，经纱趋于平直，纬纱绕经纱弯曲，形成纵凸条效应。如果伸直纱线比弯曲纱线粗，凸条效应会更加明显，而且均匀地呈现在织物的正反两面，如图3-2-5（a）所示。

若在经纱和纬纱中交替采用粗特纱和细特纱，在横凸条设计中，经密远大于纬密，粗的经纱总是跨过粗的纬纱，并且沉在细的纬纱下面；对于纵凸条来说，纬密远大于经密，粗的纬纱总是跨过粗的经纱，并且沉在细的经纱下面。凸条纹被细纱分开，仅仅在织物某一面显示凸条效应，如图3-2-5（b）所示。

经纱交替卷绕在张力不同的织轴上，也可形成凸条效应。奇数根经纱从张力大的织轴上退绕下来，经纱几乎是直的；而偶数根经纱从低张力的织轴上退绕下来，经纱受力弯曲绕着纬纱，因此布面上就形成了横向的凹凸条纹，如图3-2-5

of ends per unit space) is greater than P_w (the number of picks per unit space), picks tend to lie straight, and ends bend round them, then warp rib is formed. If P_w is greater than P_j, ends tend to lie straight, and picks bend round them, and weft rib (cord) is formed. In each case the prominence of the rib is accentuated if the straight threads are thicker than those which bend. In addition, rib lines are uniform in size appearing on both sides of the cloth as shown in Fig. 3-2-5(a).

A thick thread and a fine thread can be alternately employed in warp and weft. For warp rib, P_j is greater than P_w, the thick ends always pass over the thick picks, and under the fine picks; for weft rib where P_w is greater than P_j, the thick picks always pass over the thick ends and under the fine ends. The rib lines, which are separated from each other by fine lines, show prominently on one side of the cloth only as shown in Fig. 3-2-5(b).

Rib effects can be formed by alternately winding the warp yarn on different beams in different tensions. Odd ends are brought from one warp beam with more heavy tension, ends lie almost straight; even ends are brought from another beam with light tension, ends are compelled to bend round the fillings. Therefore, horizontal ridges and depressions are formed in the cloth as shown in Fig. 3-2-5(c). Seersucker is produced on the principle. The rib formation is quite

（c）所示，泡泡纱织物就是采用这个原理生产的。即使所有经纱粗细相同，凸条也相当明显；若低张力的经纱较粗，则凸条效应更加突出。泡泡纱织物中，织物中的经纱分为地经和泡经，呈条形相间排列。织物采用两个织轴织造，两个织轴的送经量不同，地经与泡经的张力就不同。地经送经量少，则纱线张力大，此处织物紧短；泡经送经量多，则纱线张力小，此处织物松长。泡泡纱织物常用于夏季面料和童装面料。

prominent if all the ends are equal in thickness, but it is more pronounced if the lightly tensioned ends are thicker than the others. In seersucker, a serial of heavily tensioned yarns and slacked yarns arrange alternatively. The cloth is made on a twin-beam loom which feeds the yarns at different speeds and the puckers are therefore woven in, i.e. a normally plain weave fabric is produced by having two warps, one heavily tensioned and the other comparatively slacked. The crinkled stripes are formed by the slack warp and the smooth ground by the tight warp. Seersucker is often used in summer fabrics and children's clothing.

图3-2-5 凸条效应形成原理
Rib or cord effects

3. 稀密纹织物 /Crammed Stripe Fabrics

平纹织物中，利用穿筘变化，即一部分筘齿中穿入的经纱根数多，一部分筘齿中穿入的经纱根数少，或经纱采用空筘穿法，从而改变部分经纱的密度，可获得仿纱罗效果和稀密纹织物。采用此法，可改善涤纶织物的透气性。

In plain weave fabrics, the sley is varied, that is, there are more warp threads in one part of the reed, and fewer warp threads in another part of the reed, or the warp threads are perforated by an empty reed, which changes the density of part of the warp threads, imitations of open leno effects and the thick and thin stripe fabric can be obtained. This method can be used to improve the air permeability of polyester fabric.

4. 起绉效应 /Crepe Effects

平纹起绉织物是一种在表面故意形成颗粒或者褶皱特征的织物，挺爽且具有弹性。绉效应可通过多种方式产生。常见的方式由 S 和 Z

The fabric is characterised by a crinkled surface, fairly crisp and with a springy handle. The crêpe effect may be produced in a variety of ways. The most common form of crêpe fabrics is produced by the use of hard twisted "S"

捻相间排列的强捻纱在织物后整理过程中收缩不同形成表面凹凸不平的绉效应（如双绉、乔其绉、重绉织物）。拷花绉表面的颗粒图案由热塑性织物在有花纹图案的热轧辊之间轧印定型形成，或由树脂整理形成。绉效应的持久性由所使用的纤维和轧花工艺共同决定。

还有一些绉织物是由特定的组织结构形成的，表面没有明显的组织交织循环，如可丽绉、燕麦绉、苔绒绉等，这部分内容将在第五章联合组织的绉组织一节中介绍。

and "Z" direction yarns which cause surface distortion of fabric due to differential shrinkage during finishing (e.g. Crêpe de Chine, Crêpon, Marocain, Georgette, Crêpe Suzette). Embossed crêpe is characterised with a pebbled or crinkled surface. The crêpe pattern is imparted by means of passing the fabric between heated engraved-embossing rollers, either into a softened thermoplastic fabric or in combination with resin. The permanence of the effect is governed by the fiber used and the finish accompanying the embossing process.

The crepe effects can be formed by the use of particularly constructed crêpe weave-types which break up the fabric surface into a random series of interlacings with no visible repeat (e.g. Crêpe, Oatmeal crêpe and Moss crêpe), which will be discussed in Crepe Weave of Chapter 5.

5. 烂花织物 /Burnt-out Fabrics

烂花织物由包芯纱组成，以聚酯纤维为芯纱，外面包覆棉纤维。在后整理中，按照纹样图案进行酸处理。由于聚酯纤维和棉纤维的耐酸性能不同，棉纤维被酸腐蚀水洗掉，而聚酯纤维不受破坏。因此，无棉部分织物薄、透，未被酸印过的部分保持不变。面料上形成的图案清晰，凹凸有致，风格独特，可用作服装面料或装饰面料。

The fabrics comprise of core-spun yarn, with polyester as the core yarn covered by cotton fibers. In finishing, acid treatment is implemented in the design of pattern. Because of difference of the acid resistance of polyester and cotton, the cotton fibers are corroded by the acid and polyester fibers are kept intact. Therefore, the fabric is thin and clear where the cottons are washed away while the places where no printed acid remain the same. The pattern of the fabric is clear, concave and convex, with a unique style, and can be used as apparel fabric or decorative fabric.

6. 色彩效应和花式纱 /Color Effects and Application of Fancy Yarns

当不同颜色的纬纱、经纱交织时，可以得到条格花型等色织产品。通过对坯布进行染色或印花也可以获得色彩效应。若采用花式纱线，效果更加独特。

图 3-2-6 所示为各种特殊效应的平纹织物。

When weft and warp yarns of different colors are interwoven, checks and striped products can be obtained. Color effects can also be obtained by dyeing and printing grey cloth. A special style fabric can be formed if the fancy yarns are applied.

Different fabrics of plain weave are shown in Fig. 3-2-6.

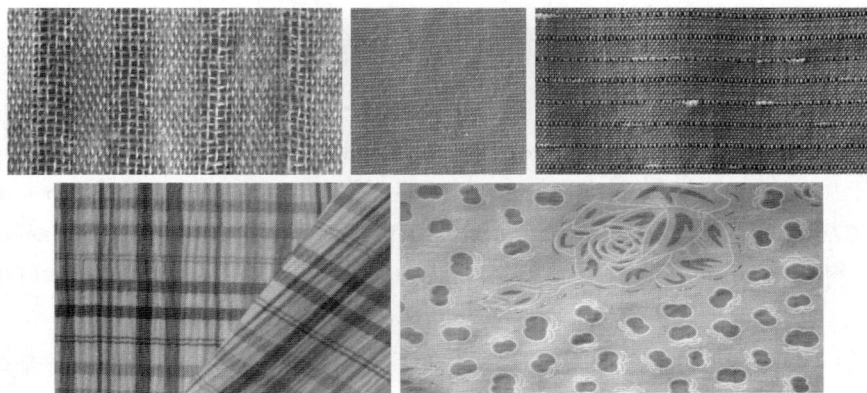

图3-2-6 特殊的平纹织物
Various fabrics of plain weave

第三节 斜纹组织 /Basic Twill Weaves

Basic Twill Weaves

当平纹组织循环扩大后，沿着单起平纹或者双起平纹的斜向，延伸经组织点，得到如图 3-3-1（a）~（d）所示的组织。这种在组织图上形成斜纹线条的组织，称为斜纹组织。若此时继续沿着经向或者纬向延伸经组织点，直到每根经纱或者纬纱方向上只有一个纬组织点，如图 3-3-1（e）~（h）所示，这些也都是经面原组织斜纹组织。需要指出的是，通过平移这些组织的起点，发现与图 3-3-1（i）~（l）是相同的组织。

By extending the marks of a plain weave diagonally in an enlarged weave repeat, weaves shown in Fig. 3-3-1(a)(b)(c)(d) are obtained. A twill line appears in these weaves, so the weaves are named. If the marks are continually extended along the warp direction or weft direction until there is only one blank in each end or pick, warp-faced basic twills are obtained as shown in Fig. 3-3-1(e)(f)(g)(h). It should be noted that the weaves shown in Fig. 3-3-1(e)(f)(g)(h)are the same as the weaves in Fig. 3-3-1(i)(j)(k)(l) respectively if the starting points are varied by transforming the first warp or weft thread of the weave.

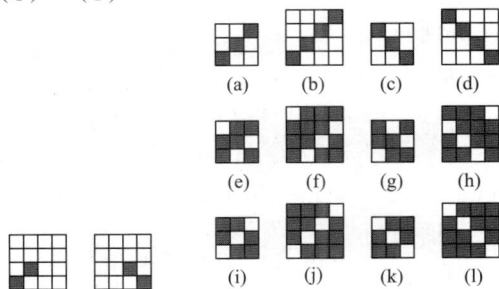

图 3-3-1 斜纹组织的形成
Formation of basic twills

一、斜纹组织斜向 /Direction of Basic Twill Weaves

在斜纹组织中，组织点以一定角度的斜向呈现。通过沿经纱方向，自下而上观察组织点的斜向来确定斜纹方向。斜线方向向右为右斜纹，斜线方向向左则为左斜纹。

There are diagonal lines constructed by the interlacing points in basic twills. The direction of a twill is generally described as a fabric is viewed looking along the warp. "Twill right" then refers to the diagonal running upwards to the right(↗), and "twill left" to the diagonal moving upwards to the left(↖).

二、斜纹组织参数与表示方法 /Parameters and Representation of Basic Twill Weaves

要构成一个原组织斜纹组织，其基本参数必须满足如下条件：

原组织斜纹的交织规律用分数表达式表示连续的经组织点个数，外加箭头表示斜纹方向。组织表示方法写作"$\frac{n}{m}$+斜纹方向箭头（↗，↖）"，读作"n上m下左（或右）斜纹"，其中n或m必有一个为1。分子是连续的经浮点个数，分母是连续的纬浮点个数。如果n>1，属于经面斜纹；如果n=1，则是纬面斜纹。组织循环根数等于分子与分母之和，即$R_j=R_w=n+m$。

对于图3-3-1（a）（b）（e）（f）中的组织来说，斜路为由左下到右上，均为右斜纹。对于图3-3-1（f），写作"$\frac{3}{1}$↗"，读作"三上一下右斜纹"。图3-3-1（c）（d）（g）（h）的组织斜路为由右下到左上，均为左斜纹。对于图3-3-1（h），表示为"$\frac{3}{1}$↖"，读作"三上一下左斜纹"。

Parameters and Representation of Basic Twill Weaves

To form a basic twill, the following parameters are required:
（1）$R_j=R_w \geq 3$;
（2）$S_j=S_w= \pm 1$.

A regular twill can be represented as a fraction like "$\frac{n}{m}$ + an arrow (for twill direction)". The fraction is read as "n up m down, left-handed (or right-handed) twill". The numerator is the number of successive marks in a vertical space while the denominator is the number of successive blanks. Either n or m is 1. If n>1, then a warp dominated twill is obtained while a weft dominated twill is obtained if n=1. The size of the weave repeat R_j or R_w is the sum of n+m.

Weaves shown in Fig. 3-3-1(a)(b)(e)(f)are all right-handed twills. In these weaves, the diagonal line points to upper right. Weave shown in Fig. 3-3-1(f) is written as "$\frac{3}{1}$↗", and read as "three up one down right-handed twill". Weaves shown in Fig. 3-3-1 (c)(d)(g)(h) are all left-handed twills. As regard to weave shown in Fig. 3-3-1(h), the diagonal line points to upper left. It is written as "$\frac{3}{1}$↖", and read as "three up one down left-handed twill".

三、斜纹织物结构特征及外观形成原理 /Appearance and Characteristics of Basic Twill Weaves

1. 外观特点 /Appearance

斜纹织物正面形成斜条纹路，如图3-3-2所示。若将斜纹织物翻面，即组织"底片翻转"，经、纬组织点根据中心对称反向变化。正面为经浮点时，反面为纬浮点；正面为纬浮点时，则反面为经浮点，如图3-3-1（a）变为图3-3-1(k)，斜纹纹路反向。一般情况下，原组织斜纹织物正面斜路效果好，但是背面斜路效果不明显，如图3-3-3所示，此时称为单面斜纹。斜路倾斜角随经纬纱密度的比例增加而增大。

Diagonal lines appear on the front side of the twill fabrics as shown in Fig. 3-3-2. If the fabric is turned over, or the weave is reversed, the weave points of the back will just reverse symmetry to the front weave. The marks on the front weave shown in Fig. 3-3-1(a) will be changed to blanks on the reversal weave shown in Fig. 3-3-1(k), and the direction of the twill line changes as well. Usually, the front side of fabrics of the basic twill shows a clear twill effect while the twill lines in the back side won't be so distinct as the front side. These fabrics are called single-side twills as shown in Fig. 3-3-3. The inclination angle of the diagonal lines on the surface of the twill fabric increases with the increase of the ratio of the warp density and weft density.

2. 形成斜路效应的关键 /Keys to the Formation of Twill Effects

织物表面的纹理效应是由于纱线在织物表面弯曲导致高低不平，反光强度不同形成的，主体组织点所在的纱线作用更重要。如果在织物表面反光强度大致相同的位置形

The texture appearance is formed by reflection the light on the bent yarns of different heights duo to interweaving. The yarn serial that dominates the fabric surface plays an important role to the appearance of the fabric. If the reflection band of the fabric forms diagonal lines, the twill

(a) 正面/Front

(b) 反面/Back

图 3-3-2　斜纹织物外观

The appearance of a twill fabric

图3-3-3　单面斜纹外观

The appearance of a single-side twill fabric

成一个斜向条带，则形成斜纹纹路。对于经面斜纹，主要看经纱反光效果；纬面斜纹则主要看纬纱反光效果。

织物表面的纱线段（浮点处）在光线照射下，其浮长高点反光部分有规律地排列成一个近似椭圆的反光带区域，其倾斜方向与纱线的纤维斜向（捻向）正交，如图3-3-4所示。浮长端点两侧是纱线弯曲部分，光照暗淡，反射光少，是阴影区。若纱线为Z捻，纤维斜向自左下至右上，呈现右斜，但是反光带的斜向为左斜；S捻纱线的反光效果则恰好相反。

effect appears. For a warp-faced twill weave, the twill effect is depended on reflection of the warp yarns while for a weft-faced twill weave, weft yarns determine the twill effect.

Under the light, the high place of the float reflects the light regularly forming a nearly elliptical reflective band. The direction of the reflective band is perpendicular to the fiber direction or twist direction of the yarn as shown in Fig. 3-3-4. The two sides of the float are the bent area receiving less light and form the dark areas. For Z twist yarn, fibers are aligned from lower left to upper right and form a right-handed twill diagonal line. However, the reflective band forms a left-handed diagonal line. The reflective band of the S twist yarn is just opposite to that of Z twist yarn.

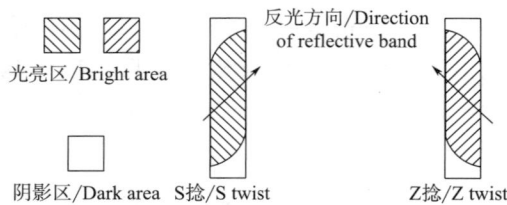

图3-3-4　不同捻向纱线的反光带方向
Direction of reflection band of yarns with different twist direction

当织物在光照下，纤维的反射光可以在浮于织物表面的纱线段上看到。纤维反射部分排列成条状，称为"反射带"（图3-3-4）。反射带的倾斜方向与纱线捻向相反，即反射光方向与纱线中纤维的排列方向垂直。当反射带的方向与织物斜向一致时，斜路清晰。

斜纹织物中，若反光带方向与斜纹方向一致、反光带连续，斜纹纹路清晰、醒目，效果好；否则，反光带被阴影区阻隔断裂，纹路不清晰，模糊不清。当经纱为S捻时，经面斜纹应为右斜纹；当经

When the fabric is exposed to light, the reflection of the fibers can be seen on each section of the yarn floating on the surface of the fabric. The reflective parts of the fibers are arranged in strips called "reflective bands". The direction of the reflective band is opposite to the twist direction of the yarn, that is, the direction of the reflective band is perpendicular to the direction of the alignment of the fibers in the yarn. The diagonal lines are clear when the direction of the reflective band is in line with the twill direction of the fabric.

If the direction of the reflection band conforms to the twill line of the weave, the reflective bands of the neighboring threads are continuous, which forms distinct and clear twill effect. Otherwise, the reflective bands of the neighboring yarns are blocked and broken, and the twill

纱为 Z 捻时，则经面斜纹应为左斜纹。对于纬面斜纹来说，当纬纱为 S 捻时，应为左斜纹；当纬纱为 Z 捻时，应为右斜纹。图 3-3-5 所示的 $\frac{2}{1}$↖ 斜纹采用 Z 捻经纱，反光带连续，斜纹纹路清晰；图 3-3-6 所示的 $\frac{2}{1}$↗ 斜纹采用 Z 捻经纱，反光带阻隔断裂，斜纹纹路模糊不清。组织点浮长线长，反光带之间容易连续，斜纹粗壮，因此 $\frac{3}{1}$ 斜纹效应要明显优于 $\frac{2}{1}$ 斜纹效应。

effect will be unclear. A warp–faced twill weave is right–handed if S twist yarns are applied to ends, or left–handed twill with Z twist ends. As regard to weft–faced twills, it is better to have S twist yarns for left–handed twills and Z twist yarns for right–handed twills. In Fig. 3–3–5, the reflective bands in a $\frac{2}{1}$↖ twill with Z twist yarns are continuously formed, which results in a clear diagonal effect. However, the reflective bands in the same weave in Fig. 3–3–6 with S twist yarns are broken by the dark area, as a result, the twill effect is damaged. Longer floats in a twill make it easy to form continuous reflective bands between the neighboring threads and make prominent twill lines, which is why a $\frac{3}{1}$ twill weave shows better twill effect than that of a $\frac{2}{1}$ twill weave.

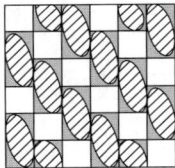

图3-3-5　Z捻左斜纹路效果
Left–handed twill with Z twist yarn

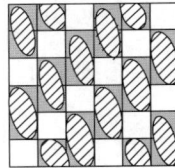

图3-3-6　Z捻右斜纹路效果
Right–handed twill with Z twist yarn

斜纹织物可密性大，增加纱线密度，有助于反光带连续。但原组织斜纹是异面组织，通常只有经纱方向密度大，纬纱方向密度小。若正面是经面组织，由经纱形成的斜纹纹路效果好；那么，反面是纬面组织，其纹路效果要看纬纱反光带的连续状态。因为纬向密度较小，反光带之间距离较大，斜纹纹路效果不明显，正反面斜纹效果差异大，形成所谓的单面斜纹效应。

一般来说，棉纱和毛纱在捻度方向上是不同的，单纱和股线也是如此。捻度的方向随着经纬方向的变化而变化，因此在设计面料时需

Twill fabrics are capable of packing threads, which is beneficial to narrow the distance between the reflective bands of the neighboring threads. A basic twill weave is an unbalanced structure, and the fabric is usually given higher setts in warp direction and lower setts in weft direction. If a warp–faced twill weave forms a good twill effect by the warp threads in face side, then, its wrong side will be a weft–faced twill, and its twill effect is determined by the status of the reflective band of the weft threads. Due to a lower weft density, the distance between the two neighboring weft threads is relatively far, which might result in unclear diagonal lines. That is why a single–side twill effect is formed.

Generally, cotton yarns and woolen yarns are different in twist direction, so do as single yarns and plied yarns.

要充分考虑。

The direction of the twist changes with the warp and weft direction, therefore, full considerations are required before designing the fabric.

3. 斜纹织物其他性能 /Other Properties of Twill Fabrics

与平纹组织相比，斜纹组织浮长线长，织物反光强，光泽好；纱线交错次数少，纱线之间可移动空间大，织物手感较柔软；可密性大，可以获得较高的纱线密度，织物厚重，悬垂性能好，强力高，坚牢。在纱线细度和织物密度相同的情况下，斜纹织物的强度低于平纹织物。

Comparing with plain weaves, twill weaves have longer float, which may better reflect the light, therefore, twill fabrics are more lustrous. There are less intersections between the weaving points, and the movement of threads is less restricted, so twill fabrics touch softer. More threads can be woven in a twill fabric, so that the fabric can be stronger, thicker, heavier, and shows better drapability. Under the condition of the same yarn thickness and fabric density, twill fabrics have less tenacity than plain weave fabrics.

Drawing of Basic Twill Weaves

四、斜纹组织图绘制 /Drawing of Basic Twill Weaves

绘制斜纹组织图的步骤如下：

（1）根据斜纹表达式 $\frac{n}{m}$，计算出组织纱线循环根数 $R_j=R_w=n+m$。

（2）确定第一根经纱和第一根纬纱在方格纸的位置，将方格纸上组织图的外框加粗。

（3）从第一根经纱最下方开始，在连续 n 个方格中填经组织点符号。

（4）根据前一纵向方格中经组织点符号和飞数，填满相邻的连续纵向方格，直至所有纵向方格都被填完。

对于右斜纹，重复上一列纵行的所有经组织点并向上平移一格，最上方的经组织点应该移到纵条格底部位置。对于左斜纹，重复上一列纵条格的所有经组织点并向下平移一格，最下方的经组织点应该移到纵行最上方位置。图 3-3-7（a）

The procedures of drawing a basic twill can be described as following:

(1) Calculate the size of weave repeat according to the twill fraction $\frac{n}{m}$. $R_j=R_w=n+m$.

(2) Fix the position of the first end and the first pick on the design paper, and draw the out frame of the weave in bold line.

(3) Fill in n marks continuously in the first vertical space.

(4) Fill the successive vertical space based on the previous vertical space and step number until all the vertical space are finished.

For a right-handed twill, copy all the marks of the previous vertical space one square ($S_j=+1$) upward. The mark above the top position will be moved to the bottom position of the vertical space. For a left-handed twill, copy all the marks of the previous vertical space one square ($S_w=-1$) downward. The mark below the bottom position will be moved to the top position of the vertical space.

Fig. 3-3-7(a) and (b) show the procedures of drawing a $\frac{1}{3}\nearrow$

图3-3-7　斜纹组织画法
The drawings of twill weaves

和（b）展示了绘制 $\frac{1}{3}$↗斜纹和

$\frac{3}{1}$↖斜纹的过程。

twill weave and a $\frac{3}{1}$↖ twill weave respectively.

五、斜纹织物上机 /Looming for Basic Twill Weaves

顺穿法适用于低密度织物，所用综页数应与组织循环数相等。一般情况下，斜纹密度较大，要采用飞穿法，如图 3-3-8 所示。根据斜纹组织的经纱循环根数，每筘齿穿入 3 ~ 4 根经纱。

经面斜纹织物可采用反织法。$\frac{3}{1}$斜纹织物采用常规织造时，织物表面容易出现缺纬、跳花、纬缩等疵点，可以及时纠偏，但是，开口装置耗电大，不易发现断经，拆坏布容易损伤经纱等。如果以 $\frac{1}{3}$ 踏盘反织，不仅省电，而且容易发现断经疵点，拆坏布方便，但不容易发现其他织疵。需要注意的是，如果使用反织法，斜纹组织必须底面翻转。例如，如果待织的是 $\frac{3}{1}$↖卡其布，那么上机图中的组织图必须改为 $\frac{1}{3}$↗，如图 3-3-8 所示。

In weaving, straight draft can be applied to the lower density twill fabric, and the number of harness shafts used equals to R (the size of the weave repeat). Skip draft as shown in Fig .3-3-8 is usually used since the twill fabrics are designed in higher warp density. 3 ~ 4 ends are placed per dent, which is depended on the size of weave repeat in warp direction.

Sometimes, face downward weaving is applied to twill weaves. When using normal weaving for a $\frac{3}{1}$ twill weave, fabric defects such as lack of weft, harness skip and weft contraction are easily found on the cloth surface, which is convenient for timely correction. However, the shedding device consumes much power, thus, it is not easy to find warp breakage, and the ends are easily damaged in raveling. If a $\frac{1}{3}$ twill cam is used, the electricity power is saved, and it is easy to find the warp breakage and dismantle the broken cloth. However, it is not easy to find other weaving defects. When the method of face downward weaving is used, the weave in looming plans must be reversed. For example, if a khaki of $\frac{3}{1}$↖ twill is woven, then the design must be changed as $\frac{1}{3}$↗ twill as shown in Fig. 3-3-8.

(a) 正常织造/Normal weaving   (b) 反织法 /Face downward weaving

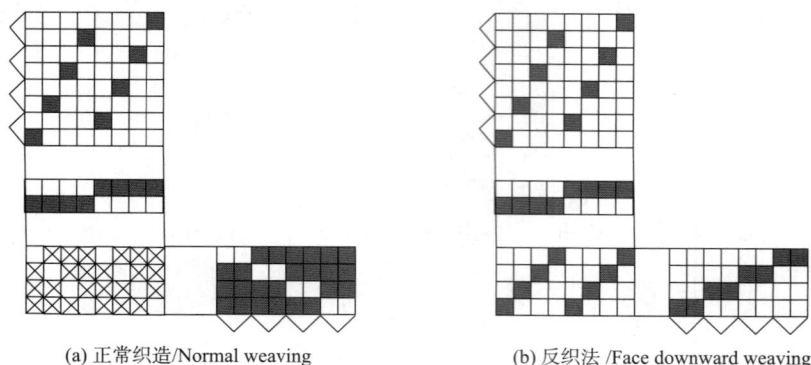

图3-3-8 斜纹组织上机图
The looming plans for basic twills

The Application of Basic Twill Weaves

六、斜纹组织的应用 /The Application of Basic Twill Weaves

原组织斜 3 页和 4 页纹织物应用广泛。例如，采用 $\frac{2}{1}$↖ 的有斜纹布、牛仔布、棉单面华达呢，采用 $\frac{3}{1}$↖ 的有纱卡、劳动布，采用 $\frac{2}{1}$↗ 的有毛单面华达呢，采用 $\frac{3}{1}$↗ 的有毛单面华达呢、里子绸、线卡等。

3-harness twill weave and 4-harness basic twill weave are widely used. For example, $\frac{2}{1}$↖ twill is applied to drill, jeans, denim and single-sided cotton gabardine; $\frac{3}{1}$↖ is used for khaki and drill; $\frac{2}{1}$↗ is used for single-sided worsted gabardine, and $\frac{3}{1}$↗ is applied to single-sided worsted gabardine, foulard and khaki of folded yarn, etc.

第四节 缎纹组织 /Satin/Sateen Weaves

缎纹组织的显著特点是浮长线长，织物正面富有光泽。经纱低捻度，浮长线至少为 4，再配以合适的纤维，织物反光强。缎纹分经面缎纹和纬面缎纹，如图 3-4-1 所示。

Satin weave is characterized by floating yarns, used to produce a high luster on one side of the fabric. Warp yarns of low twist float or pass over four or more filling yarns. Low twist and floating of warp yarns, together with fiber content, give a high degree of light reflection. There are satins and sateens as shown in Fig. 3-4-1 respectively.

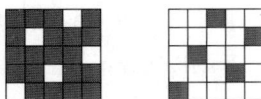

图3-4-1 经面缎纹和纬面缎纹
Satin and sateen weaves

一、缎纹组织形成条件 /Prerequisites of Satin/Sateen Weaves

缎纹组织必须满足如下条件：

（1）组织循环纱线数 $R \geqslant 5$（6除外）。

（2）$1<S<R-1$，飞数为常数。

（3）S 与 R 必须互为质数。

由于 S 是常数，所以原组织经（纬）面缎纹也被称为正则经（纬）面缎纹。经向飞数用于经面缎纹，纬向飞数用于纬面缎纹。

To form a satin or sateen weave, the following require-ments must be met:

(1) $R \geqslant 5$ (with the exception of 6).

(2) $1<S<R-1$. Step number S must be a constant number.

(3) Step number S has no common factor with R.

Since step S is a constant number, simple satin/sateen is also called regular satin /sateen. For warp effects, the step is counted in the warp direction while for weft effects, the step is counted in the weft direction.

二、缎纹组织表示方法 /Representation of Satin/Sateen Weaves

经面 / 纬面缎纹可以用"分数式 + 缎纹类型"表示，如" $\dfrac{R}{S}$ 经面（纬面）缎纹"，读作" R 枚 S 飞经（纬）面缎纹"，分子式中的分子表示组织循环数，分母表示飞数。若是经面缎纹，该参数是经向飞数 S_j；若是纬面缎纹，则该参数是纬向飞数 S_w。图 3-4-1 中的缎纹分别是 $\dfrac{5}{3}$ 经面缎纹和 $\dfrac{5}{3}$ 纬面缎纹。

Representation of
Satin/ Sateen Weaves

Satins and sateens can be represented as a fraction with the type like " $\dfrac{R}{S}$ Satin/Sateen". In the fraction, the numerator R is the size of the weave repeat, and the denominator S is the step number (or count /base for satin/sateen). S is counted as S_j in warp direction for satins or S_w in weft direction for sateens. The weaves in Fig. 3-4-2 can be represented as $\dfrac{5}{3}$ satin and $\dfrac{5}{3}$ sateen weave respectively.

三、缎纹组织的特征 /Characteristics of Satin/Sateen Weaves

缎纹组织有如下特点：

（1）组织中有长浮长线，交织次数相对少。

（2）缎纹组织是对组织循环较

Satin/sateen weave has the following characteristics:

(1) There are long floats and less intersections in satin/sateen weave compared with plain weave.

(2) Satins or sateens are actually rearranged at certain

大的原组织斜纹按照一个给定的间隔规律，重新排列纱线顺序得到的，如图 3-4-2 所示，所以组织点飞数不为 1。

（3）缎纹组织的交织点相距较远。由于单个浮长的组织点均匀分布，避免了斜纹效应的出现。

intervals from a simple larger regular twill as shown in Fig. 3-4-2. Therefore, satins or sateens differ from twills by having a step number different from 1.

(3) Intersection points are separated in satins/sateens. Due to the distribution of interlacing points, all emphasized diagonal effects are avoided.

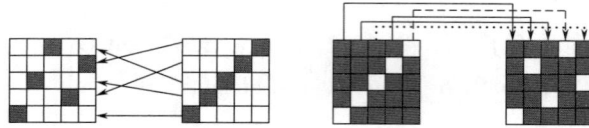

图3-4-2　重新排列斜纹组织的纱线次序得到缎纹组织

Obtain satin and sateen weaves by rearranging the threads in regular twills

四、缎纹织物外观与性能 /Appearance and Properties of Satin/Sateen Weaves

缎纹织物具有如下特性：

（1）缎纹织物布面光滑，光泽丰富，质地柔软。在一个完整的组织循环中，缎纹织物的交织次数最少。织物表面几乎全部由经纱或纬纱的浮长线组成。浮长线紧密排列在织物表面，与相互垂直的另一系统纱线很少交织。长浮长线上的反光，形成了缎纹织物最主要的特征——光泽好。

（2）单个交织点彼此分散，织物表面基本不可见。单独组织点高度较低，形成凹陷，被其相邻长浮长线上较高的组织点遮盖，因此斜纹纹路中断，模糊不清。缎纹组织的斜路并不明显，这一点与斜纹织物不同。由于交织点斜路被故意打乱，形成所需的平整、光滑、光泽好的表面。

（3）结构不紧密。交织会导致纱线弯曲，限制相邻纱线的运动。

Satin/sateen fabrics have the following properties:

(1) Satin/sateen fabrics are smooth, rich in luster and soft in texture. In a complete weave repeat, there are the least intersections. The surface of the fabric is almost entirely composed of the floats of warp or weft. The floats lie compactly on the surface with very little interruption from the yarns going at right angles to them. Reflection of light on the floats gives satin/sateen fabric its primary characteristic of luster.

(2) The individual intersection points are separated and invisible on the surface. The depressions or furrows that are formed by the single interweaving points are interrupted and hidden by their higher neighboring intersecting points of the long float, thus, the diagonals are indistinct. It differs in appearance from the twill weave because the diagonal line of the satin/sateen weave is not visible; it is purposely interrupted in order to contribute to the flat, smooth, lustrous surface desired.

(3) Satin/sateen fabrics are less firm in structure. Intersection results in the bending of the threads, blocking the movements of the neighboring threads. Adjacent

交织点靠近可使纱线从两个系统都获得支撑，织物的结构更紧密。但缎纹织物交织点之间远离，纱线从同一系统相邻纱线所获得的支撑较少，因此，它不如同等循环大小斜纹组织紧密。

（4）织物密度高，悬垂性好。因为交织少，组织结构相对松散，只能加大密度，以提高同一系统相邻纱线之间的支撑作用。如果缎纹织物密度小，织物容易纰裂，影响服用性能。为了突出经面效应和经面缎纹织物的光泽，经面缎纹组织的经纱密度比纬纱密度更高，并且经纱的捻向应该与交织点主斜向相配合。织物结构越紧密，其孔隙率就越小，织物的保暖性就更好。纬面缎纹的纬密远大于经密。

（5）缎纹织物正反面差异大。考虑到捻度效应，纱线捻向应与正面交织点的主斜向相配合。织物两面有明显差异。由于交织原因，组织上仍然有一些斜纹效果。有些经面缎纹与纱线捻向配合，在一定程度上有意显示出斜纹纹路。除非另有说明，纬面缎纹没有明显的斜纹纹路。

（6）缎纹织物耐磨性能较差。缎纹织物通常需要5～12页综框，这些浮长线会导致潜在缺陷。为了增加织物平滑度和光泽，纱线捻度低，强力相对较弱，耐磨性不良，容易起毛。浮长线越长，织物越有可能发生钩丝现象，显得粗糙，也易磨损。

intersections make the texture of the fabric firmer duo to support from both directions. The interweaving points are separated in satin/sateen fabric, so that threads in satin/sateen fabric get less support from their neighbors of the same series sateen fabric, thus, satin/sateen fabric is lesser firmer than the twills of same size.

(4) Satin or sateen fabrics have high thread density and good draping property. The fabrics are relatively loose due to less intersections, and the thread density has to be increased to improve the mutual support from the neighbor threads of the same series. If the fabric is low in thread density, slippage is liable to happen and damage the performance of the fabric. In order to enhance the appearance effect and luster of the satin fabric, the warp density of the satin is higher than the weft density, and the twist direction of the warp threads should be consistent with the texture direction of the weave points. The compactness gives the fabric more body as well as less porosity, which makes the fabric warmer. The weft density of the sateen fabric is higher than the warp density.

(5) Satin/sateen fabrics are irreversible. Considering the twist effect, the twist direction of the threads should be consistent with the texture direction of the weave points on the face side. It is natural that the two sides of the fabric have significant difference. There is still some twill effect based on the interlacing. Some satins show twill lines purposely to some extent with the combination of the direction of the yarn twist. Unless otherwise stated, sateens have no distinct twill effect.

(6) Satin/sateen fabrics are poor in wearing resistance. From five to as many as twelve harnesses are used to produce the satin/sateen fabric. The longer floats may result in a potential weakness in the fabric. Furthermore, to increase the smoothness and luster of the fabric, the threads are given a minimum of twist and are therefore relatively weak. The longer the floats, the greater the chance the fabric will snag, roughen, and show signs of wear.

五、缎纹组织图绘制 /Drawing of Satins/Sateen Weaves

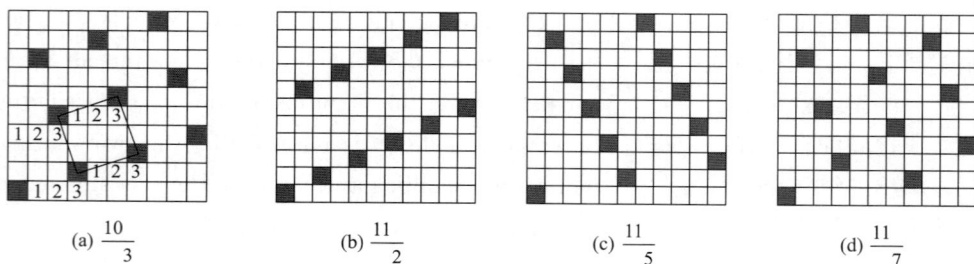

设计并绘制缎纹组织图的步骤如下：

（1）基于组织循环数 R 确定方格纸上完整组织的大小。

（2）选择恰当的飞数。确保组织点在组织图中均匀分布，相邻的 4 个组织点最好形成方形，飞数 S 应该远离 2、R-2 和 R/2 这三个值。图 3-4-3（a）所示的组织（十枚纬面缎纹）中，飞数为 3，组织点分布非常均匀。在十一枚纬面缎纹中，飞数有 2、3、4、5、6、7、8 和 9 共 8 种选择。2（9）、5（6）几种飞数的缎纹组织的斜路明显 [图 3-4-3（b）（c）]，3（8）、4（7）几种飞数的缎纹组织点较为均匀 [图 3-4-3（d）]。

The followings are the procedures of designing and drawing of a satin/sateen weave.

(1) Define the range of the weave repeat unit on the design paper based on R.

(2) Select a suitable step for the satin/sateen fabric. Since all the intersection points are expected to be evenly distributed in the design, it is better for the 4 neighboring interweaving points to form a square, therefore, step number S is recommended to be far away from 2, R-2, R/2. In a 10-shaft sateen shown in Fig. 3-4-3(a), the marks are evenly distributed in the weave. For a 11-shaft sateen, the possible step number is 2,3,4,5,6,7,8 and 9. It is not suitable to choose the design of step number 2 (9), 5(6) as shown in Fig. 3-4-3(b)(c), instead, step number of 3(8), 4(7) are better, which is demonstrated in Fig. 3-4-3(d).

(3) Fill in the marks in squares. Satin weave is started

(a) $\frac{10}{3}$ (b) $\frac{11}{2}$ (c) $\frac{11}{5}$ (d) $\frac{11}{7}$

图3-4-3 不同飞数的缎纹组织点分布

The distribution of the interweaving points

（3）填绘组织点。经面缎纹在起始点处绘纬组织点，根据经向飞数 S_j 找到其余各纵行上的纬组织点。其方法是，自起始点向右移一行纵格，向上数 S_j 格，得到第二根经纱上的纬组织点，依次类推，直到所有纵行的纬组织点查找完毕。再将图中余下的空格全部绘上经组

from the blank at the bottom-left corner, and all the rest blanks on the other vertical space are found according to S_j. From the commencing point, S_j squares are counted and the blank at the successive vertical space is located, and so on until all the blanks of the next vertical spaces are determined. Then, the marks are drawn on all the rest squares. Sateen weave is started from the mark at the bottom-left corner, and all the rest marks on the next

织点符号。纬面缎纹在起始点处绘经组织点，按纬向飞数 S_w 绘其余横行上的经组织点。其方法是，自起始点向上移一行横格，向右数 S_w 小格，得到下一横行上的经组织点，其余横行上的经组织点依次类推绘出。

图3-4-4是根据该法绘制的 $\dfrac{5}{3}$ 经面缎纹和 $\dfrac{5}{3}$ 纬面缎纹组织图。

horizontal spaces are located according to S_w. From the commencing point, S_w squares are counted and the mark at the successive horizontal space is located, and so on until all the marks of the next horizontal spaces are determined.

Fig. 3-4-4 shows how a $\dfrac{5}{3}$ satin weave and a $\dfrac{5}{3}$ sateen weave are drawn respectively.

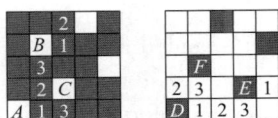

图3-4-4　缎纹组织的绘制
The drawing of a satin/sateen weave

六、缎纹织物上机 /Looming for Satin/Sateen Weaves

制织缎纹织物所用综框数为一个完全缎纹组织循环的经纱数，采用顺穿法穿综。当经密过高时，可用复列式综框，用飞穿法穿综，以减少断经。经面缎纹也常用反织法织造。每筘穿入经纱数一般为 2 ~ 4 根。

In weaving a satin or sateen fabric, the number of the heals used is the size of the weave repeat in warp direction. The ends are straight drafted. If the fabric is high in warp density, the ends are skip drafted to reduce the breakage. Satins can also be woven by face downword weaving. In denting, 2 ~ 4 ends are usually spaced in a split of reed.

七、纱线捻向对缎纹织物风格的影响 /Effects of Yarn Twist on Satin/Sateen Weaves

尽管组织点在缎纹组织中均匀分布，但组织点仍然存在某种程度上的斜纹纹路。为了满足特定的织物风格，纱线的捻向应与缎纹主斜纹线的方向配合。缎纹主斜纹线方向就是相邻纱线上具有最长的共同浮长形成的纹路方向。

在图 3-4-5（a）$\dfrac{5}{2}$ 经面缎纹

Although the weaving points are designed evenly in satins/sateens, there are twill lines to some extent. In order to realize a special style desired, the direction of thread twist should be in coincide with the direction of the main twill lines, which is the direction formed by the longer common float between the neighboring yarns.

In the $\dfrac{5}{2}$ satin weave shown in Fig. 3-4-5(a), the common floats of the float A and its neighboring float B

中，浮长 A 与相邻浮长 B、C 的共同浮长分别为 1 和 2，因此 A 与 C 形成的纹路是主浮长纹路；在图 3-4-5（b）所示的 $\frac{5}{3}$ 经面缎纹中，浮长 A 与相邻浮长 B、C 的共同浮长分别是 2 和 1，因此 A 与 B 形成的纹路则是主浮长纹路。有些经面缎纹需要贡子清晰或斜纹效应的效果，织物表面被经浮长线覆盖，若经纱为 Z 捻，则主浮长线左斜与纱线斜向 Z 捻一致，是合适的设计。即 $\frac{5}{3}$ 经面缎纹符合设计要求，$\frac{5}{2}$ 经面缎纹则不符合要求。

and C are 1 and 2 respectively, therefore, the common floats between warp float A and float C will form the main twill line of the weave. In the $\frac{5}{3}$ satin weave shown in Fig. 3-4-5(b), the common floats of the warp float A and its neighboring float B and C are 2 and 1 respectively, therefore, the common floats between float A and float B will form the main twill line of the $\frac{5}{3}$ satin weave. For some satins, the twill effect is desired. The fabric is covered with warp float, and the ends are Z twisted. Therefore, a left-handed direction of twist is desirable. Thus, $\frac{5}{3}$ satin weave conforms to it, and $\frac{5}{2}$ satin weave doesn't.

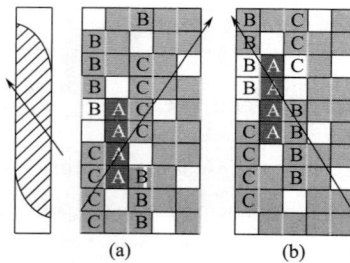

图3-4-5　经面缎纹中纱线捻向与主斜向的配合
The relationship of direction of the main twill lines and direction of thread twist in satins

对于纬面缎纹来说，需要光滑、有光泽的效果，不显斜向。织物表面被纬浮长线覆盖，因此要根据纬浮长判断主纹路方向。在图 3-4-6（a）所示的 $\frac{5}{2}$ 纬面缎纹中，浮长 A 与相邻浮长 B、C 的共同浮长分别为 1 和 2，因此 A 与 C 形成的纹路是主浮长纹路；在图 3-4-6（b）所示的 $\frac{5}{3}$ 纬面缎纹中，浮长 A 与相邻浮长 B、C 的共同浮长分别是 2 和 1，故 A 与 B 形

For sateens, the smooth and lustrous effect is desired, and a twill effect is undesired. Since the fabric is covered with weft float, it is necessary to judge the main twill line of a sateen weave according to weft floats. In the $\frac{5}{2}$ sateen weave shown in Fig. 3-4-6(a), the common floats of the weft float A and its neighboring float B and C are 1 and 2 respectively, therefore, the common floats between float A and float C will form the main twill line of the sateen weave. In the $\frac{5}{3}$ sateen weave shown in Fig. 3-4-6(b), the common floats of the weft float A and its neighboring float B and C are 2 and 1 respectively, therefore, the common floats

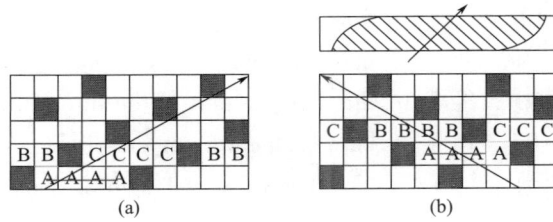

图3-4-6 纬面缎纹中纱线捻向与主斜向的配合
The relationship of direction of the main twill lines and direction of yarn twist in sateens

成的纹路则是主浮长纹路。若纬纱为 Z 捻，则主斜纹线向左斜的方向是合理的。即 $\frac{5}{3}$ 纬面缎纹符合设计要求，$\frac{5}{2}$ 纬面缎纹则不符合设计要求。

between float A and float B will form the main twill line of $\frac{5}{3}$ sateen weave. If the picks are Z twisted, a left-handed twill is desirable. That is to say, $\frac{5}{3}$ sateen weave conforms to it, but $\frac{5}{2}$ sateen weave doesn't.

The Application of Sateen/Satin Weaves

八、缎纹组织的应用 /The Application of Satin/Sateen Weaves

经面 / 纬面缎纹组织主要用于真丝织物，如素缎、横贡缎等，常用的有 5 枚缎纹、8 枚缎纹、10 枚缎纹、12 枚缎纹等。在棉织物中，缎纹组织通常与其他组织结合制成各种花式织物，如缎条府绸、缎条手帕、缎条床单等。精纺毛织物中有直贡呢、横贡呢等。

The satin/sateen weaves are mostly used in silk fabrics, such as plain soft satin, figured soft satin, etc., commonly used are 5-shaft satin, 8-shaft satin, 10-shaft satin, 12-shaft satin, etc. In the cotton fabrics, satin weaves are usually combined with other weaves to make a variety of fancy fabrics, such as satin poplin, satin handkerchief, satin sheets, etc. In the wool fabrics, there are venetian, dice venetian, etc.

第五节 原组织对比 /Comparison Among Elementary Weaves

一、平均浮长与飞数 /Average Length of Float and Step Number

组织规律不同，各个组织点的浮长长度与交错点分布就不同，体现在平均浮长与飞数不同。平均浮长小，交错次数多，纱线弯曲次数多，纱线之间的支撑作用大。飞数小，交织点之间距离小，相邻的弯

The length of float of each interlacing and the distribution of intersection points differ in different weave, which is reflected in the average length of floats and step number. The less the average length of floats, the more the intersections, the more the bending of threads in fabric, and the more the support between the warp threads and the weft

曲部分距离小，纱线之间的支撑作用也大。

平纹组织平均浮长最小，经纬向飞数均为 ±1；斜纹组织平均浮长比平纹大，经纬向飞数也均为 ±1；缎纹组织平均浮长最大，飞数变化也最大。

由于平均浮长与飞数的不同，导致平纹、斜纹和缎纹织物的外观与性能明显不同。

threads. The less the step number, the shorter the distance between the intersection points, the less the distance between the bending part of the threads, and the more the support between the warp threads and the weft threads as well.

The average length of floats of the plain weave is the shortest, and the step number in both warp direction and weft direction is ±1. Twill weaves have a longer average length of floats, but the same step number in both directions. Satins or sateens have the longest average length of floats and greatest change in step number.

The appearance and properties of plain weave, twill weave and satin/sateen weave fabrics are obviously different due to the difference of average float length and step number.

二、织物正反面外观特征 /Appearance of Both Sides

1. 密度 /Density

平纹织物平均浮长最小，交织最频繁，可密性最差；缎纹织物平均浮长最大，交织最弱，可密性最大；斜纹织物介于两者之间。为了体现或弱化斜纹纹路效应，斜纹组织和缎纹组织仅仅在一个方向上纱线密度大。

The plain weave fabrics have the shortest average float, the largest intersections, and the lowest fabric density. However, the satin/sateen weave fabrics have the longest average float, the least intersections, and the highest fabric density. The twill weave fabrics fall somewhere in between. Twill and satin/sateen weave fabrics have a high fabric density in only one direction in order to reflect or weaken the twill effect.

2. 厚度 /Thickness

一般来说，织物某一方向纱线密度越大，该方向的缩率就越大，纱线屈曲波高就越大，织物厚度就可能越大，因此平纹织物最薄，缎纹织物最厚。从另一个角度来说，纱线在平纹织物中弯曲最多，那么每个弯曲的波高也最小，可见，平纹织物仍然是最薄的。

Generally speaking, the greater the fabric density in one direction, the greater the contraction in that direction, and the higher the crimp height, the greater the thickness of the fabric is likely to be, so fabrics of plain weave are the thinnest and fabrics of satin weave are the thickest. On the other hand, the thread in fabrics of plain weave that bends the most has the smallest wave height per bend, therefore, plain weave fabrics are still the thinnest.

3. 光泽与斜向 /Luster and Direction of Diagonal Lines

平纹组织浮长短，织物表面最高点的面积最小，反光强度最小，光泽暗淡；缎纹组织浮长表面最高处面积最大、最平整，故反光最

The surface area of the highest place of the plain fabric is the smallest, and the reflective intensity is the smallest, thereby forms a dim surface. However, in fabrics of satin/sateen weave, the longest float makes the largest level area

强，光泽明亮。

平纹每一个组织点与相邻组织点没有共同浮长，因此，没有斜纹纹路产生，正反面没有区别。斜纹织物只在一个方向有明显的斜路。缎纹织物在两个方向都有斜路，但是需要根据与相邻浮长之间的共同浮长线的长度来判断主斜路方向。织物斜纹效果是否明显，还与纱线捻向和相邻纱线之间的距离相关。因此，斜纹与缎纹织物正反面斜向效果差异较大。

in highest place, so more light is reflected, and a lustrous surface is formed.

Each interweaving point in a plain weave has no common float length with its neighbor, therefore, no twill lines are produced, and there is no difference between front and back. Twill fabrics have distinct diagonal lines only in one direction. Satin/sateen fabrics have twill lines in both directions, but the main diagonal direction needs to be determined by the length of the common float between the adjacent floats. The effect of fabric twill is also related to the direction of the twist of the threads and the distance between adjacent threads. Accordingly, there might be a huge difference of diagonal effects of the face side and the back side for twill fabrics and satin /sateen fabrics.

三、织物性能 /Properties of Elementary Weave Fabrics

1. 强度与耐磨性 /Tenacity and Resistance to Abrasion

在其他条件相同的情况下，平纹织物经纬纱之间弯曲多，摩擦大，相互支撑作用强，强度优于斜纹与缎纹织物。但是斜纹、缎纹织物密度大，单位长度内承受载荷的纱线根数多，一般情况下，斜纹与缎纹织物的强力要高于平纹织物。

Other conditions equaling, there are more intersections, larger friction and mutual support in fabrics of plain weave, and the tenacity of the fabric is higher than that of fabrics of twills or satins. However, the fabric density of the twills or satins is usually higher than that of the plain fabric, so there are more yarns bearing the load per unit space, therefore, the tensile strength of the fabrics of twill or satin/sateen weave is higher than that of plain ones.

平纹织物（府绸除外）一般是同支持面结构，经纬纱都同时与外界摩擦，共同承担磨损，耐磨性能较好。原组织斜纹（异面）与缎纹织物由于经纬向密度差异大，屈曲波高差异大，与外界接触时，主要是一个系统纱线与外界摩擦，容易出现单向磨损。

Plain fabric (except poplin) is generally balanced structure, both the warp yarns and weft yarns rub against external body and bear wear together, therefore, the abrasive resistance is good. Due to the great difference in warp and weft density and the great difference in crimp height of warp-wise and weft-wise, fabrics of twill or satin/sateen weave are prone to show unidirectional wear due to the friction only takes place between one serial of yarns and the outside object.

2. 手感与悬垂性 /Handle and Drapeability

平纹织物中，交织多，导致纱线之间弯曲多，阻碍纱线相互移动，织物硬挺，手感粗糙，悬垂性

There are more bends in plain fabric, which prevents yarn from moving freely each other, therefore, the fabric is stiff, feels rough and shows poor drape performance. The

差。缎纹织物中，纱线之间弯曲次数相对少，纱线之间运动较自由，因此织物手感平滑、柔软、悬垂性好。斜纹织物手感与悬垂性介于两者之间。

number of yarn bends between satin/sateen fabrics is the minimum, and the movement between yarns is relatively free, so the fabric feels smooth, soft, and has the best drape performance. The handle and the drapeability of the twill fabrics are just between the two kinds of fabrics.

习题 /Questions

1. 已知棉平布规格为 58tex × 58tex，181 × 141.5，在踏盘织机上织造，绘制其上机图。

2. 已知派力司织物规格为 (16.9tex × 2) × 25.6tex，232 × 225，其纬向缩率是 15%，在踏盘织机上织造，绘制其上机图。

3. 已知全毛平纹花呢织物规格为 [(26.3tex × 2) × (26.3tex × 2)，217 × 184]，其纬向缩率是 15%，在踏盘织机上织造，绘制其上机图。

4. 分别绘制在踏盘织机和多臂织机上织造府绸织物的上机图。

5. 绘制线卡织物反织法织造的织物上机图。

6. 绘制纱卡织物反织法织造的织物上机图。

7. 绘制 $\frac{2}{1}$ 左斜的斜纹布织物的上机图。

8. 绘制采用 Z 捻单纱的直贡织物反织法织造的织物上机图。

9. 绘制采用 S 捻股线为经纱的直贡织物采用反织法织造的织物上机图。

10. 绘制 7 枚缎纹构成的所有可能的缎纹组织。

1. Draw the looming plans for a calico (58tex × 58tex, 181 × 141.5) fabric on a tappet loom.

2. Draw the looming plans for a palace fabric [(16.9tex × 2) × 25.6tex，232 × 225] on a tappet loom. The contraction rate of weft direction is 10% in total.

3. Draw the looming plans for a worsted fancy suiting [(26.3tex × 2) × (26.3tex × 2)，217 × 184] on a tappet loom. The contraction rate of weft direction is 15% in total.

4. Draw the looming plans for a poplin fabric on a tappet loom and a dobby loom respectively.

5. Draw the looming plans for a khaki of folded yarns by face downward weaving.

6. Draw the looming plans for a khaki of single yarns by face downward weaving.

7. Draw the looming plans for a $\frac{2}{1}$ twill weave.

8. Draw the looming plans for a venetian fabric of single cotton yarn (Z twist) by face downward weaving.

9. Draw the looming plans for a venetian fabric of folded cotton yarn (S twist) by face downward weaving.

10. Draw all the possible 7-shaft satin/ateen weaves.

第四章　变化组织 /Derivatives of Elementary Weaves

变化组织

Derivatives of
Elementary Weaves

在原组织的基础上，通过改变浮长、飞数，得到原组织的变化组织。变化组织的组织循环大小将随之改变，但是仍然保留原组织的基本特征。变化组织包括平纹变化组织、斜纹变化组织和缎纹变化组织。

The derivatives of elementary weaves are mainly obtained by modifying the floats and step number. Although the size of the weave repeat may change, the derivatives keep the main characteristics of elementary weaves. These derivatives include plain weave derivatives, twill weave derivatives and satin/sateen weave derivatives.

第一节　平纹变化组织 /Plain Weave Derivatives

以平纹为基础，沿着经纱方向延长组织点（即同一种组织点连续），得到经重平组织，如图 4-1-1（a）（b）所示；沿着纬纱方向延长组织点所形成的组织，称作纬重平组织，如图 4-1-1（c）（d）

By extending the intersection points of the plain base vertically, grouping together several picks in the same shed, the warp rib weaves as shown in Fig. 4-1-1(a)(b) are formed; or horizontally, several ends are lifted together in tandem, and the weft rib weaves as shown in Fig. 4-1-1(c) (d) are produced; or in both directions, the basket (matt or

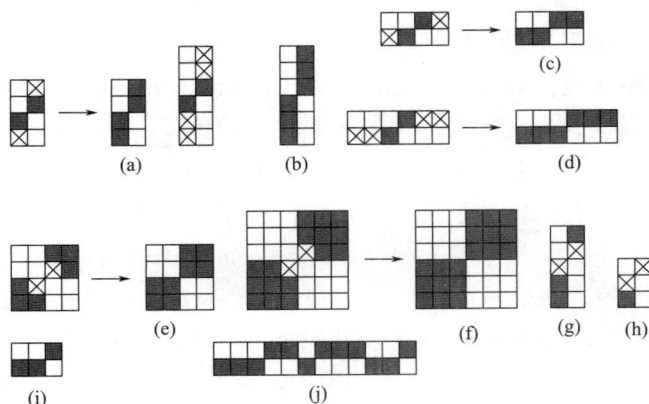

图4-1-1　平纹变化组织的形成
The formation of plain weave derivatives

103

所示；同时沿经纬方向延长组织点形成的组织，称为方平组织，如图 4-1-1（e）（f）所示。平纹变化组织只用 2 片综框即可织造。

hopsack) weaves as shown in Fig. 4-1-1(e)(f) are resulted. All these plain weave derivatives can be woven on two healds.

一、经重平组织 /Warp Rib Weaves

一般情况下，沿经纱两侧延伸的组织点个数相同。若不相等或者仅仅单侧延伸，则得到变化经重平组织，如图 4-1-1（g）（h）所示。

Generally, the numbers of the interlacing points extended in both sides of the plain base are equal. However, if it is not the case, irregular warp weaves are formed as shown in Fig. 4-1-1(g)(h).

1. 表示方法 /Representation

经重平组织用分子式和文字说明表示，如 $\dfrac{n}{m}$ 经重平组织，分子 n 表示第一根经纱上的经组织点个数，分母 m 表示第一根经纱上的纬组织点个数。 图 4-1-1（a）的组织表示为 $\dfrac{2}{2}$ 经重平组织，读作"二上二下经重平组织"；图 4-1-1（h）的组织为 $\dfrac{2}{1}$ 变化经重平组织。

Warp rib weaves can be represented as a fraction with literal expression, such as $\dfrac{n}{m}$ warp rib weave. The numerator n is the number of marks in first vertical space while the denominator m is the number of blanks in first vertical space. The weave as shown in Fig. 4-1-1(a) is represented as $\dfrac{2}{2}$ warp rib weave, and read as "two up two down warp rib weave". The weave as shown in Fig. 4-1-1(h) is represented as $\dfrac{2}{1}$ irregular warp rib weave.

2. 经重平组织图的绘制 /Drawing of Warp Rib Weaves

经重平组织图的绘制步骤如下：

（1）首先确定组织循环经纬纱数 R_j 与 R_w，然后勾画图的范围。经重平组织的组织循环经纱数 R_j = 2，组织循环纱数 R_w = 分子 n+ 分母 m。n 为第一根经纱的连续经组织点个数，m 为第一根经纱的连续纬组织点个数。

（2）从左下角开始，在第一根经纱上按分式所示的交织规律填绘连续 n 个经组织点。

（3）在第二根经纱上填绘与第一根经纱上相反的组织点。

The procedures of drawing a warp rib weave can be described as following:

(1) Calculate the size of the weave repeat R_j and R_w, and draw the weave range in bold lines. For a warp rib weave, R_j=2, and R_w= numerator n+ denominator m, where n is the number of the successive marks that are drawn on the first end, and m is the number of the successive blanks that are drawn on the first end.

(2) Draw n successive marks on the first vertical space from the bottom-left corner of the design.

(3) Opposite intersective points are drawn on the second end.

3. 经重平组织的结构特点 /Structure of Warp Rib Weaves

经重平组织中，若干根纬纱成组地在一个梭口中织入，相当于在平纹组织中织入几根纱线并成的粗纬，因此，经纱弯曲程度大，在织物表面形成横凸条效应。为了突出该效应，通常采用高密、线密度低的经纱，使经纱弯曲程度大；同时，采用低纬密、粗支纬纱，使纬纱在织物结构中呈现更加平直的状态。

Several picks are grouped to be inserted into the same shed in weaving, which is equivalent to insert a thick pick by doubling yarns in a plain weave. Therefore, the ends bend more and form horizontal lines on the surface. The effects can be accentuated by applying high warp density and coarse picks. The picks almost lay straight in the fabric structure.

4. 经重平组织上机图 /Looming Plans for Warp Rib Weaves

尽管理论上经重平织物可以用2页综框织造，但因经密较大，宜用飞穿法，或者采用4页综框顺穿。每筘齿穿入2 ~ 4根经纱。上机图如图4-1-2所示。

Due to its high warp density, the fabric of warp rib weave should be woven in skip draft or straight draft with 4 shafts although 2-shaft is feasible theoretically. 2 ~ 4 ends are threaded per dent split. Fig. 4-1-2 shows two forms of the looming plans for a warp rib weave.

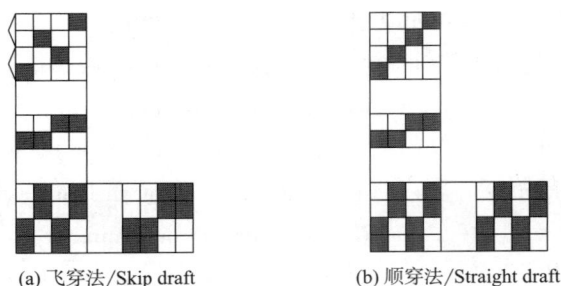

(a) 飞穿法/Skip draft　　　(b) 顺穿法/Straight draft

图4-1-2　经重平组织上机图
The looming plans for the warp rib weave

5. 经重平组织的应用 /Applications of Warp Rib Weaves

除服用和装饰织物外，经重平组织可用作毛巾组织的基础组织和各种织物的布边，以调整织物结构，改善织造工艺。

Besides for apparel and decorative uses, warp rib weaves can be used as the ground of a terry weave and the selvedges of various fabrics for its even-sided weave to adjust the structure of the fabric for improving the weaving process.

二、纬重平组织 /Weft Rib Weaves

若平纹组织沿纬纱两侧延伸的

A weft rib weave is obtained if the numbers of the

组织点个数相同，得到纬重平组织。若不相等或者仅仅单侧延伸，则得到变化纬重平组织，如图 4-1-3（i）（j）所示为 $\dfrac{2}{1}$ 变化纬重平组织。

1. 表示方法 /Representation

纬重平组织也用分子式和文字说明表示，如 $\dfrac{n}{m}$ 纬重平组织，分子 n 表示第一根纬纱上的经组织点个数，分母 m 表示第一根纬纱上的纬组织点个数。图 4-1-1（c）的组织表示为 $\dfrac{2}{2}$ 纬重平组织，读作"二上二下纬重平"组织。图 4-1-1（i）的组织是 $\dfrac{2}{1}$ 变化纬重平组织，也称麻纱组织。

2. 纬重平组织图的绘制 /Drawing of Weft Rib Weaves

纬重平组织图的绘制步骤如下：

（1）首先确定组织循环经纬纱数 R_j 与 R_w，然后勾画图的范围。纬重平组织循环经纱数 R_j = 分子 n + 分母 m，组织循环纬纱数 R_w =2。n 为第一根纬纱连续经组织点的个数，m 为第一根纬纱连续纬组织点个数。

（2）从左下角开始，在第一根纬纱上按分式所示的交织规律填绘连续 n 个经组织点。

（3）在第二根纬纱上填绘与第一根纬纱上相反的组织点。

3. 纬重平组织的结构特点 /Structure of Weft Rib Weaves

纬重平组织中，若干根经纱同时沉浮，相当于在平纹组织中将几根经纱并成单根粗经，因此，纬纱弯曲程度大，在织物表面形成纵凸条效应。为了突出该效应，通常采

interlacing points extended in weft direction of the plain base are equal. However, if it is not the case, irregular weft weaves are formed, such as weave in Fig. 4-1-3(i)(j).

Weft rib weaves can also be represented as a fraction with literal expression, such as $\dfrac{n}{m}$ weft rib weave. The numerator n is the jenumber of marks in first horizontal space while the denominator m is the number of blanks in first horizontal space. The weave as shown in Fig. 4-1-1(a) is represented as $\dfrac{2}{2}$ weft rib weave, and read as "two up two down weft rib weave". The weave as shown in Fig. 4-1-1(i) is represented a $\dfrac{2}{1}$ irregular weft rib weave, which is also called hair cords.

Drawing of Weft Rib Weaves

The procedures of drawing a weft rib weave can be described as following:

(1) Calculate the size of the weave repeat R_j and R_w and draw the weave range in bold frame. For a weft rib weave, R_j =numerator n+ denominator m, and R_w =2, where n is the number of the successive marks that are drawn on the first pick, and m is the number of the successive blanks that are drawn on the first pick.

(2) Draw n successive marks on the first horizontal space from the bottom-left corner of the design.

(3) Opposite intersective points are drawn on the second horizontal space.

3. 纬重平组织的结构特点 /Structure of Weft Rib Weaves

Several ends are grouped to be lifted and depressed simultaneously in weaving, which is equivalent to have a thick end by doubling yarns in a plain weave. Therefore, the picks bend more and form vertical lines on the surface. The effects can be accentuated by applying high weft density

用高密、细度低的纬纱，使纬纱弯曲程度大；同时，采用经密较低，稍粗的经纱，使经纱在织物结构中呈现更加平直的状态。

一个较好的纵凸条效应举例如下：采用 $\dfrac{2}{1}$ 变化纬重平组织，经纱排列是 2 根粗经纱（ 2×20tex），1 根细经纱（15tex），经密为 20 根 /cm；纬纱为 30tex，纬密为 42 根 /cm。

需要指出的是， $\dfrac{2}{1}$ 变化纬重平并不绝对形成纵条纹，其结构或外观取决于织造条件。下面的织造条件就可形成横凸条效应：经纱为 18tex；经密为 52 根 /cm；两根经纱张力小，一根经纱张力大；纬纱为 38tex，纬密为 16 根 /cm。

4. 纬重平组织上机图 /Looming Plans for Weft Rib Weaves

纬重平组织当经密不大时，可以采用 2 页综的照图穿法。当经密较大时，可以采用飞穿法，或者增加综框后顺穿。在织造中，同一筘齿中组织规律相同的经纱往往会相互纠缠，使织造产生困难，影响织物外观。需要将交织规律相同的经纱穿在不同的筘齿中。图 4-1-3 所示为纬重平组织分别采用 4 页综框的顺穿法和照图穿法的上机图。

and coarse ends. The ends almost lay straight in the fabric structure.

For example, the looming parameters for making the vertical lines by a $\dfrac{2}{1}$ irregular weft rib weave: 2 coarse ends, 2×20 tex; 1 fine end, 15 tex; 20 ends per cm; weft, 30tex; 42 picks per cm.

It should be noted that an irregular $\dfrac{2}{1}$ weft rib doesn't necessarily make vertical lines, its structure or appearance depends on the looming conditions of the fabric. The looming conditions for making the horizontal lines: warp, 18tex; 52 ends per cm; two ends working together slack, single end tight; weft, 38tex; 16 picks per cm.

When in low warp density, the fabric of weft rib weave can be woven in 2-shaft loom. However, when in higher warp density, the fabric can be skip drafted or straight drafted with 4 or more healds. The ends which work together tend to twist or roll round each other as the cloth is woven, and if this takes place the cloth suffers in appearance, while the weaving process is made more difficult. The twisting of the ends can be prevented by denting them in such a manner that those which work alike are separated by the wires of the reed. Fig. 4-1-3 shows the looming plans for weft rib weave in straight draft and broken draft with 4 shafts.

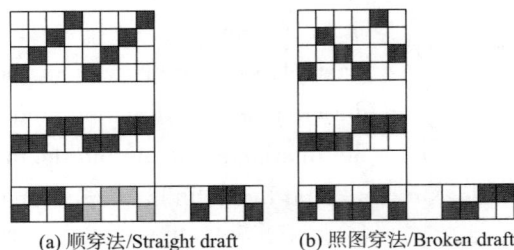

(a) 顺穿法/Straight draft   (b) 照图穿法/Broken draft

图 4-1-3 纬重平组织上机图
The looming plans for the weft rib weave

5. 纬重平组织应用 /Applications of Weft Rib Weaves

除了夏季衬衣、裙、裤面料和窗帘等装饰织物外，纬重平组织也可用作各种面料的布边。

Besides for summer shirts, skirts, trousers and curtains, weft rib weaves can also be used as the selvedges of various fabrics.

三、方平组织 /Basket Weaves

在平纹组织基础上，经、纬方向延伸相同数量的组织点，形成普通方平组织，如图4-1-1（e）和图4-1-4所示；如果经、纬方向延伸相同的变化重平组织或者多个浮长，则形成变化方平组织，如图4-1-5（a）所示；如果经、纬方向延伸不同的变化重平组织，则形成复杂变化方平组织，如图4-1-5（b）所示。

A basket (matt, hopsack) weave is obtained if the numbers of the interlacing points extended in both direction of the plain base are equal, as shown in Fig. 4–1–1(e) or Fig. 4–1–4. If the same interlacing is extended in both directions, irregular basket weaves are formed, such as weave as shown in Fig. 4–1–5(a). If different interlacing is extended warp–wise or weft–wise, a fancy irregular basket weave is formed as shown in Fig. 4–1–5(b).

图4-1-4 方平组织
A basket weave

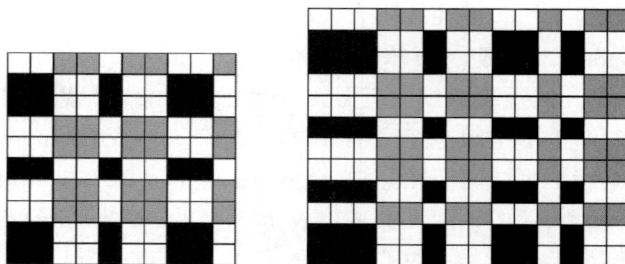

(a) (b)

图4-1-5 变化方平组织
Irregular basket weaves

1. 表示方法 /Representation

方平组织用组合分子式和文字说明表示，如 $\dfrac{n_1\,n_2\,n_3\cdots n_i}{m_1\,m_2\,m_3\cdots m_i}$ 方平组织，每组分子式 $\dfrac{n_i}{m_i}$ 表示沿经向或者纬向的延伸规律，i 是延伸规律（浮长）的顺序。若经纬方向相同，只需写一个分子式。图 4-1-1(e) 的组织表示为 $\dfrac{2}{2}$ 方平，读作"二上二下方平"。图 4-1-5

Basket weaves can also be represented as a combination of fractions with literal expression, such as $\dfrac{n_1\,n_2\,n_3\cdots n_i}{m_1\,m_2\,m_3\cdots m_i}$ basket weave. Each fraction dictates the principle how the interweaving points are extended in warp direction or weft direction, and i denotes for the index of the interlacing. If the interlacing in warp direction is the same as the weft direction, a group of the fraction is enough to represent the basket weave. The weave as shown in Fig. 4–1–1(e) is represented as $\dfrac{2}{2}$ basket weave, and read as

（a）的组织是$\frac{2\ 1\ 2}{2\ 2\ 1}$变化方平组织，而图 4-1-5（b）所示为经向按照$\frac{2\ 1\ 1\ 2}{1\ 2\ 2\ 1}$规律变化、纬向按照$\frac{3\ 1\ 2\ 1}{2\ 2\ 1\ 2}$规律变化的复杂方平组织。

"two up two down basket weave". The weave as shown in Fig. 4-1-5(a) is represented a $\frac{2\ 1\ 2}{2\ 2\ 1}$ irregular basket weave. The weave as shown in Fig. 4-1-5(b) is obtained by extending the weaving points as the principle $\frac{2\ 1\ 1\ 2}{1\ 2\ 2\ 1}$ in warp direction, and $\frac{3\ 1\ 2\ 1}{2\ 2\ 1\ 2}$ in weft direction.

2. 方平组织图的绘制 /Drawing of Basket Weaves

方平组织图的绘制步骤如下：

（1）首先确定组织循环经纬纱数 R_j 与 R_w，然后勾画图的范围。方平组织循环经纱数 R_j = 沿纬向延伸规律分式的分子 n+ 分母 m，组织循环纬纱数 R_w = 沿经向延伸规律分式的分子 n+ 分母 m。

（2）分别根据经向的分数式和纬向的分数式绘制第一根经纱和第一根纬纱的组织点。

（3）凡与第一根纬纱交织有经组织点的经纱，均按照第一根经纱规律绘制组织点。

（4）其余经纱上按照相反组织点绘制。

在设计图 4-1-5（b）所示的组织时，其组织循环根数的计算方法为：

The procedures of drawing a basket weave can be described as following:

(1) Determine the range of the weave. The size of the weave repeat R_j in warp direction equals to the sum of the numerators and the denominators of the fraction in weft direction, and the size of the weave repeat R_w in weft direction equals to the sum of the numerators and the denominators of the fraction in warp direction.

(2) Draw the marks on the first end and first pick according to the fraction for the warp direction or weft direction respectively.

(3) The end whose interweaving point with the first pick is a riser or "warp over weft" should be drawn as the mode for the first end.

(4) The other ends should be drawn otherwise.

In designing the weave as shown in Fig. 4-1-5(b), the size of the weave repeat is calculated as :

$$R_j=3+2+1+2+2+1+1+2=14$$

$$R_w=2+1+1+2+1+2+2+1=12$$

该组织设计的步骤如图 4-1-6 所示。

The procedures of drawing the design are illustrated as Fig. 4-1-6.

3. 方平组织的结构与外观特征 /Structure and Appearance of Basket Weaves

方平组织是由平纹组织变化而来的，相邻的经纱都当成一组经纱，并在每个梭口中以两根或两根纬纱以上为一组引入纬纱。由于浮长线均匀、长度适中，织物表面平整、光泽较好。

In basket weaves, groups of adjacent warps are each woven as one and picks are inserted in groups of two or more in each shed. Due to uniform and moderate floats distribution in the structure, the fabric is flat and lustrous.

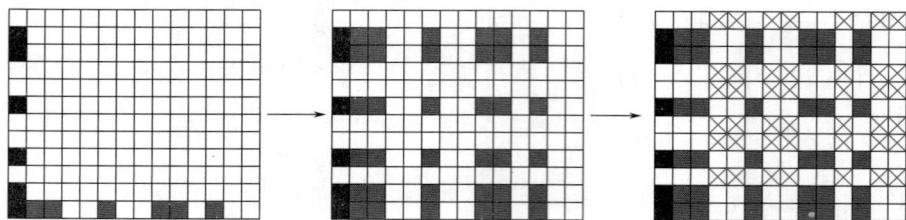

图4-1-6 复杂变化方平组织的绘制过程
The steps to draw an irregular basket weave

但是变化方平组织双、三经纬并列成组，相当于纱线加粗，刚性变大，不易弯曲，织物表面具备粗细宽窄不一的纵横交叉的凸条纹，外观粗犷、自然。由于组织中上下、左右组织点均相反，浮长线起始点相同，浮长聚集和分离作用明显，组外经纬纱均不易靠拢，形成小空隙，织物透气性好。

However, the appearance of the irregular basket weaves is quite different. Several ends and picks are grouped together, which is equivalent to use thick threads. The threads are rigid, stiff, and the vertical lines and horizontal lines of different sizes appear on the surface. Thus, the fabric is sturdy, rough and natural. Because the interlacing points of one group are opposite to their neighbors around, floats have functions of distinctive aggregation and separation, and the threads in different groups are separated and small interstices form in the structure. Therefore, the fabrics of irregular basket weave show better permeability.

4. 方平组织上机图 /Looming Plans for Baske Weaves

方平织物的上机类似于纬重平组织，穿综可以采用顺穿法、2页综的照图穿法，或者2页复列式飞穿法。根据组织的经纱循环根数，确定每筘穿入数。为了避免同规律的两根经纱在穿入同一筘齿中的产生相互纠缠现象，可以将其穿在不同的筘齿中，如图4-1-7所示。若要形成一些孔洞效应，则同规律的两根经纱可以穿入同一筘齿中。

Like weft rib weaves, basket weaves can be drafted by design of a 2-shaft, straight draft, or skip draft. The number of ends that are dented per split depends on the size of the warp repeat. The ends of same interlacing can be separated by the dent wire to avoid turning over as shown in Fig. 4-1-7. Sometimes, the ends of same interlacing are threaded in the same dent split to form hole effect purposely.

5. 方平组织的应用 /Applications of Basket Weaves

方平组织因为平整，光泽较好，可用于各类服装，如板司呢、牛津纺、学生服。因为有孔洞效应，可用作银幕布。也常用作布边组织，如图4-1-8所示，用作布边

Basket weaves are suitable for apparel, such as Basket, Oxford, Harvard shirtings. The weave can be used as screen cloth due to the hole effect. When used as selvage, the weaves should be stagger over for one pick, and the proper picking order is required for the first picking for a shuttle

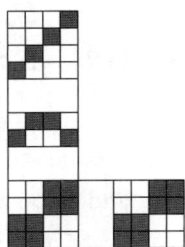

图4-1-7　方平组织上机图
The looming plans for a basket weave

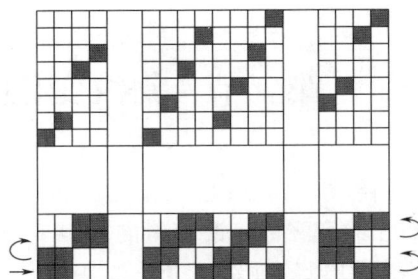

图4-1-8　方平组织作为布边
Basket weave works as selvedge

时，两边要错开一纬，并注意第一纬的投梭方向。变化方平组织常用于制织仿麻织物。

6. 花式方平组织 /Fancy Irregular Basket Weaves

为了增加织物的牢固程度，在变化方平组织的基础上，部分区域修改一些交织规律，可形成其他的精美图案，如图 4-1-9（a）（b）（c）（e）和（f）所示的鸟眼组织，（d）和（g）所示的麦粒组织，但这些组织不能用 2 页综框织造。

loom as shown in Fig. 4-1-8. Irregular basket weaves are used for imitating cambric.

In order to improve the firmness of the weave, fancy irregular basket weaves are formed by modifying some interweaving points at local area on the basket base or warp rib weaves. The weaves produce exquisite patterns. The typical weaves are bird-eye weaves shown in Fig. 4-1-9(a) (b)(c)(e) and (f), and barley-corn weaves shown in Fig. 4-1-9(d) and (g).

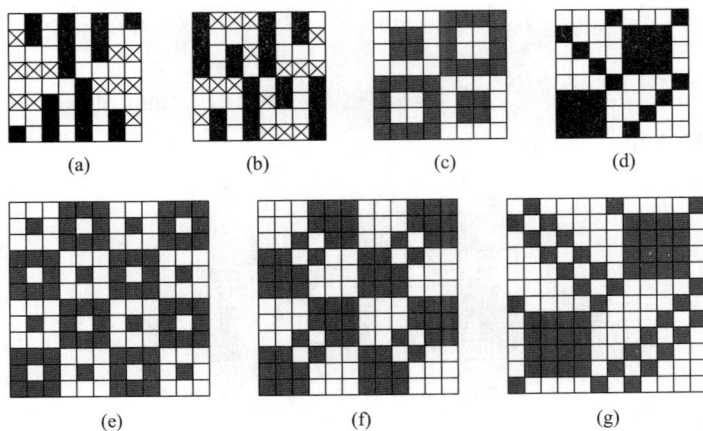

(a)　　(b)　　(c)　　(d)

(e)　　(f)　　(g)

图 4-1-9　各种花式方平组织
Various fancy irregular basket weaves

第二节　斜纹变化组织 /Twill Weave Derivatives

在斜纹组织基础上，通过延伸组织点、改变飞数、改变斜向和斜条数量，得到各种斜纹变化组织，广泛地应用在各类织物上。

By extending the interweaving points, modifying the step number, changing the direction and the number of the diagonal lines, twill kindred weaves are obtained and applied to various applications.

一、加强斜纹组织 /Double Twills

1. 组织构成 /Construction

加强斜纹组织是以原组织的斜纹组织为基础，在其组织点旁（经向或纬向）延长组织点而成（$R_j=R_w \geqslant 4, S=\pm 1$）。在加强斜纹的组织图中不存在单独的组织点，组织可用分数式 $\dfrac{n}{m}$ 表示，$\dfrac{n}{m}$ 含义与原组织斜纹中的意思相同。图 4-2-1（a）（b）所示组织分别为 $\dfrac{2}{2}\nearrow$ 加强斜纹和 $\dfrac{3}{3}\nearrow$ 加强斜纹，是双面组织；（c）（d）所示组织分别为 $\dfrac{2}{3}\nwarrow$ 加强斜纹和 $\dfrac{3}{2}\nwarrow$ 加强斜纹，是异面组织。

Double twills are formed by extending the single interweaving points horizontally or vertically in a basic twill weave (with $R_j=R_w \geqslant 4, S=\pm 1$). There is no single interweaving point in a double twill weave. Double twills are also represented by a fraction $\dfrac{n}{m}$. n and m have the same meaning as in a basic twill weave. The even-sided double twills as shown in Fig. 4-2-1(a) and (b) are $\dfrac{2}{2}\nearrow$ twill weave and $\dfrac{3}{3}\nearrow$ twill weave respectively, and single twills as shown in Fig. 4-2-1(c) and (d) are $\dfrac{2}{3}\nwarrow$ twill weave and $\dfrac{3}{2}\nwarrow$ twill weave respectively.

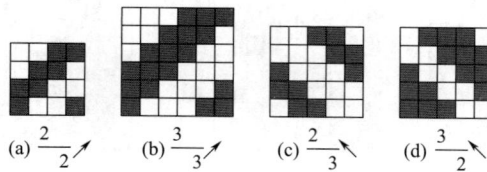

(a) $\dfrac{2}{2}\nearrow$　(b) $\dfrac{3}{3}\nearrow$　(c) $\dfrac{2}{3}\nwarrow$　(d) $\dfrac{3}{2}\nwarrow$

图 4-2-1　加强斜纹组织
Double twills

2. 加强斜纹组织的结构特点 /Structure of Double Twills

加强斜纹正反面都没有单独的组织点存在，斜纹线反光带正反面

There is no single interlacing point on both the face and the back of the twill fabric, so the reflective bands keep

都可以连续。因此，正反面斜纹效应明显，称为双面斜纹。最常见的是 $\dfrac{2}{2}$ 加强斜纹，浮长不长，但可密性比平纹大，织物紧密厚实，适合中厚型棉、毛精纺、丝等各类织物。由于有一定浮长，交织较为紧密，加强斜纹织物还可以缩绒、起毛，粗纺织物呈现丰厚、柔软的风格。采用该组织的哔叽织物因为经密稍大于纬密，斜条宽、浅，斜纹线角度为 50° 左右；华达呢的经密远大于纬密，斜条细、深，斜纹线角度为 63° 左右。

continuous and the distinctive diagonal effect shows in both the face and the back, which is the reason why the fabric is called even-sided twill. For most widely used $\dfrac{2}{2}$ double twill, the floats are moderate, a thread density superior to plain weave can be achieved, therefore, the fabric is firm and thick and is suitable for the medium-weight fabrics of cotton, worsted and silk fibers. For the same reason, $\dfrac{2}{2}$ double twill is also suitable for felting, raising and pulling to show bulk, full and soft style. For serge, the warp sett is little higher than the weft sett, the diagonal line is shallow and wider, and the angle of the diagonal line is about 50°. However, for gaberdine, the warp sett is far higher than weft sett, the diagonal line is finer and deeper, and the angle of the diagonal lines is about 63°.

3. 加强斜纹组织图的绘制 /Drawing of Double Twills

一个 $\dfrac{4}{2}\nearrow$ 加强斜纹组织图的绘制步骤如图 4-2-2 所示，描述如下：

（1）计算 R_j 和 R_w。$R_j=R_w=$ 基础组织循环 $R_b=$ 分子 $n+$ 分母 $m=4+2=6$。

（2）确定组织的方格纸范围。

（3）根据分式在第一根经纱上画出经组织点符号（连续 4 个）。

（4）复制前一列经纱的经组织符号点并分别根据 $S_j=+1$、$S_w=+1$ 的规律将其在第二至第六根经纱每次向上平移一方格。

The steps of drawing $\dfrac{4}{2}\nearrow$ double twill weave are described in Fig. 4-2-2 as following:

(1) Calculate R_j and R_w: $R_j = R_w = R_b=$numerator $n+$ denominator $m=4+2=6$. Here, R_b is the size of the base weave repeat.

(2) Define the square range of the weave on the design paper.

(3) Draw the marks (4 successive marks) on the first vertical space by the fraction.

(4) Copy the marks of the previous end a square higher to the next end from 2nd to 6th respectively according to $S_j=+1$, $S_w=+1$.

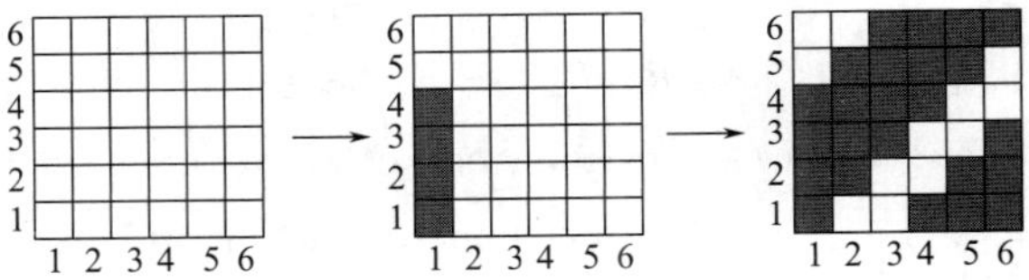

图 4-2-2 $\dfrac{4}{2}\nearrow$ 加强斜纹组织的绘制步骤

The steps of drawing $\dfrac{4}{2}\nearrow$ a double twill weave

4. 加强斜纹组织上机图 /Looming Plans for Double Twills

因为 $R \geqslant 4$，当织物经纱密度小时，织造时可以用顺穿法穿综；但此类织物一般经纱密度较大，故采用飞穿法穿综。棉、丝织物中每筘齿穿 2 ~ 4 根经纱，毛织物中每筘齿穿入数可达 6 ~ 8 根经纱。

In weaving fabrics of double twill, the ends are straight drafted when with a small warp sett due to $R \geqslant 4$. However, skip draft is frequently used since the fabrics are likely designed with a higher warp sett. For cotton and silk fabrics, 2 ~ 4 ends are threaded-in per split of reed; for worsted fabrics, as high as 6 ~ 8 ends are placed in per dent.

5. 加强斜纹组织的应用 /Applications of Double Twills

加强斜纹组织织物外观，纹路清晰，广泛用于棉、毛、化纤及混纺面料。常用于棉织物中的哔叽、双面卡其等，精纺毛织物中的哔叽、华达呢和啥味呢等，粗纺织物中的麦尔登、海军呢、制服呢、海力斯、大衣呢、毛毯等，丝织物中的绫类、斜纹绸等。$\dfrac{2}{2}$加强斜纹还可用作斜纹织物的布边。

The fabrics of double-twill weave have a clear appearance with distinct ridges and furrows and are widely used in cotton, wool, chemical fiber and blended fabrics. Cotton fabrics of double twill includes serge, double-sided khaki, etc. The weaves are also found in worsted fabrics like serge, gabardine and worsted flannel, woolen fabrics like Melton, pilot cloth, uniform cloth, Helis, coat, silk fabrics like silk twills and surah, etc. $\dfrac{2}{2}$ twills can be used as the selvedge in weaving fabrics of twill weave.

Composed Twills

二、复合斜纹组织 /Composed Twills

1. 组织构成 /Construction

在组织循环小的斜纹组织中，浮长线较短，斜纹线细，纹路不够明显。如果增加组织循环大小，使浮长线变长，则斜纹效果突出，但坚牢度降低，因为纱线交织次数少，布面柔软，纱线间的摩擦阻力较小，纱线将有滑脱的倾向。解决措施是在浮长线内增加交织点。

复合斜纹组织中的斜条纹粗细不同，其特点是 $R_j = R_w \geqslant 5$，$S = \pm 1$。可用复合分数式表示，意义与斜纹或方平组织类似。图 4-2-3（a）所示组织是 $\dfrac{3\quad2}{2\quad1}\nearrow$ 复合斜纹组织，读作"三上二下二上一下右斜复合斜纹组织"。在每组分子式中，

In twills of small size with short floats, the twill lines are fine but not distinct. It will be more accentuated if the length of float is increased by enlarging the size of the weave unit, but this more marked diagonal will be at the expense of solidity. As the threads are less interlaced, the cloth will be softer, and as they offer less resistance, the threads tend to slip. One way of remedying this disadvantage is to insert intersections within the floats to form composed twill.

Composed twill weaves have unequal wales of warp and weft appeared on the surface. For a composed twill, $R_j = R_w \geqslant 5$, $S = \pm 1$. Like a double twill weave or an irregular basket weave, composed twills can be represented by a compound fraction and an arrow denoting the direction of the diagonal lines. The weave as shown in Fig. 4-2-3(a) is represented as $\dfrac{3\quad2}{2\quad1}\nearrow$ composed twill, read as "three-

分子表示连续的经组织点数，分母表示连续的纬组织点数，箭头表示斜纹方向。分子式的组数就是组织的斜纹条数量。

up, two-down, two-up, one-down right composed twill". In each pair of the fraction, the numerator is the number of the successive marks while the denominator is the number of the successive blanks in design. The arrow denotes the direction of the twill. The number of the pairs is the number of the diagonal lines in the weave.

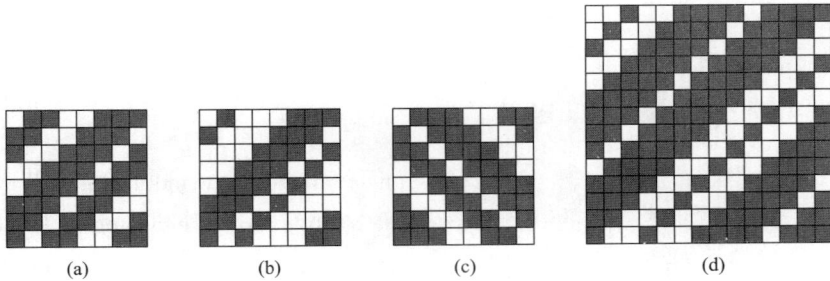

图4-2-3　各种复合斜纹组织
Various composed twills

因此，图4-2-3（b）～（d）所示的组织分别为$\dfrac{3\quad1}{3\quad1}\nearrow$、$\dfrac{3\quad2}{2\quad1}\nwarrow$和$\dfrac{4\quad4\quad1}{1\quad2\quad2}\nearrow$复合斜纹组织。

The weave as shown in Fig. 4-2-3(b)(c) and (d) are represented as $\dfrac{3\quad1}{3\quad1}\nearrow$, $\dfrac{3\quad2}{2\quad1}\nwarrow$ and $\dfrac{4\quad4\quad1}{1\quad2\quad2}\nearrow$ composed twills respectively.

2. 复合斜纹组织图的绘制 /Drawing of Composed Twills

下面以$\dfrac{3\quad2}{2\quad1}\nearrow$为例，介绍复合斜纹组织图的绘制过程：

（1）首先确定组织循环经纬纱数 R_j 与 R_w，然后勾画图的范围。复合斜纹组织循环经纱数 $R_j=R_w=$分子之和＋分母之和。因此，该组织 $R_j=R_w=3+2+2+1=8$。

（2）根据分数式绘制第一根经纱上的组织点。

（3）如果箭头朝右上方，复制前一根经纱的组织点向右上方移到下一根经纱，相反，方格向下平移。

（4）重复步骤（3），直至所有

Take composed twill $\dfrac{3\quad2}{2\quad1}\nearrow$ as an example, the drawing of various composed twill weaves can be described as following:

(1) Determine the range of the design. The size of the weave repeat unit equals to the sum of all the numerators and the denominators of the fraction. So, the size of the weave is calculated as $R_j=R_w=3+2+2+1=8$.

(2) Draw the marks on the first vertical space on design paper according to the fraction.

(3) Copy the marks of the previous vertical space to the succeeding vertical space on design paper one square upward if the arrow is up-right direction, otherwise, one square downward.

(4) Repeat the step (3) until all the vertical spaces are

经纱绘制完成。

复合斜纹一般采用顺穿法穿综，每筘穿入数随织物密度不同而改变，棉织物每筘齿穿入 2 ~ 4 根经纱，毛织物每筘齿穿入 4 ~ 6 根经纱。复合斜纹可用于棉彩格女线呢、粗花呢，还可以作为急斜纹的基础组织。

finished.

As a routine, compose twills are straight drafted. The number of the ends threaded varies with the warp density, say, 2 ~ 4 ends per dent for cotton fabrics and 4 ~ 6 ends per dent for worsted fabrics. Composed twills can be used in lady's colored check, tweed, and as the foundation of the steep twill weave.

Elongated Twills

三、角度斜纹组织 /Elongated Twills

斜纹组织的斜路在布面形成的角度取决于：①经纱与纬纱密度的比值；②前后两根纱线对应点之间的距离，即飞数。组织中斜纹线的角度通常是45°。为了增加角度，可以增加 P_j 和 P_w 的比值。但是，比值过大，会损害织物的力学性能和外观。改变斜纹线角度的另外方法是改变飞数 S，这就是角度斜纹，如图4-2-4所示。

The angle formed in the cloth by a twill weave depends upon: ① the relative ratio of ends and picks per unit space; ② and step, or the rate of advance of one interlacing in respect of the following one. Usually, the angle of twill line in a weave is 45°. To increase the angle, the ratio of P_j and P_w can be increased. However, too large the ratio will damage the mechanical properties and appearance of the cloth. Another method to change the angle of the twill line is completed by changing the step number S, and elongated twills are formed as shown in Fig. 4-2-4.

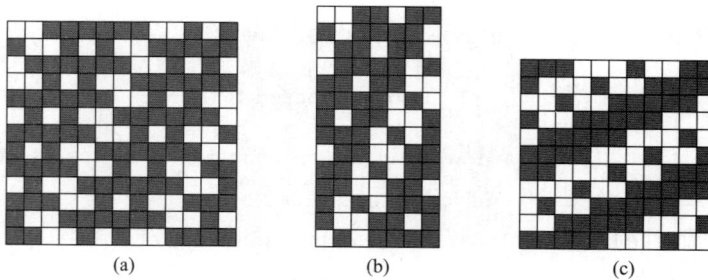

(a)　　　　(b)　　　　(c)

图 4-2-4　角度斜纹
Elongated twills

1. 角度斜纹分类 /Classifications

从图 4-2-4 可知，若 S_j >1，斜纹角度 θ > 45°，则组织称为急斜纹组织；若 S_w >1，斜纹角度 θ < 45°，则组织称为缓斜纹组织。

It can be seen from Fig. 4-2-4 that if S_j>1, the twill angle θ > 45°, and the twill weave is called steep twill weave (high-angle twill weave). If S_w>1, then θ < 45°, and a reclined twill weave (low-angle twill or flat twill weave) is formed.

角度斜纹需要用分式、箭头和飞数联合表示。图4-2-4（a）~（c）中的角度斜纹组织分别表示为 $\frac{3\ 3\ 1\ 1}{1\ 1\ 2\ 1}\nearrow$, $S_j=2$; $\frac{4\ 4\ 1}{1\ 2\ 2}\nearrow$, $S_j=2$; $\frac{5\quad 1}{3\quad 2}\nearrow$, $S_w=2$。

An elongated twill weave must be represented by the combination of fraction, arrow and step. The elongated twill weaves as shown in Fig. 4-2-4(a)(b) and (c) are $\frac{3\ 3\ 1\ 1}{1\ 1\ 2\ 1}\nearrow$, $S_j=2$; $\frac{4\ 4\ 1}{1\ 2\ 2}\nearrow$, $S_j=2$; $\frac{5\quad 1}{3\quad 2}\nearrow$, $S_w=2$ respectively.

2. 斜纹角度计算 /Calculation of Twill Angle

斜纹角度与飞数的关系如图4-2-5所示。经向飞数越大，斜纹角度越陡峭；纬向飞数越大，则斜纹角度越平缓。如果再考虑织物密度，则斜纹角度的计算式为：

The relationship between the step number and the angle of twills is demonstrated in Fig. 4-2-5. Obviously, the larger the value of S_j, the steeper the diagonal lines; the larger the value of S_w, the flatter the diagonal lines. If the fabric density is also considered, the twill angle θ can be calculated as :

$$\tan\theta=\frac{P_j\times S_j}{P_w\times S_w}$$

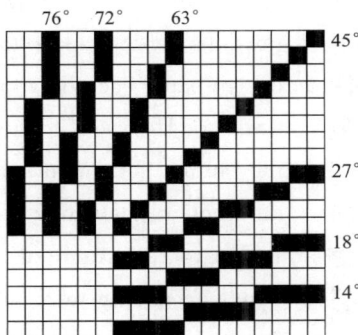

图4-2-5　飞数与斜纹线角度的关系
The relationship between the step number and the angle of twills

3. 角度斜纹组织图的绘制 /Drawing of Elongated Twills

角度斜纹组织图的绘制步骤如下：

（1）选择基础组织，常用加强斜纹、复合斜纹。

（2）确定飞数。急斜纹组织要求 $|S_j|>1$，缓斜纹组织要求 $|S_w|>1$，但应小于基础组织的最长浮长。飞数一般取 2 或 3。

The procedures of designing elongated twills can be described as following:

(1) Choose the foundation twill weave for the elongated twill weave. Double twills and composed twills are the better choices.

(2) Determine the step number. Steep twills require $|S_j|>1$ while reclined twills require $|S_w|>1$. To get a better effect, $|S|$ should be less than the longest float in the

（3）计算基础组织循环根数 R_b。R_b 等于分式中所有分子与分母之和，即：

（4）计算角度斜纹组织的循环根数，并在意匠纸上确定组织范围。

对于急斜纹组织：

对于缓斜纹组织：

lcm 表示最大公约数。

（5）绘制组织图。急斜纹组织先按照基础组织规律，从左边第一纵行绘图，然后按照 S_j 依次完成其他纵行；但是缓斜纹组织要从最下第一横行开始，按照基础组织规律绘图，然后按照 S_w 依次完成其他横行。

图 4-2-6 展示了两个角度斜纹绘制的例子。图 4-2-3（a）为急斜纹组织 $\frac{5\quad4}{2\quad1}\nearrow$，$S_j$=2，其基础循环根数 R_b =5+2+4+1=12，故 R_j =12/2=6，R_w =12。图 4-2-6（b）为缓斜纹组织 $\frac{5\quad1}{3\quad3}\nwarrow$，$S_w$=−3 其基础循环根数 R_b =5+3+1+3=12，故 R_j =12，R_w =12/3=4。

4. 角度斜纹组织的应用 /Applications of Elongated Twills

急斜纹组织织物的斜纹线角度远大于45°，外观粗糙，斜纹线明显，立体感强，常用于制织棉纺和精纺织物，常用作各种外穿制服的

foundation weave, therefore, $|S|$ is usually 2 or 3.

(3) Calculate the size of foundation weave R_b。R_b equals to the sum of all the numerators and the denominators of the fraction.

$$R_b=\sum(n_i+m_i)$$

(4) Calculate R_j and R_w of the elongated twill weave and define the square range of the weave on design paper.

For a steep twill weave:

$$R_j=\frac{R_b}{lcm(R_b,|S_j|)}\ ,\ R_w=R_b$$

For a reclined twill weave:

$$R_w=\frac{R_b}{lcm(R_b,|S_w|)}\ ,\ R_j=R_b$$

Here, lcm is the largest common factor.

(5) Draw the design. For designing a steep twill, draw the marks on the first vertical space on design paper according to the fraction, then copy the marks of the previous vertical space to the succeeding vertical space on design paper according to S_j until all the vertical spaces are finished. However, for reclined twills, draw the marks on the bottom horizontal space on design paper according to the fraction, then copy the marks of the previous horizontal space to the succeeding horizontal space on design paper according to S_w until all the horizontal spaces are finished.

Fig. 4-2-6 demonstrates how to draw elongated twills. Weave as shown in Fig. 4-2-6(a) is $\frac{5\quad4}{2\quad1}\nearrow$，$S_j$=2，$R_b$=5+2+4+1=12，thus，$R_j$=12/2=6，$R_w$ =12。Weave as shown in Fig. 4-2-6(b) is $\frac{5\quad1}{3\quad3}\nwarrow$，$S_w$=−3，thus，$R_b$ = 5+3+1+3=12，and R_j =12，R_w =12/3=4。

The twill lines in steep twill weaves are greater than 45°. The fabric is of greater thickness, rough appearance, steep inclination, and strong three-dimensional sense. Steep twill weaves are widely used in cotton and worsted

面料。图 4-2-7 是马裤呢的常用组织，高经密低纬密配置，飞数为 2，斜条陡峭，多用于军服。图 4-2-8 是毛直贡呢常用的组织，风格类似缎纹，但斜纹线角度可达 75°。图 4-2-4（a）所示的组织用于棉克罗丁（缎纹卡其）织物，图 4-2-4（b）所示的组织则用于精纺巧克丁织物，布面上均有 2 条明显并列的斜纹条。缓斜纹组织并不常见，偶尔用于粗纺毛织物。

fabrics, especially in uniform for outwears. Weave in Fig. 4-2-7 is used in whipcord in green color as army uniform. The prominent, indented, steep twill weave is produced by having the warp closely sett and the weft more open, and a special weave in which the twill interlacings are "stepped-up" two weft yarns to give a steeper twill line. The weave in Fig. 4-2-8 is used in worsted Venetian fabric which has the style of satin but the twill angle can be as high as 75°. Weave as shown in Fig. 4-2-4(a) is used in keluotine or satin kahki, and weave as shown in Fig. 4-2-4(b) is applied to worsted Tricotin. Both the fabrics have two clear diagonal lines on the surface. Flat twills are occasionally used in woolen fabrics.

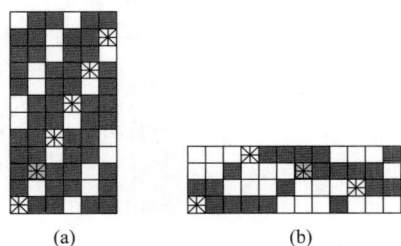

(a)　　　　　(b)

图 4-2-6　角度斜纹绘制
The drawing of reclined twills

图 4-2-7　马裤呢
Whipcord

图 4-2-8　毛直贡呢
Worsted Venetian

四、曲线斜纹组织 /Curved Twills

曲线斜纹是曲线效应的斜纹组织，通过连续变化飞数构成急 / 缓斜纹，产生波浪斜纹线。斜纹可以只沿一个方向［图 4-2-9（a）］，也可以不断反转方向［图 4-2-9（b）］。经向飞数不断变化，可得到经向曲线斜纹（图 4-2-9）；若纬向飞数不断变化，则得到纬向曲线斜纹（图 4-2-10）。

Curved twill is a class of twill weave producing wavy twill lines by continuously changing of steps to form the steep and/or reclining twill to obtain fine curved effects. The weaves may run in only one direction［Fig. 4-2-9(a)］or may reverse［Fig. 4-2-9(b)］.

1. 曲线斜纹组织构成 /Construction of Curved Twill

斜纹基础组织保持不变，但是飞数 S 不断变化。图 4-2-9（a）（b）

The curved twill is based on a twill foundation with changing steps. The weaves as shown in Fig. 4-2-9(a)

(a)

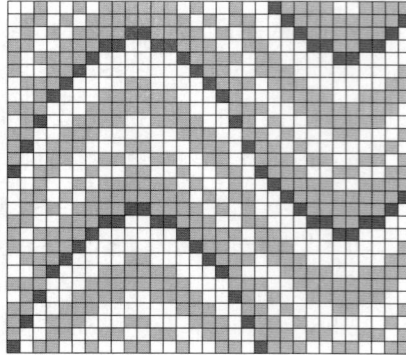
(b)

图 4-2-9　经向曲线斜纹

A warp curved twill

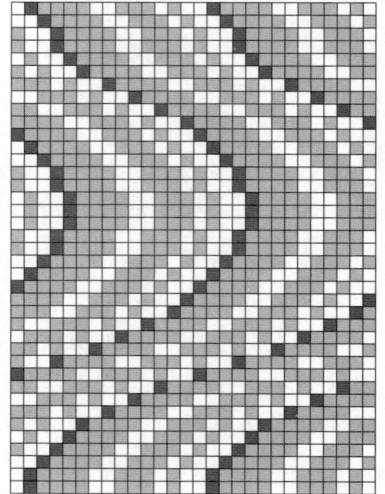

图 4-2-10　纬向曲线斜纹

A weft curved twill

所示经向曲线斜纹的基础组织分别是 $\dfrac{4\ 1\ 1}{2\ 2\ 3}$ 复合斜纹、$\dfrac{4\ 1\ 1}{1\ 1\ 3}$ 复合斜纹，图 4-2-10 纬向曲线斜纹的基础组织是 $\dfrac{5\ 3\ 1}{2\ 1\ 2}$ 复合斜纹。

　　飞数之和等于基本组织的循环倍数，或者为 0。图 4-2-9（a）所示组织的经向飞数序列是 2、2、2、2、1、1、1、0、1、0、1、1、1、1、2、2、2、2、1，经向飞数之和 26 是基础组织循环根数 13 的倍数；图 4-2-9（b）所示组织的经向飞数序列是 2、2、2、1、1、1、1、0、1、0、-1、0、-1、-1、-1、-2、-2、-2、-1、-1、-1、-1、0、-1、0、1、0、1、0、1、1、1，经向飞数之和为 0；图 4-2-10 的纬向飞数序列为 0、1、1、1、1、2、2、2、3、3、3、2、2、2、1、1、1、1、1、0、1、0、0、-1、0、-1、-1、-1、-1、-2、-2、-2、-2、-1、-1、

(b) are based on $\dfrac{4\ 1\ 1}{2\ 2\ 3}$ and $\dfrac{4\ 1\ 1}{1\ 1\ 3}$ respectively while weave in Fig. 4-2-10 is based on $\dfrac{5\ 3\ 1}{2\ 1\ 2}$ composed twill weave.

　　The sum of the steps is zero or the multiple of the size of the base weave repeat . The sequence of S_j for weave as shown in Fig. 4-2-9(a) is: 2, 2, 2, 2, 1, 1, 1, 1, 0, 1, 0, 1, 1, 1, 1, 2, 2, 2, 2, 1. Thus, the sum of the steps is 26, a multiple of 13, the size of the base weave. For weave as shown in Fig. 4-2-9(b), the sum of sequence of the S_j (2, 2, 2, 1, 1, 1, 1, 0, 1, 0, -1, 0, -1, -1, -1, -1, -2, -2, -2, -1, -1, -1, -1, 0, -1, 0, 1, 0, 1, 0, 1, 1, 1) is zero. For weave in Fig. 4-2-10, the sequence of S_w is 0, 1, 1, 1, 1, 2, 2, 2, 3, 3, 3, 2, 2, 2, 1, 1, 1, 1, 1, 0, 1, 0, 0, -1, 0, -1, -1, -1, -1, -2, -2, -2, -2, -1, -1, 0, -1, -1, 0, and the sum is 14.

0、−1、−1、0，纬向飞数之和为 14。

2. 曲线斜纹织物特点 /Features of Curved Twills

（1）曲线斜纹飞数不断变化，但是基础组织保持不变。

（2）在斜纹线的不同位置，经、纬浮长可能变化，织物结构紧密度可能变化。

(1) The step number S changes from one place to another while the twill basis keeps constant.

(2) The length of the weft /warp float and the firmness of the cloth may vary in different parts of the twill line.

3. 曲线斜纹组织图的绘制 /Drawing of Curved Twills

曲线斜纹组织图的绘制步骤如下：

（1）确定基础组织。任何斜纹组织都可以作为基础组织，但在整个组织循环中必须使用相同的基础组织。一般选用复合斜纹组织。

（2）先确定基本曲线形状。曲线形态可以任意设计，但要达到较好的效果，曲线首尾要连续，因此必须满足以下条件：$\sum S_i = A \times R_b$，$A=0,1,2,\cdots$，i=j 或 w。故 $\sum S_j$ 或 $\sum S_w$ 等于 0 或为基础组织循环纱线数的整数倍。故设计中，经向曲线斜纹起点和终点在同一水平线上，纬向曲线斜纹起点和终点在同一垂直线上。

（3）确定飞数序列。最大飞数必须小于基础组织中最长的浮线长度，以保证曲线的连续。

（4）根据基础组织 R_b 和飞数序列，确定组织循环和绘制范围。经向曲线斜纹的 R_j 为一个循环内经向飞数 S_j 序列的个数，纬纱循环根数 R_w 是基础组织循环根数；纬向曲线斜纹的 R_j 是基础组织循环根数，纬纱循环根数 R_w 而为一个循环内纬向飞数 S_w 序列的个数。

（5）根据基础组织分式，绘制第一根纱线（经曲线斜纹：纵行；纬曲线斜纹：横行）的组织点。根

The drawing of curved twills can be described as following:

(1) Select the twill base. Any twill weave may be used as a basis but the same base must be used throughout the design. Usually, a composed twill is a good choice.

(2) Design the shape of the curve twill. The shape of the undulating can be designed arbitrarily, however, the following conditions must be met to have a better effect: for warp curved twills, $\sum S_i = A \times R_b$, $A=0,1,2,\cdots$, which means the commencing point of the curve must locate at the same level with the final point; for weft curved twills, $\sum S_w = A \times R_b$, $A=0,1,2,\cdots$, which means the commencing point of the curve must locate at the same vertical line with the final point.

(3) The sequence of S_j or S_w is determined after the curve is designed. To make a continuous curve, the largest step number must not exceed the length of the longest float.

(4) Determine the size of the weave repeat according to the composed base and the sequence of the steps. For a warp curve twill, R_j equals to the number of sequence S_j, and R_w equals to R_b, the size of the base weave repeat. For a weft curve twill, R_w equals to the number of sequence S_w, and R_j equals to R_b.

(5) For designing a warp curved twill weave, draw the marks on the first vertical space on design paper according to the fraction of the composed twill base, then copy the marks of the previous vertical space to the succeeding vertical space on design paper according to the sequence of warp steps S_j until all the vertical spaces are finished. However, for a weft twill weave, draw the marks on the

据飞数序列，依次填绘下一根纱线组织点，直至完成所有纱线。

图 4-2-11 所示为一个经向曲线斜纹组织图绘制的例子。

bottom horizontal space on design paper according to the fraction, then copy the marks of the previous horizontal space to the succeeding horizontal space on design paper according to the sequence of weft steps S_w until all the horizontal spaces are finished.

Fig. 4-2-11 demonstrates a good example of designing a warp waved twill.

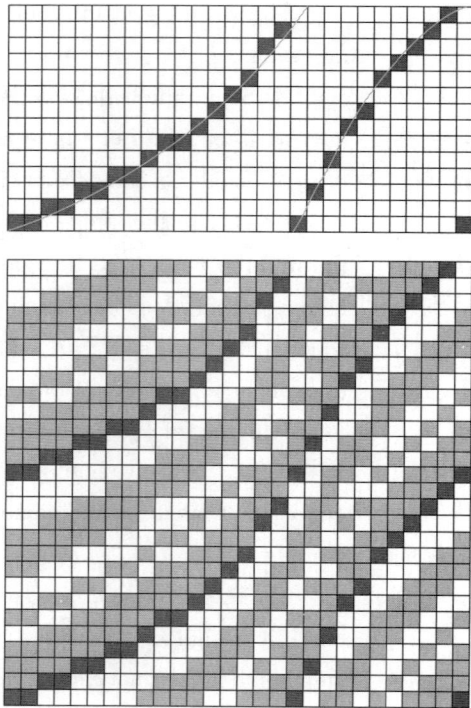

图 4-2-11　经向曲线斜纹组织图绘制
The drawing of a warp curved twill

4. 曲线斜纹组织上机 /Looming for Curved Twills

曲线斜纹穿综所需综页数与基础斜纹组织所需综页数相同。经向曲线斜纹采用照图穿法，而纬向曲线斜纹一般采用顺穿法。每筘齿穿入 2 ~ 4 根经纱。

The number of heals for weaving the fabric of a curved twill needed is the same as the base twill weave. The warp curved twill is drafted by design while the weft curved twill is usually straight drafted. 2 ~ 4 ends are dented-in per split of reed.

5. 曲线斜纹组织的应用 /Applications of Curved Twills

曲线斜纹用途不太广泛，因为有长浮长线，且不断变化，牢

This class of design is only use to a limited extent, as there is the disadvantage that the length of the weft float and

固程度不足。曲线斜纹可用于花式条纹棉织物、窗帘、影院银幕、床罩等。

the firmness of the fabric vary in different parts of the twill line. Cured twills usually used for the fancy striped cotton fabrics, curtains, film screen, bed coverings, etc.

五、山形斜纹组织 /Waved Twills

斜纹组织每隔一段距离翻转斜纹线方向，产生山峰状或 Z 字形的山形斜纹。在经纱方向翻转斜纹方向，形成水平波纹或山峰状的经山形斜纹（图 4-2-12）；在纬纱方向翻转斜纹反向，形成垂直波纹或 Z 字形的纬山形效果（图 4-2-13）。山形斜纹组织的缺点是在变向时会出现过长的浮长线。

The waved twill is a twill weave with a wave or zig-zag design produced by reversing the direction of the twill at intervals. The reversal can occur either upon a warp end, in which case a horizontal wave as shown in Fig. 4-2-12 is produced, or upon a weft pick which results in a vertical wave or a zig-zag effect as shown in Fig. 4-2-13. The disadvantage of the weave is that overlong float may appear at the reversal.

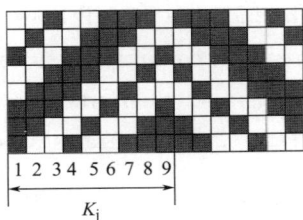

图4-2-12　经山形斜纹组织
The horizontal waved twill

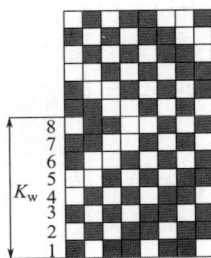

图4-2-13　纬山形斜纹组织
The vertical waved twill

1. 山形斜纹组织特点 /Features of Waved Twills

在经（纬）山形斜纹中，若组织斜向翻转的间隔距离为 K_j (K_w)，那么第 K_j (K_w) 根经（纬）纱将是组织的对称中心。因此，第一根经（纬）纱到对称中心经纱的距离决定了组织循环根数。若经（纬）山形斜纹的基础组织为复合斜纹 $\frac{n_i}{m_i}$，组织循环为 R_b，为了避免转向处产生双经（纬），则组织循环数计算式为：

In a horizontal/vertical waved twill, if the length of the interval to reverse the direction of the twill is K_j (K_w), the end/pick No. K_j (K_w) will be the symmetrical center of the weave. So, the length between the first end/pick to the symmetrical center determines the size of the warp/weft repeat of the weave. If the composed twill base is $\frac{n_i}{m_i}$, then the size of the base weave is R_b. In order to avoid doubling threads happen in the turning of the twill direction, the size of the waved twill is calculated by:

$$R_b = \sum(n_i + m_i)$$

经山形斜纹组织：

纬山形斜纹组织：

For horizontal waved twills:

$R_j=2K_j-2$, $R_w=R_b$

For vertical waved twills:

$R_w=2K_w-2$, $R_j=R_b$

2. 山形斜纹组织图的绘制 /Drawing of Waved Twills

山形斜纹组织图的绘制步骤如下：

（1）选定基础组织，确定 K_j (K_w) ，计算 R_b ，计算 R_j 和 R_w ，确定组织图范围。

（2）从第一根经（纬）纱到第 $K_j (K_w)$ 根经（纬）纱，按顺序填绘基础组织，要求 S_j =+1 和 S_w =+1。

（3）从第 $K_j+1(K_w+1)$ 根经（纬）纱到最后一根经（纬）纱，斜纹线方向翻转，仍根据基础组织，但按照斜向相反的方向（即经山形斜纹 S_j =−1 或者纬山形斜纹 S_w =−1）填绘组织点。

图 4-2-12 是按照该步骤绘制的经山形斜纹组织图，基础组织为 $\frac{3\ 1}{2\ 2}\nearrow$ ， $K_j=9$ ；图 4-2-13 所示纬山形斜纹组织图的基础组织为 $\frac{1\ 2\ 2}{1\ 1\ 1}\nearrow$ ， $K_w=8$ 。需要说明的是，①即使基础斜纹起点不同，产生的山形斜纹也可以是相同的；②设计纬山形斜纹时，不要从第一根经纱开始填绘组织点，而是应该根据基础组织表达式，从第一根纬纱开始填绘组织点。以 $\frac{2\ 1}{2\ 1}$ 为基础组织， $K_w=6$ 的纬山形斜纹组织不能沿经向绘制成图 4-2-14（a）所示组织，而是要沿着纬向绘制成图 4-2-14（b）所示组织。实际上，图 4-2-14（a）所示组织是以 $\frac{2\ 1}{1\ 2}$

The procedures of drawing a waved twill can be described as following:

(1) Choose the foundation twill for the horizontal/vertical waved twill, determine the length of interval $K_j (K_w)$, calculate the size of foundation weave R_b , then calculate R_j and R_w of the waved twill and define the range of the weave on design paper.

(2) From the end (or weft) No .1 to No. $K_j (K_w)$, the marks are drawn according to the base weave, requiring S_j =+1 and S_w =+1.

(3) From the end (pick) No. K_j+1 (K_w+1) to the final one, the direction of the twill lines are reversed, and the marks are drawn according to the base weave. For horizontal waved twills, S_j =−1, or for vertical waved twills, and S_w =−1.

Fig. 4-2-12 demonstrates a horizontal waved twill which is based on $\frac{3\ 1}{2\ 2}\nearrow$, $K_j=9$ while the vertical waved twill in Fig. 4-2-13 is based on $\frac{1\ 2\ 2}{1\ 1\ 1}\nearrow$, $K_w=8$. It should be noted that ① different commencing point in a base twill won't result in a different horizontal curved twill; ② do not draw the marks from the first end when design a vertical curved twill. The vertical waved twill based on $\frac{2\ 1}{2\ 1}$ twill and $K_w=6$ can't be drawn as weave as shown in Fig. 4-2-14(a) by drawing the first thread along the vertical space on design paper. Instead, the first thread should be marked on the horizontal space on design paper and weave as shown in Fig. 4-2-14(b) is the correct one. Actually, weave as shown in Fig. 4-2-14(a) is based on $\frac{2\ 1}{1\ 2}$ composed twill weave.

为基础组织，K_w=6 的纬山形斜纹组织。

3. 山形斜纹组织上机图 /Looming Plans for Waved Twills

经山形斜纹组织采用照图穿法，因为组织对称，穿综图也如山峰一样对称，故此种穿法也称山形穿法［图4-2-15（a）］，所用综页的数目取决于基础组织的组织循环经纱数。纬山形斜纹组织采用顺穿法［图4-2-15（b）］。通常，每筘齿穿入 2 ~ 4 根经纱。

Horizontal weave twills are drafted by design. Since the weave is symmetrical about the center point, the draft is also called pointed drafting shown in Fig. 4-2-15(a). The first half is straight drafted, and the next half is reversal. The number of the hearlds used is the same as the base twill. However, vertical waved twills are straight drafted ［Fig. 4-2-15(b)］. 2 ~ 4 ends are threaded-in per dent.

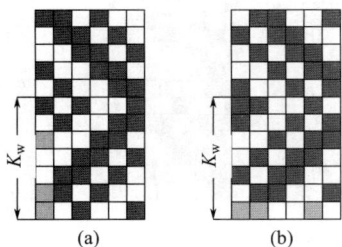

图 4-2-14　纬山形斜纹组织的起点
The commencing point of the vertical waved twills

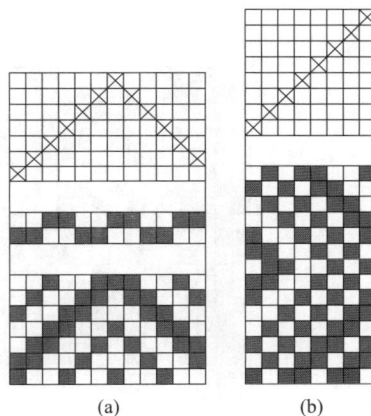

图 4-2-15　山形斜纹组织上机图
The looming plans for waved twills

4. 山形斜纹组织的应用 /Applications of Waved Twills

通过改变山形斜纹转向处之间的间隔K_j或者K_w值，形成一些变化，如图4-2-16所示。山形斜纹组织常用于棉型和毛型织物。

By irregularly changing K_j or K_w, variation of waved twills can be obtained, as shown in Fig. 4-2-16. Waved twills are widely used in cotton-type, worsted-type or woolen-type fabrics.

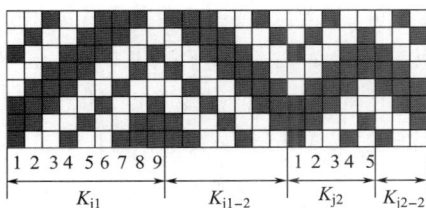

图 4-2-16　变化山形斜纹组织
The variation of a horizontal waved twill

六、锯齿斜纹组织 /Zigzag Twills

在设计变化山形斜纹组织时，若斜纹一侧间隔总比另一侧大，则形成锯齿斜纹组织，如图 4–2–17 所示。斜向翻转后的两侧斜纹线长度不同，对应的组织点既不在同一水平线上，也不在同一垂直线上，而是在一条斜线上，形成了锯齿台阶（飞数）。

In designing the variation of the waved twill, if the interval in one direction of twill is always longer than the reversal direction, a zigzag twill is formed as shown in Fig. 4–2–17. The difference is that the length of each side of the reversal is not equal. The corresponding intersection points are not at the same horizontal or vertical line, and form a serpentine step.

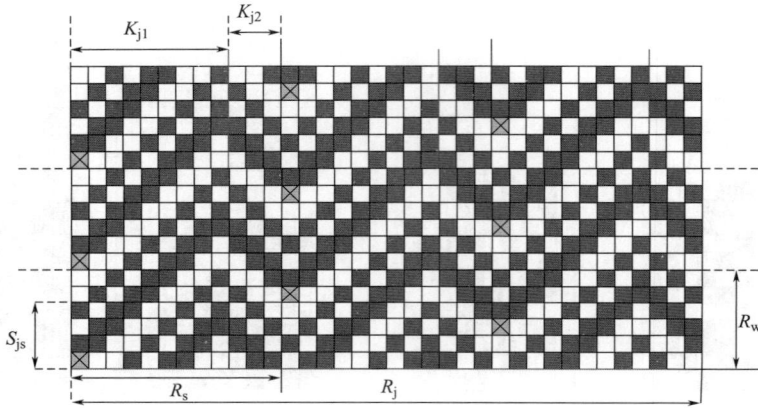

图 4–2–17　锯齿斜纹组织
A zigzag twill

1. 锯齿斜纹组织特征 /Features of the Zigzag Twills

以图 4–2–17 经向锯齿斜纹组织为例，锯齿斜纹组织有如下特点：

（1）若两个相邻锯齿单元的起点之间的垂直距离是经向锯齿飞数 S_{js}。有如下等式成立：$K_{j1}=K_{j2}+2+S_{js}$。若 $S_{js}=0$，则是普通经山形斜纹。对于锯齿斜纹，锯齿飞数取值范围为 $1 \leqslant S_{js} \leqslant K_{j1}-2$。

（2）每个经向锯齿单元长度 $R_s= 2K_{j1}-2-S_{js}$。

（3）每个经向锯齿斜纹的锯齿

The horizontal zigzag twill in Fig. 4–2–17 is taken as an example to elaborate the construction of the weave.

(1) Serpentine step S_{js} in a horizontal zigzag twill is defined as the number of the squares along vertical space between two neighboring serpentine units. Thus, $K_{j1}=K_{j2}+2+S_{js}$. If $S_{js}=0$, the weave is the average horizontal waved twill. For a horizontal zigzag twill, S_{js} is defined within the range as $1 \leqslant S_{js} \leqslant K_{j1}-2$.

(2) $R_s= 2K_{j1}-2-S_{js}$. Here, R_s is the number of threads in a serpentine unit (zigzag unit).

(3) The number of the serpentine units n is calculated by the following formula:

数 n 计算式为：

$$n=\frac{基础斜纹组织的组织循环纱线数 R_b}{基础组织的组织循环纱线数 R_b 和锯齿飞数 S_{js} 的最大公约数}=\frac{R_b}{\mathrm{lcm}(R_b,\ S_{js})}$$

$$n=\frac{\text{Size of base twill repeat } R_b}{\text{The largest common factor of } R_b \text{ and } S_{js}}=\frac{R_b}{\mathrm{lcm}(R_b,\ S_{js})}$$

（4）经向锯齿斜纹组织的循环根数计算式为：

(4) The size of the whole weave repeat is calculated by the following formula:

$$R_j=R_s\times n$$

$$R_w=R_b$$

纬向锯齿斜纹的计算式以此类推。

The same principle can be used to construct a transverse zigzag twill.

2. 锯齿斜纹组织图的绘制 /Drawing of Zigzag Twills

下面以 $\dfrac{2\quad 1}{1\quad 2}$ 为基础组织，$K_j=9$，$S_{js}=4$ 的经向锯齿斜纹组织为例，介绍组织图的绘制方法（图 4-2-18），其步骤如下。

The followings describe how a horizontal zigzag twill (Fig. 4-2-18) is drawn. The zigzag twill is based on a $\dfrac{2\quad 1}{1\quad 2}$ twill, $K_j=9$, $S_{js}=4$.

（1）计算组织循环根数 R_j 和 R_w。

(1) Calculate the size of the weave repeat R_j and R_w.

$$R_b=6，R_s=2\times 9-4=12$$

$$n=\frac{6}{\mathrm{lcm}(6,4)}=3,\ R_j=3\times 12=36,\ R_w=6$$

（2）在方格纸上画出组织图的范围及每个锯齿的范围，并按照锯齿飞数画出每个锯齿第一根经纱的起始组织点，标记出斜向转折点位置。

(2) Delineate the scope of the weave pattern and the scope of each zigzag unit, draw the starting point of the first warp of each zigzag unit according to the serpentine step number, and mark the position of the turning of the twill direction within each serpentine unit.

（3）在当前锯齿单元中，从第 1 根到第 K_j 根经纱按顺序填绘基础组织。然后从第 K_j+1 根经纱开始，按与基础组织相反方向的斜纹线填绘组织点，直至本锯齿单元填完。

(3) In the current zigzag unit, fill in the twill line according to the foundation weave to the end No. K_j. After the end No. (K_j+1), fill in the twill line in the opposite direction until the zigzag unit is finished.

（4）重复（3）步骤，直至所有锯齿单元的组织点绘制完毕，形成完整的锯齿斜纹图案。

(4) Repeat the step (3) until the other zigzag units are finished to form a complete zigzag twill pattern.

3. 锯齿斜纹组织上机 /Looming for Zigzag Twills

锯齿斜纹组织的上机方法与山形斜纹组织相同。

Looming for zigzag twills is similar to that for waved twills.

Looming for Zigzag Twills

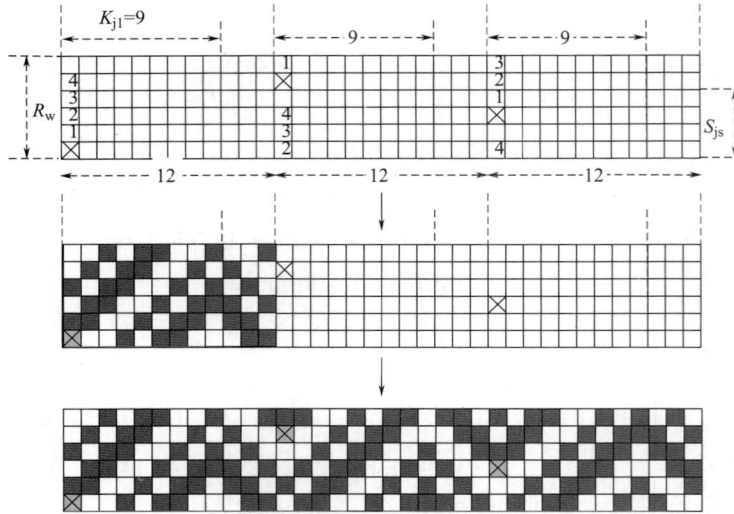

图4-2-18 锯齿斜纹组织图的绘制
The drawing of a zigzag twill pattern

七、破斜纹组织 /Broken Twills

破斜纹是斜纹组织由于一定间隔或者不定间隔的纱线跳跃导致斜纹纹路中断形成的，具有纵向或者横向的条带效应。定间隔跳跃形成等宽条格，不定间隔则导致条格宽度不断变化。条格效应取决于基础斜纹组织和形成破界的方式。形成条格的斜纹斜向可能总是相同，也可能不断变化。广义上说，有三类斜纹组织，可以在组织图上形成破界，分别是人字形破斜纹、变位破斜纹（狭义）和飞断斜纹。

Broken twills are produced by breaking the continuity of any continuous twill weave at either regular or irregular intervals of threads. Such a weave somewhat tends to the formation of stripes, either in the direction of warp or of weft, according to the direction of the twill that is broken lengthwise or crosswise respectively. If the twill is broken at regular intervals of threads, the stripes will be of uniform width; but if broken at irregular intervals, the stripes will be variegated. In either case the stripes will be more or less pronounced according to the character of twill employed, and the manner in which it is broken. The twill may incline in one direction throughout or it may be reversed in alternate stripes, or in any other manner, to emphasise the striped effect. In its broad sense, distinct breaking/cut can be formed in three types of the twills, i.e., herringbone twills, broken twills (in narrow sense), and skip twills.

（一）人字形破斜纹组织 /Herringbone Twills

1. 人字形破斜纹组织构成 /Construction of Herring Bone Twills

类似经山形斜纹，人字形破斜

Like horizontal waved twills, a herringbone twill is

纹组织也是由左右斜纹纵条交替构成。但是在斜纹线发生改变的地方，经纬组织点相反，组织点不连续，且呈间断状态，一般称此为"断界"或"破界"。破界左右两边的组织点完全相反，并且斜纹线从破界处的组织点开始翻转斜纹方向。由于组织点不同，产生交错，使相邻纱线彼此分离，导致斜纹转向处纵条纹效应非常明显，如图4-2-19所示，而且可以避免像山形斜纹那样在斜纹转向处产生长浮长线。

因为人字形破斜纹需要用两条斜纹来形成，织造时需要较多综框。为了减少综框用量，基础斜纹常用双面斜纹。为了使经纬向都有良好的条带效应，经纬纱的细度和密度一般相同。如果斜纹翻转方向是在纬纱方向上进行的，则产生的横向人字形破斜纹组织如图4-2-20所示。

composed of vertical sections that are alternately righthand and lefthand in direction, resembling a fish backbone. However, the twill line does not come to a point where it changes the direction, but instead one twill line is said to "cut" into the other at the point of the reversal. The marks on each side of the cut are exactly opposite to each other, and commencing to run the twill line from this point down in the reverse direction. The effect of the diametrically opposite lifts at the reversal point is to throw the neighboring ends apart from one another to produce "break" or "cut". The method of joining the twills tends to produce a distinct stripe effect as shown in Fig. 4-2-19, and prevents the formation of an extended float where the twill line turns, which is the drawback of waved twills.

Since two twills are used to form a herringbone twill, a large number of the healds will be necessary to produce the fabric. To save the healds, an even-sided twill weave is usually chosen as the base. In order to achieve a better stripe effect in both warp and weft direction, the fabric is woven in a square sett. Transverse or cross-over stripes as shown in Fig. 4-2-20 can be produce if the reversal of the direction is made upon a pick of weft.

2. 人字形破斜纹组织图的绘制 /Drawing of Herringbone Twills

人字形破斜纹组织图的绘制步骤如下：

（1）选定基础组织$\frac{n_i}{m_i}$，确定K_j

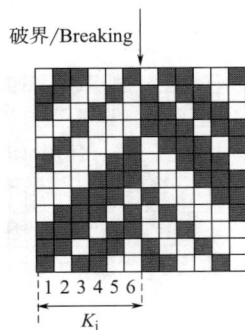

The procedures of drawing a herringbone twill can be described as following:

(1) Choose the foundation twill $\frac{n_i}{m_i}$ for the herringbone

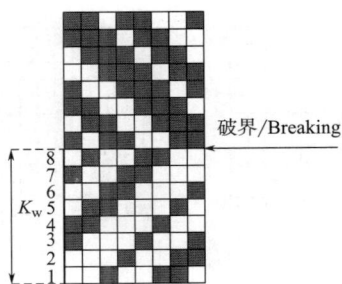

图4-2-19 人字形破斜纹
The herringbone twill

图4-2-20 横向人字形破斜纹
The transverse herringbone twill

(K_w)，计算 R_b，计算 R_j 和 R_w，确定组织所用方格纸的大小。

对于纵向人字形破斜纹：

对于横向人字形破斜纹：

（2）从第一根经（纬）纱到第 $K_j(K_w)$ 根经（纬）纱，按顺序填绘基础组织，要求 $S_j=+1$ 和 $S_w=+1$。

（3）从第 $K_j+1(K_w+1)$ 根经（纬）纱到最后一根经（纬）纱，斜纹线方向翻转，但按照斜向相反的方向（即纵向人字形破斜纹 $S_w=-1$ 或者横向人字形破斜纹 $S_w=-1$）用"底片翻转法"填绘对应的组织点。

图4-2-19 所示人字形破斜纹组织以 $\dfrac{3\ 1\ 2}{3\ 2\ 1}$ 复合斜纹组织为基础，$K_j=6$。故 $R_j=2K_j=12$，$R_w=R_b=$ 3+1+2+3+2+1=12。图4-2-21 所示是该组织的绘制过程。

3. 人字形破斜纹组织的上机和应用 /Looming and Applications of Herringbone Twills

人字形破斜纹一般采用顺穿法。如果是双面组织，可以采用照

twill, determine the length of interval $K_j(K_w)$, calculate the size of foundation weave R_b, then calculate R_j and R_w of the herringbone twill and define the square range of the weave on design paper.

$$R_b=\sum(n_i+m_i)$$

for a horizontal herringbone twill:

$$R_j=2K_j \quad R_w=R_b$$

for a transverse herringbone twill:

$$R_w=2K_w \quad R_j=R_b$$

(2) From the end (or weft) No. 1 to No. $K_j(K_w)$, the marks are drawn according to the base weave, requiring $S_j=+1$ and $S_w=+1$.

(3) From the end (pick) No. K_j+1 (K_w+1) to the end (pick) No. $2K_j$ $(2K_w)$, the direction of the twill lines are reversed, and the marks are drawn according to the principle of herringbone reverse. For horizontal herringbone twills, $S_j=-1$, or for transverse herringbone twills, $S_j=-1$.

In designing the herringbone weave in Fig. 4-2-19, select the base twill $\dfrac{3\ 1\ 2}{3\ 2\ 1}$, and let $K_j=6$. Therefore, $R_j=2K_j=12$, $R_w=R_b=$ 3+1+2+3+2+1=12. Fig. 4-2-21 demonstrates the procedures of drawing.

Herringbone twills are usually straight drafted. If a herringbone twill is derived from an even-sided base,

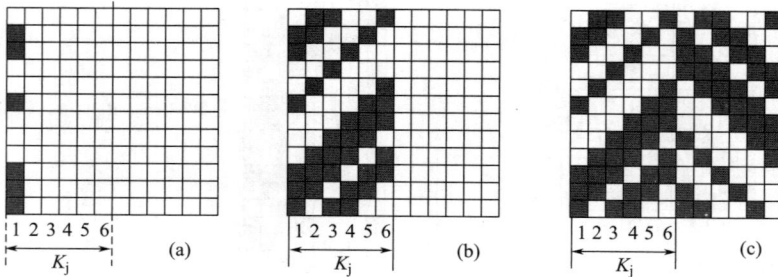

图4-2-21　人字形破斜纹组织图的绘制过程
The drawing of a herringbone twill

图穿法。通常每筘齿穿入 2 ~ 4 根经纱。

　　该组织织物广泛用于床单、大衣、西装、运动套装和女士套装，有时也用来作壁挂和软家具。

broken draft can be used. 2 ~ 4 ends are usually threaded-in per dent. Fabrics of this weave can be extensively used in sheeting, overcoating, suiting, sport coats and dress goods, occasionally employed for hangings and soft furnishings.

（二）变位破斜纹 /Broken Twills（In Narrow Sense）

　　通过改变正则复合斜纹组织穿综方式，使斜纹方向反向，产生类似人字形的条纹组织，称变位破斜纹，如图 4-2-22 所示。

　　在图 4-2-22 所示的组织中，正则斜纹的连续斜纹线被不断改变的穿综方向所终止。按照原来穿综顺序进行到一半后，另一半反向穿综的方法得到图 4-2-22（b）（d）和（f）所示的组织，这样后一半的经纱斜向反向。双面斜纹得到的组织双面平衡 ［图 4-2-22（d）（f）］，产生的破界与人字纹一样。但是，图 4-2-22（b）所示的组织由 $\frac{1}{3}$ 斜纹变化而来，也称四枚缎纹，并不

In broken twills, the direction of the regular twill is broken (usually by drafting) to produce stripes resembling "herringbones", as shown in Fig. 4-2-22.

As regarding the weaves in Fig. 4-2-22, the continuity of the regular twills is stopped by the frequent reversals of the direction in drafting. The weaves as shown in Fig. 4-2-22(b)(d) and (f) are obtained by stopping the orderly progression of the regular twill half-way through the repeat and running the ends in the second half the repeat in reverse order. Even-sided twills as shown in Fig. 4-2-22(d) (f) result in well balanced effects upon "break", which is just the same as the herringbone twill. However, weave as shown in Fig. 4-2-22(b), derived from $\frac{1}{3}$ twill and known as satinette, doesn't show twill effect.

The above weaves are derived by break and reversal

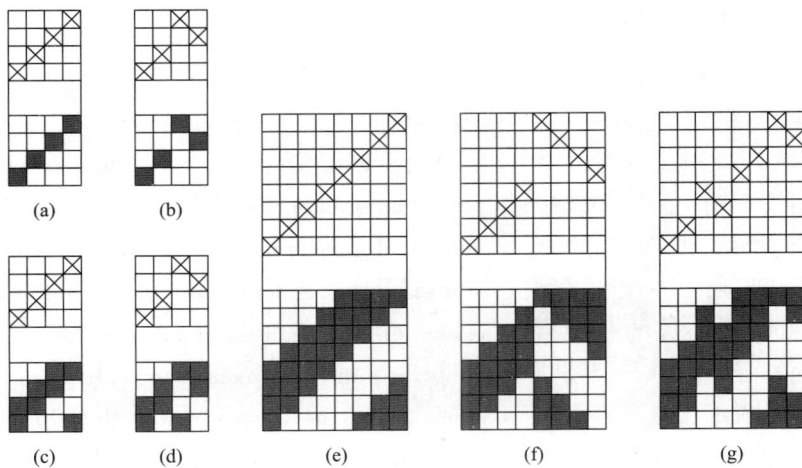

图4-2-22　变位破斜纹
Broken twills (in narrow sense)

产生斜纹效应。

上述组织是通过穿综到一半后，中断斜纹纹路后再反向得到整个循环。实际上，该方法还可以多次中断斜纹纹路，得到图4-2-22（g）所示的组织和图4-2-23所示的组织。

从图4-2-22（b）（g）所示的组织和图4-2-23（b）所示的组织可以看出，采用移位方式得到破斜纹的破界不一定明显。

half-way through the repeat. Actually, the methods can be used by breaks of the draft at more frequent intervals as shown in Fig. 4-2-22(g) and Fig. 4-2-23.

From weaves as shown in Fig. 4-2-22(b)(g) and Fig. 4-2-23(b), the breaking of the broken twill weave derived by transposing might not be as distinct as that of herringbone twills.

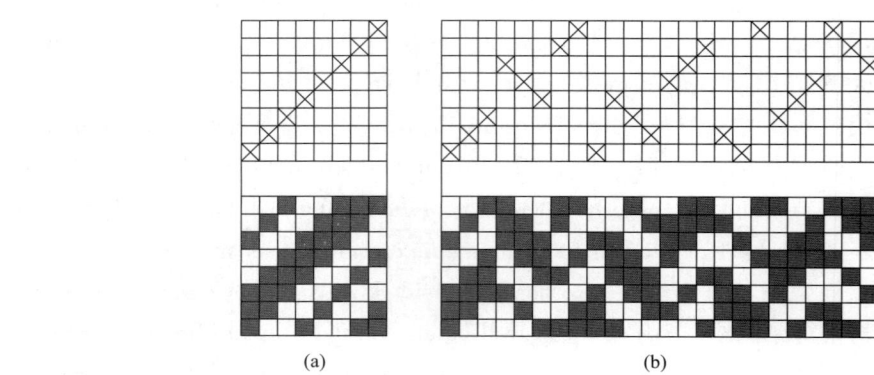

(a)　　　　(b)

图4-2-23　多重变位破斜纹
Transposed twills

（三）飞断破斜纹 /Skip Twills

飞断破斜纹是通过对正则斜纹进行"取纱飞跳"得到的，如图4-2-24所示，特别适合等经纬浮长的加强斜纹织物。正则斜纹中任何数量的连续纱线和飞跳数值都可以得到飞断破斜纹，但是飞跳数值最好是组织循环数的一半再减一。在双面加强斜纹中，经纬浮点恰好相对，形成细小破界，斜纹的纹路中断。与人字形破斜纹不同的是，飞断破斜纹的破界两侧斜向不是相反的，而是完全相同，如

Skip twill construction as shown in Fig. 4-2-24 consists of "entering and skipping" the threads of an ordinary twill. The manner is particularly applicable to twills that are composed of equal warp and weft float. Any number of threads may be entered and skipped respectively at a place, but generally the most suitable number to skip is one less than half the number of threads in the repeat of the twill. If the latter condition is observed in certain even-sided twills, the warp and weft floats oppose each other, and a fine line or "cut" is made where the twill is broken. As will be seen from the examples given in Fig. 4-2-24 the "broken" portions of the twills are not in this method

图 4-2-24 所示。若在纬纱方向取纱填入并飞跳，则形成横向飞断破斜纹，如图 4-2-25 所示。

以图 4-2-24 为例介绍"取纱飞跳"原理。图 4-2-24（a）所示组织是由一个 $\dfrac{3}{3}$ 斜纹按照连续取 5 根经纱，再飞跳 2 根经纱得到的。图 4-4-24（b）所示组织则连续取 4 根经纱，飞跳 2 根经纱得到。而图 4-2-26（a）所示组织则同样在 $\dfrac{3}{3}$ 斜纹中连续取 4 根经纱，飞跳 2 根经纱，连续取 4 根经纱，再飞跳 2 根经纱得到，图 4-2-26（b）所示组织的取纱和飞跳次序更为复杂。

alternately reversed as herringbone twill, but run in the same direction. If designs are constructed by filling and skipping the picks, the cross-over skip twills are formed as shown in Fig. 4-2-25.

The method of "entering and skipping" is illustrated in Fig. 4-2-24, at weave as shown in Fig. 4-2-24(a) the threads are arranged in the order of 5 entered and 2 skipped, while at weave as shown in Fig. 4-2-24(b) the threads are arranged in the order of 4 entered and 2 skipped. The designs can also be formed by entering unequal numbers of threads, as shown in Fig. 4-2-26. In Fig. 4-2-26(a), the $\dfrac{3}{3}$ twill is arranged as 4 entered, 2 skipped, 2 entered and 2 skipped; while at weave as shown in Fig. 4-2-26(b) the weave is arranged more complicatedly.

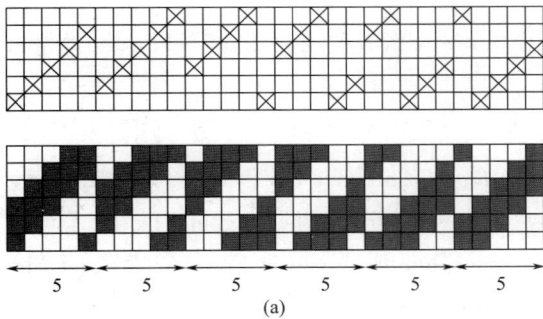

(a)

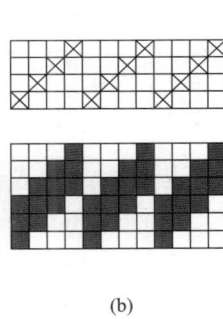

(b)

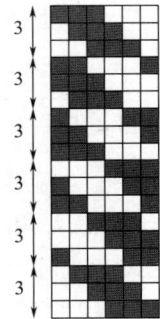

图4-2-24 经向飞断破斜纹
Vertical skip twills

图4-2-25 横向飞断破斜纹
The cross-over skip twill

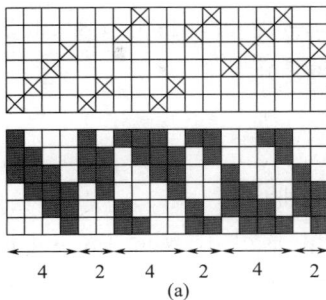

(a)

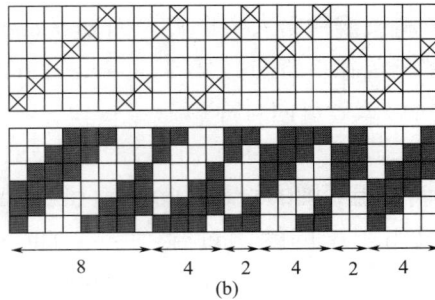

(b)

图4-2-26 不规则间断经向飞断破斜纹
Skip twills with irregular intervals

这种"取纱飞跳"法不局限于等经纬浮长的加强斜纹，如果恰当安排，任何斜纹都能得到较好的破界效果，且不改变原有组织结构的牢固程度。

The system of "entering and skipping" is by no means limited to twills which are composed of equal warp and weft float, but can be used with good results in re-arranging the threads of almost any type of twill. An advantage of the construction has little or no effect upon the firmness of the fabric structure.

Diamond Twills

八、菱形斜纹组织 /Diamond Twills

1. 组织构成 /Construction

若同时在经纱和纬纱方向上改变斜纹纹路的方向，则得到菱形斜纹。通常采用如下两种方式：组合经山形斜纹和纬山形斜纹得到菱形斜纹，如图 4-2-27 所示；组合纵向人字形破斜纹和横向人字形破斜纹则得到破菱形斜纹，如图 4-2-28 所示。

图 4-2-27 是以 $\frac{2\quad 1}{2\quad 2}$ 为基础组织、K_j=7、K_w=7 的菱形组织；图 4-2-28 是以 $\frac{2\quad 1}{2\quad 2}$ 为基础组织、K_j=6、K_w=6 的破菱形组织。这里的基础组织、K_j 和 K_w 的含义与山形斜纹或者人字形破斜纹完全一致。从图 4-2-27 和图 4-2-28 可以看出，这两种菱形斜纹都由 4 部分组成，如图 4-2-29 和图 4-2-30 所示。第 1、第 2 部分形成经山形斜纹或人字形破斜纹，第 1、第 3 部分形成纬山形斜纹或横向人字形破斜纹。第 4 部分与第 2 部分或第 3 部分对称（菱形）或者反向对称（破菱形）。

The weaves are the collection of lozenge-shaped designs obtained by reversing a twill in both warp and weft. The design is obtained by either combination of horizontal waved twills and vertical waved twills or combination of herringbone twills and transverse herringbone twills. The former is a diamond as shown in Fig. 4-2-27, and the latter is a diaper as shown in Fig. 4-2-28.

The diamond weave in Fig. 4-2-27 is based on $\frac{2\quad 1}{2\quad 2}$ twill weave with K_j=7 and K_w=7 while the diaper weave in Fig. 4-2-28 is based on $\frac{2\quad 1}{2\quad 2}$ twill weave with K_j=6 and K_w=6. The base twill weave, K_j and K_w are the same as for waved twills or herringbone twills. As will be seen from the examples given in Fig. 4-2-27 and Fig. 4-2-28, both a diamond and a diaper weave comprise of 4 quarters, as shown in Fig. 4-2-29 and Fig. 4-2-30. Part 1 and Part 2 form a horizontal waved twill or herringbone twill while Part 1 and Part 3 form a vertical waved twill or transverse herringbone twill. Part 4 is just symmetrical (for a diamond weave) or diametrically opposing (for a diaper weave) to Part 2 or Part 3.

2. 菱形斜纹组织图的绘制 /Drawing of Diamond Twills

菱形斜纹组织图的绘制过程如下：

（1）选择基础组织，一般是双面复合斜纹。确定 K_j 和 K_w，若 K_j

The procedures of drawing a diamond weave are described as following:

(1) Select a suitable base twill, usually an even-sided composed twill, and determine K_j and K_w. It is better that K_j

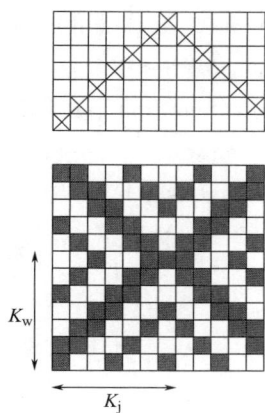

图4-2-27　菱形斜纹
The diamond weave

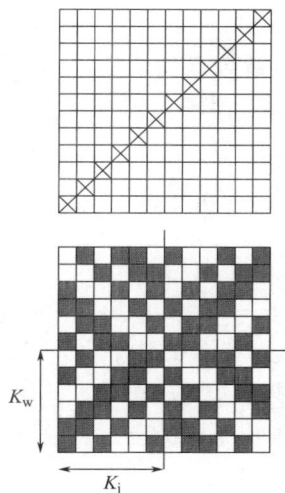

图4-2-28　破菱形斜纹
The diaper weave

图4-2-29　菱形斜纹组成
Components of the diamond weave

图4-2-30　破菱形斜纹组成
Components of the diaper weave

和 K_w 与基础组织循环根数 R_b 相同，效果更好。

（2）计算组织循环数，确定组织循环范围。

对 于 菱 形 斜 纹：$R_j=2K_j-2$，$R_w=2K_w-2$；对 于 破 菱 形 斜 纹：$R_j=2K_j$，$R_w=2K_w$。

（3）填绘第1部分组织点。在 K_j 和 K_w 的范围内，根据基本斜纹填绘组织点。它是菱形斜纹或菱形花纹的基础部分。

如果 K_j、$K_w \le R_b$，直接填绘基础斜纹。

如果 K_j、$K_w > R_b$，将基础斜纹延伸几次，以确保它可以覆盖菱形斜纹或菱形花纹。根据不同的起点，分步填绘菱形斜纹基本组织或

and K_w are just the size of the base twill repeat R_b.

(2) Calculate the weave repeat.

For diamonds: $R_j=2K_j-2$, $R_w=2K_w-2$;

For diapers: $R_j=2K_j$, $R_w=2K_w$.

(3) Fill the marks in Part 1. Fill the marks within the range of K_j and K_w, and draw the marks according to the base twill. It is the base part of the diamonds or diapers.

If K_j or $K_w \le R_b$, fill the base twill directly.

If K_j or $K_w > R_b$, extend the base twill several times to ensure it can cover the diamond or diaper base, then take some part (different commencing point) to fill the diamond or diaper base.

(4) Obtain Part 2, Part 3 and Part 4 of the diamond or diaper weave by the principle of drawing the waved twill or herringbone twill.

In drawing the diamond weave based on $\dfrac{2\quad 1}{2\quad 2}$ as

菱形花纹基本组织。

（4）通过山形斜纹或破斜纹的对称规律，绘制第2、第3、第4部分的组织点。

在绘制图4-2-27以 $\frac{2\ \ 1}{2\ \ 2}$ 为基础的菱形斜纹组织时，$R_j=2K_j-2=2\times7-2=12$，$R_w=2K_w-2=12$。若 $K_j=13$，$R_w=9$，则得到图4-2-31所示的菱形斜纹组织。需要注意的是：①起点不同会形成不同的菱形斜纹组织，如图4-2-32（a）（b）所示（起点用交叉线表示）；②第一根纱线绘制从经纱开始与从纬纱开始也会形成不同的效果，如图4-2-32（a）（c）所示。

shown in Fig. 4-2-27, $R_j=2K_j-2=2 \times 7-2=12$, $R_w=2 K_w -2 = 12$. If $K_j=13$ and $R_w=9$, then the weave will be drawn as shown in Fig. 4-2-31. It should be noted that ① different commencing point will result in a different weave, which is shown at weaves as shown in Fig. 4-2-32(a)(b); ② different commencing direction（warp-wise, or weft-wise）also results in difference, which is observed by the weaves as shown in Fig. 4-2-32(a)(c).

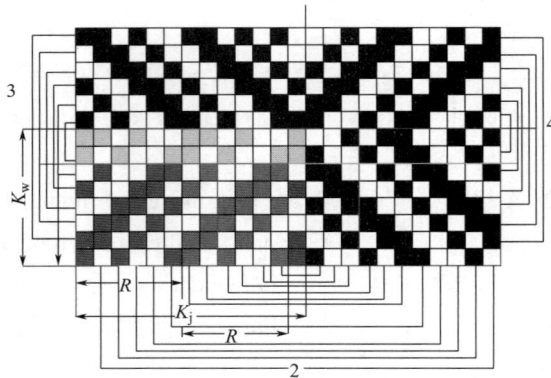

图4-2-31　菱形斜纹组织图的绘制过程
The drawing of a diamond weave

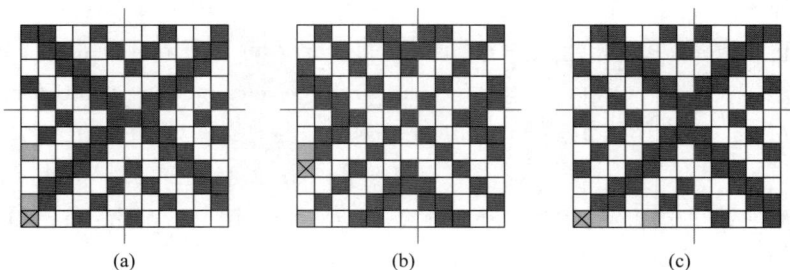

图4-2-32　不同方式绘制的菱形斜纹组织
The drawing of diamond weaves with different order

3. 菱形组织上机及应用 /Looming and Applications of Diamond Weaves

菱形斜纹组织中间可能会形成长浮线，采用山形穿综，使用的综框数量少；破菱形中间没有长浮长线，交界处纹路清晰，但是必须采用顺穿法，使用综框多。

菱形斜纹组织花型对称，变化繁多，图案精美，在各类织物中都有应用。

Looming and Application of Diamond Weave

Long floats may appear in a diamond weave. The point drafting is suitable for diamond weave and less healds may be used. However, less long floats and more distinct cut split effect are possible to achieve by using a diaper weave which must be straight drafted.

The diamonds and diapers are used in all kinds of fabrics due to their characteristics that are symmetrical in pattern, varied and exquisite in pattern.

九、芦席斜纹组织 /Entwining Twills

芦席斜纹由右斜和左斜两个区域的斜纹条采用正交方式组合而成，图案类似菱形。芦席斜纹用基础斜纹组织和在每个区域上的斜纹条的数量来表征。此类组织的织物表面斜条相互垂直。尽管有些异面斜纹也能产生良好效果，但采用双面斜纹的效果更好。

Entwining twills are constructed from regular twills by running sections of twill lines both to the right and to the left so that each section meets other sections at right angles to form diamond-shaped patterns. Based on the ordinary twill, such weaves are described with reference to the foundation twill and the number of twill lines that run in each section. The effects produced by these twills have an entwined or interlaced appearance. The more perfect ones are obtained when the separate sections are composed of even-sided twills, although in some cases uneven-sided twills give good results.

1. 芦席斜纹组织构成 /Construction of Entwining twills

图 4-2-33 中（a）是以 $\frac{2}{2}$ 加强斜纹为基础组织、同一方向为两条平行斜纹线的芦席组织。图 4-2-33（b）是以 $\frac{2}{2}$ 加强斜纹为基础组织、同一方向为四条平行斜纹线的芦席组织。图 4-2-33（c）为以 $\frac{3}{3}$ 加强斜纹为基础组织、同一方向为三条平行斜纹线的芦席组织。

The weave as shown in Fig. 4-2-33 shows an entwining twill constructed by running two twill lines to the right and two to the left, with $\frac{2}{2}$ twill base. The weave as shown in Fig. 4-2-33(b) is an entwining twill with four twill lines of $\frac{2}{2}$ twill in each section while the weave as shown in Fig. 4-2-33(c) with three twill lines of $\frac{3}{3}$ twill in each section.

2. 芦席斜纹组织图的绘制 /Drawing of Entwining Twills

下面以 $\frac{2}{2}$ 加强斜纹作为基础

Take $\frac{2}{2}$ twill base as the example to describe how to

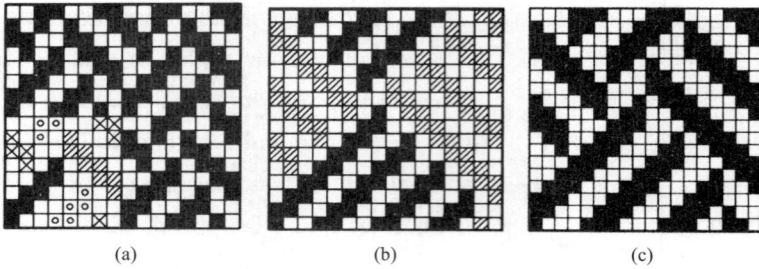

图4-2-33　芦席斜纹
Entwining twills

组织，设计同一方向为四条平行斜纹线的芦席组织。组织图的绘制过程如图 4-2-34 所示，步骤如下：

（1）计算组织循环根数 R_j、R_w。

$R_j = R_w =$ 基础组织的组织循环纱线数 $R_b \times$ 同一方向的平行斜纹线的条数 $= (2+2) \times 4=16$

（2）把组织循环沿经向分为相等的两部分，然后在左半部分，从左下角开始，按基础组织绘制第一条右斜的斜纹线。

（3）在右半部，从第一根斜纹线的顶端向上移动基础组织的连续组织点数，以此作为起点，向下画出相反方向的斜纹线（左斜线）。

（4）画出其他各条右斜的斜纹线，其长度与第一条斜纹线一样长。按基础组织的组织点规律，先在前一条斜纹线向右下方移动两

design an entwining twill weave with 4 twill lines running on the surface of the pattern. The steps are shown in Fig. 4-2-34, and the procedures are explained as following:

(1) Calculate R_j and R_w.

$R_j = R_w =$ Repeat of foundation twill $R_b \times$ number of twill lines= $(2+2) \times 4=16$

(2) Draw a vertical line and cut the weave in two halves in tandem. From the bottom left corner of the weave, draw the first right-handed twill line to the vertical line according to the foundation twill.

(3) At the right half, from the top of the first right-handed twill line, move to the square that is just over the riser of the first left-handed twill, and draw a reversal twill line to the border of the weave.

(4) Draw the other right-handed diagonal lines, the length of which is the same as the first diagonal line. Take any mark on the first right-hand twill line, move it 2 squares to the lower and obtain a starting point, and draw the twill line outward and upward according to the interlacement of

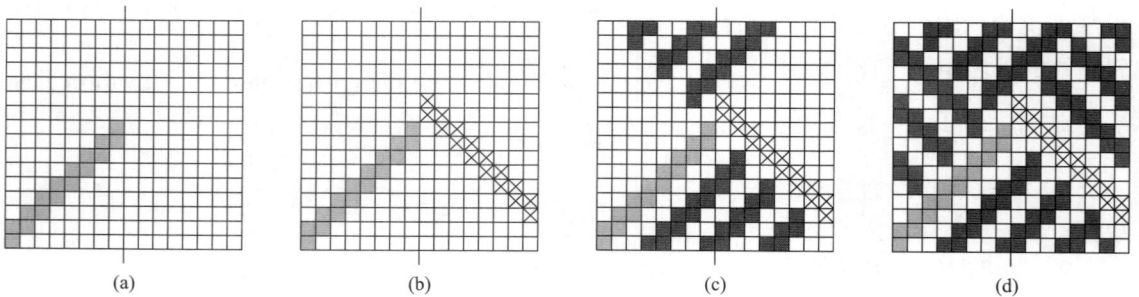

图4-2-34　芦席斜纹组织图的绘制过程
The steps to draw an entwining twill

（浮长长度）根纬纱，根据基础组织的交织情况在上方和右侧绘制斜纹线，直到斜纹线刚好到达左斜的斜纹线为止。然后回到起始点，将斜纹线向左下方绘制，直到斜纹线长度相同为止。如果有第三条右斜纹，那么找到第二根右斜纹的起始点，并画出第三条右斜纹。直至所有斜纹线全部画完。

（5）同样，绘制其他左侧斜纹线。取第一个左斜纹上的任意组织点，并将其向上移动两个方格，即可获得起点。然后画出左斜纹线，直到所有的斜纹线都画完。

3. 芦席斜纹组织上机和应用 /Looming and Application of Entwining Twills

芦席组织一般采用顺穿法或者照图穿法。因为图案精美，可用于各类花呢、女衣呢。

除了上面介绍的斜纹变化组织之外，还有螺旋斜纹（捻斜纹）、夹花斜纹、阴影斜纹，每种斜纹还可以有多种变化。

the ground weave until the twill line just reaches the first left-handed twill line. Then go back to the starting point, draw the twill line downward and leftward until the twill line of the given length is finished. If there is a third right-handed twill, find the starting point based on the second right-handed twill, and draw the third right-handed twill until all the right-handed twill lines are finished.

(5) Similarly, draw the other left-handed twill lines. Take any mark on the first left-handed twill and move it 2 squares upward and obtain the starting point. Then draw the left-handed twill lines until all the lines are finished.

Entwining twills are usually straight drafted or broken drafted. The weaves are found their uses in fancy suiting or lady's dressing.

In addition to the twill derivatives described above, there are also corkscrew twills (twist twills), figured twills, shaded twills. Each sub-class can also have a variety of variations.

第三节　缎纹变化组织 /Satin/Sateen Weave Derivatives

Satin/Sateen Weave Derivatives

通过改变缎纹组织点的个数、飞数，经纬向延伸组织点，扩大组织循环，组合经面纬面缎纹在一起的方式，得到各类缎纹变化组织。缎纹变化组织包括加强缎纹、变则缎纹、重缎纹和阴影缎纹四大类。

Satin/sateen weave derivatives are formed by adding or subtracting risers, modifying skip numbers, extending the interlacing points in both horizontal and vertical direction and enlarging the weave repeat, and combining several satins/sateens together. There are double satins /sateens, irregular satins /sateens, extended satins/sateens and shaded satins/sateens.

一、加强缎纹组织 /Double Satins/Sateens

1. 结构特征 /Characteristics

在缎纹组织的单独组织点周围

Double satins/sateens are derived from satin /sateen

添加一个或多个组织点，形成加强缎纹，如图4-3-1所示。加强缎纹组织仍然保持缎纹的基本特征，组织循环纱线数保持不变，坚牢性提高，并可能增加某些风格。在增加组织点时，确保不产生连续的直线，以免破坏缎纹效应。

2. 常见加强缎纹组织 /Common Double Satins/Sateens

图4-3-1是几种常见的加强缎纹组织。其中图4-3-1（a）所示组织是$\dfrac{5}{3}$或（$\dfrac{5}{2}$）加强缎纹，在单个组织点的上方添加了2个组织点形成，也可认为是$\dfrac{3}{2}$、飞数S_j为2的急斜纹。由于缎纹组织是重构的斜纹组织，急斜纹组织可以认为是加强缎纹组织。图4-3-1（b）（c）（d）所示组织均由八枚纬面缎纹增加经组织点得到，图4-3-1（b）所示组织常用于棉、毛、丝的起绒织物，增加经组织点可以避免纱线移位。图4-3-1（c）所示组织增加了交织次数，常用于被面的花纹部分组织。图4-3-1（d）所示组织为花岗石组织，因纹路如花岗石得名，是$\dfrac{8}{3}$加强缎纹。因为该组织兼备平纹变化、斜纹变化组织特点，故又称斜纹板司呢，其外观平整，常用于毛织物军服。图4-3-1（e）所示组织为$\dfrac{11}{7}$加强缎纹组织，经纱密度高，正面如华达呢，背面如缎纹，故称为缎背华达呢组织。图4-3-1（f）所示组织为$\dfrac{13}{4}$纬面加强缎纹，表面斜纹陡直但

by adding one or more interlacing points (risers or sinkers) around the individual interlacing points of the satin/sateen weave. Therefore, the size of the weave remains unchanged, firmness is increased, and some new style is added, the basic features of satins/sateens are still maintained, however. When adding or subtracting the marks, make sure not form a continuous straight line to weaken the satin/sateen effect.

Some common double satins/sateens are shown in Fig. 4-3-1. The weave as shown in Fig. 4-3-1(a) , a $\dfrac{5}{3}$ (or) $\dfrac{5}{2}$ double satin weave, is derived by adding 2 marks on the original mark in a 5-shaft sateen. The weave can also be regarded as a steep twill of $\dfrac{3}{2}\nearrow$ with S_j=2. Since satin weave is a rearranged twill, therefore, many steep twills can be regarded as double satins. Weaves as shown in Fig. 4-3-1 (b)(c) and (d) are all derived from an 8-shaft sateen weave. The weave as shown in Fig. 4-3-1(b) is usually used in the napped or elysian fabrics from cotton, wool and silk because the added marks will prevent the threads slipping in pulling or napping process. The weave as shown in Fig. 4-3-1(c) is used in the figure part of the quilt covering fabrics due to the intersection is increased by adding the marks. The weave as shown in Fig. 4-3-1(d) is called $\dfrac{8}{3}$ Granite named by the texture. The weave is a combination of plain, twill, and satin, and is also called twilled basket. Barathea, a worsted uniform cloth is made from the weave due to its flat and smooth appearance. The weave as shown in Fig. 4-3-1(e) is derived from an $\dfrac{11}{7}$ satin weave. Usually, the fabric of the weave is highly sett in warp, the face looks like a gaberdine while the back looks as a satin, therefore, the weave is called satin-back gaberdine. The weave as shown in Fig. 4-3-1(f), a $\dfrac{13}{4}$ double sateen, is used in doeskin with steep but faint twill lines on the surface. Weaves as

不明显，常用于毛驼丝锦织物。图 4-3-1（g）（h）所示组织均为 $\dfrac{10}{3}$ 纬面加强缎纹，表面光滑细洁，手感厚实柔软，用于毛色子贡织物。图 4-3-1（i）所示组织为 $\dfrac{10}{7}$ 纬面加强缎纹，每个基础组织点上下左右各增一个经组织点，形成交叉十字线，经纱微微凸起；同时，对角则是十字形纬组织点交叉，纬纱稍微凸起。一个循环内经纬纱线交错次数不多，手感柔软，经纬交替凹凸，呈海绵状，称为海绵组织。

shown in Fig. 4-3-1(g)(h)are both derived from a $\dfrac{10}{3}$ sateen, and used for dice venetian. The fabrics are smooth, fine and clean, touch soft and firm. The weave as shown in Fig. 4-3-1 (i) is derived from a $\dfrac{10}{7}$ sateen, and 4 marks are inserted around each individual mark to form a cross of marks, therefore, the ends are slightly raised. Meanwhile, cross of blanks is also formed at the diagonal positions for each cross of mark, and the weft threads are raised slightly as well. Since the intersections in each weave repeat are low, the threads are slightly raised alternately, and the fabric touches soft and has good absorbency, hence, the weave is named sponge weave.

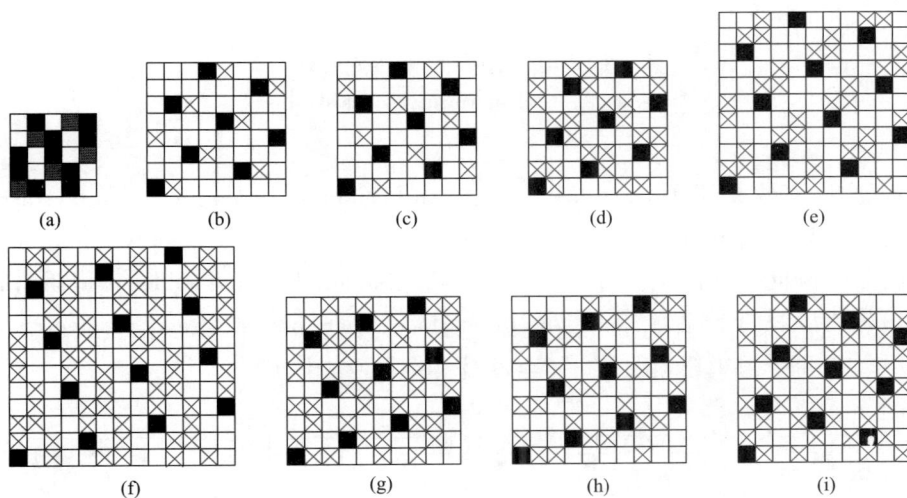

(a)　(b)　(c)　(d)　(e)

(f)　(g)　(h)　(i)

图4-3-1　几种典型的加强缎纹组织
Typical double satins/sateens

二、变则缎纹组织 /Irregular Satins/Sateens

如果一个缎纹组织循环中飞数是变数，称为变则缎纹（有两个以上的飞数值）。

Irregular satins/sateens are those weaves that the step number changes from one place to another. There are at least 2 different step numbers in a weave repeat unit.

1. 使用变则缎纹的原因 /Reasons of Using Irregular Satins/Sateens

（1）不能形成正则缎纹。规则

(1) Regular satins/sateens are unavailable. Regular

缎纹组织纱线循环根数不能是 4 或 6，因为没有比 5 小的整数与 4 或 6 互为质数。由于组织循环数为 4 和 6 的缎纹大小适中，且是偶数，经常要与平纹等组织联合使用。为了减少织造困难，只能采用变则缎纹。图 4-3-2 是四枚和六枚变则缎纹的例子。四枚变则缎纹更多情况下称其为破斜纹，除了单独使用外，经常作为绉组织和复杂组织的基础组织。

satins/sateens cannot be constructed on four or six threads, because no number can be counted which has the common factor with 4 or 6. However, the size of the weave is especially suitable for fabrics, and it is even number which makes it convenient to combine with a plain weave. To reduce the problems in weaving, 4-shaft and 6-shaft irregular satins/sateens are designed as shown in Fig. 4-3-2. 4-shaft irregular satins/sateens or satinettes, but more frequently, known as broken twills, are usually used as the base weaves of crepe weaves or compound weaves as well as use alone.

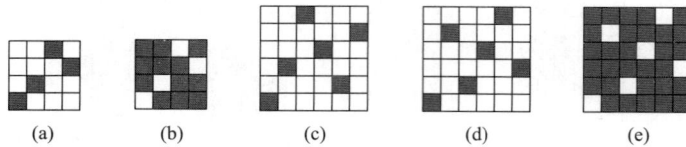

(a)　　(b)　　(c)　　(d)　　(e)

图4-3-2　四枚和六枚变则缎纹
4-shaft，6-shaft irregular satins/sateens

（2）正则缎纹效果不好。七枚正则缎纹，无论如何选择飞数，都有斜路的倾向，如图 4-3-3（a）~（d）所示。为了改善效果，使用图（e）所示的变则缎纹。

(2) Unpleasant effects with a regular satin/sateen. 7-shaft regular weaves, such as weaves as shown in Fig. 4-3-3(a)(b)(c)(d), have the tendency to show twill lines on the surface. Therefore, an irregular sateen as shown in Fig. 4-3-3(e) is introduced.

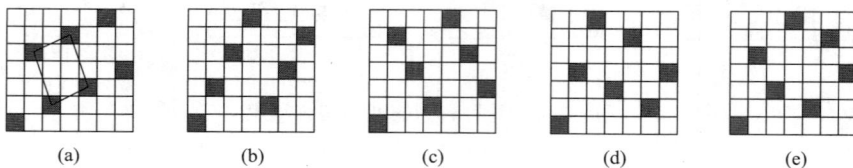

(a)　　(b)　　(c)　　(d)　　(e)

图4-3-3　七枚缎纹
7-shaft sateens

（3）特殊要求。有些缎纹组织点分布较均匀［图 4-3-4（a）］，但为了故意获得不均匀的效果，使用变则缎纹，如图 4-3-4（b）（c）（d）所示的变则八枚缎纹。

(3) Special requirements. Sometimes, in order to obtain an irregular effect, the irregular marks in the weave ［Fig. 4-3-4(a)］ are modified purposely to form irregular sateens as shown in Fig. 4-3-4(b)(c)(d) for an 8-shaft sateen.

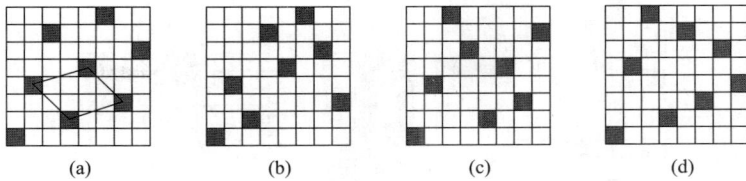

图4-3-4　八枚缎纹
8-shaft sateens

2. 变则缎纹飞数设计原则 /Principles of Modifying the Step Numbers

在设计变则缎纹的飞数时，必须符合如下原则：

（1）任意一个飞数仍要满足 $1<S_i<R-1$ $(0<i<R)$。

（2）各飞数之和应为组织循环根数的整数倍。图 4-3-2（c）（d）中的六枚缎纹纬向飞数 S_w 序列分别为：{4，3，2，2，3，4} 和 {2，3，4，4，3，2}，故飞数之和均为 24，是 6 的倍数。

（3）通常情况下，组织点要分布均匀。

The principles below should be followed in designing the sequence of step number S_i $(0<i<R)$ in an irregular satin/sateen weave:

(1) $1<S_i<R-1$

(2) The sum of S_i should be the multiple of the size of the repeat R. S_w of weaves as shown in Fig. 4-3-2(c)(d) are {4,3,2,2,3,4} and {2,3,4,4,3,2} respectively. The sums of the two sequences are both 24, a multiple of 6, the size of the weave repeat.

(3) As a routine, the marks or blanks should be arranged uniformly.

三、重缎纹组织 /Extended Satins/Sateens

重缎纹组织延长了缎纹组织的经/纬向组织循环根数，也就是延长了组织点的经/纬浮长。因此，面料更加松软。采用三种方法延伸组织点：①沿着纬纱方向，R_j 加倍，S_w 加倍，如图 4-3-5（c）所示；②沿着经纱方向，R_w 加倍，S_j 加倍，如图 4-3-5(d) 所示；③同时沿着经纬纱方向，R_j 和 R_w 加倍，S_j 和 S_w 加倍与 0 交替出现，如图 4-3-5(e) 所示。重缎纹多应用于粗花呢、粗纺女式呢和手帕织物中。

Extended satins/sateens are constructed by doubling the weave repeat and extending the floats. Therefore, the fabrics are softer and even loose. There are 3 ways to extend the floats: ①both R_j and S_w are doubled cross-wise as shown in Fig. 4-3-5(c); ②both R_w and S_j are doubled warp-wise as shown in Fig. 4-3-5(c); ③both R_j and R_w are doubled two-way as shown in Fig. 4-3-5(e), meanwhile, both S_j and S_w double and alternate with zeros. The weaves are applied to tweed, lady's woolen fabric and handkerchief.

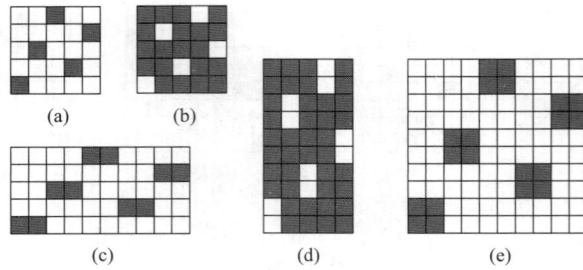

图4-3-5　重缎纹
Extended satins/sateens

四、阴影缎纹组织 /Shaded Satins/Sateens

阴影缎纹组织是由经面缎纹逐渐过渡到纬面缎纹，或者由纬面缎纹逐渐过渡到经面缎纹的变化组织。由于浮长线长度的逐渐变化，在织物表面形成由明到暗或者由暗到明的阴影变化。若阴影由明到暗或者由暗到明仅变化一次，称为单过渡阴影缎纹，如图4-3-6所示；若阴影由明到暗再到明，或者由暗到明再到暗变化两次，称为对称过渡阴影缎纹，如图4-3-7所示。若从经纱方向添加组织点，则称经纱阴影缎纹（图4-3-6）；若从纬纱方向添加组织点，则称纬纱阴影缎纹（图4-3-7）。过渡组织沿经向并列的称为经向阴影组织（图4-3-6，图4-3-7）；过渡组织沿纬向并列的称为纬向阴影组织（图4-3-8）；过渡组织同时沿经纬向的称为双向阴影组织（图4-3-9）。

阴影缎纹组织的循环根数取决于基础组织循环、过渡方向、过渡方法，由于组织循环大，交错复杂，表现影光效果好，多用于棉、毛、丝的提花织物。

Shaded satins/sateens are those satin derivatives in which satin changes to sateen gradually or from sateen to satin. The diagonal shading effect of light and dark changing is produced by increasing or decreasing the float length in the weave gradually. If the shadow changes only from light to dark or from dark to light, then the weave is called single transition shaded satin as shown in Fig. 4-3-6. If the shadow changes from light to dark then to light again, or from dark to light to dark, then the weave is called symmetrical transition shaded satin as shown in Fig. 4-3-7. End shaded satins as shown in Fig. 4-3-6 applied to those that the marks/blanks are inserted on ends while filling shaded satins as shown in Fig. 4-3-7 for inserting marks/blanks on the filling yarn. If the transition weaves are arranged in tandem, then warp-wise shaded satin forms as shown in Fig. 4-3-6; or stacked up and down to form a weft-wise shaded satin as shown in Fig. 4-3-8, or along both directions form a two-way shaded satin as shown in Fig. 4-3-9.

The size of the shaded satin/sateen repeat depends on the base satin, direction of transition, and the way of transition. The weave repeats on many ends and picks, interlaces in complex way, demonstrates a better shadow effect, and is used for jacquard fabrics of cotton, wool and silk.

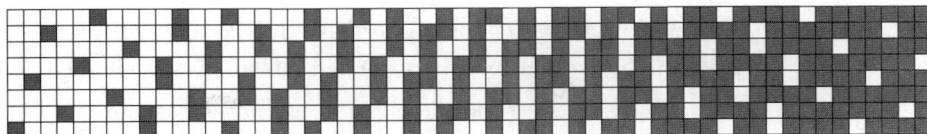

图4-3-6 单过渡经向经纱阴影缎纹
Single transition, warp direction end shaded satin

图4-3-7 对称过渡经向纬纱阴影缎纹
Single transition, warp direction filling shaded satin

图4-3-8 单过渡纬向纬纱
阴影缎纹
Single transition, weft
direction filling shaded satin

图4-3-9 单过渡双向纬纱阴影缎纹
Symmetrical transition, two-way direction filling shaded satin

第四节 传统布边 /Traditional Selvage

Traditional Selvage

在织物两端最边部的织物部分称为布边，布边在织造中作为纬纱的支撑点，在染整中保持恰当的幅宽。故要求布边坚牢、外观平整、边道平直。在精纺毛织物上，往往

The two edge portions of the woven fabric are called selvages (selvedge). Selvages serve as a binding point for the weft when weaving and as an aid to maintain proper width when dyeing, printing, and finishing fabric. The selvage of the cloth should be strong, flat, and parallel to the warp. The

在布边上提花织字，不仅可提高织物的外观质量，而且兼具织物商标的宣传作用。

机织物的布边有多种形式，最常见的是传统普通布边，在有梭织机上不需要特殊装置即可织造。但是新型织机并不形成普通布边。无梭织机的各种布边使边部织物并不磨损，其染整、缝纫效果很好，甚至最终产品不需包边也可得到满意的效果。

这些布边包括折入边、双纬边、双纬针织边、热熔边、绳状边、纱罗边等。在本节中，仅仅介绍普通布边的结构要求。

contraction of a selvage is usually in a firmer construction than body. In worsted fabrics, fancy characters are often woven at the edge, which not only improves the aesthetic effects of the fabric, but also serves as an advertisement of the fabric trademark.

There are many types of selvages. The most common selvage is traditional conventional selvedge, which is formed without making special provision on a shuttle loom weaving plain cloth. However, fabrics woven by new fast looms do not produce conventional selvedges. On shuttleless looms, the cloth is provided with special selvedges that will not allow the fabric to fray at the edges, but will enable it to perform satisfactorily during finishing and making-up, or even give satisfaction in the final article with-out the need for hemming.

The selvages for shuttleless looms include tuck-in selvedge, double-pick interwoven selvedge, double-pick knitted selvedge, fused selvedge, helical selvedge, leno selvedge. In this section, only traditional selvage is introduced.

一、传统布边设计原则 /Principle of Designing Traditional Selvage

1. 布边经纱选择 /Selection of Warp Threads

由于布边在织造、后整理中所承受的机械摩擦力比布身要大得多，故布边的经纱应选用布身中强度和耐磨性好的一组经纱为原料，并注意保持布边与布身的收缩性一致。

Since the mechanical friction of selvage received in weaving and finishing is much greater than that of the body of the fabric, the warps of the selvage should use a group of yarns with good strength and wear resistance. The contraction of the fabric selvage is expected to be consistent with the body.

2. 布边的宽度与密度 /Width and Thread Density

在保证布边作用的前提下，布边宽度以窄为宜。布边的宽度一般为布身幅宽的0.5%，在0.5 ~ 2cm。为使布边平挺，布边的经密应略大于布身的经密，或与布身经密相同。对于高经密高紧度的织物，布边经密与布身经密相同；对于低经密低紧度的织物，布边经密

Under the premise of ensuring the effect of cloth selvage, the width of selvage should be narrow and is generally 0.5% of the width of the body, say, between 0.5cm and 2cm. In order to make the selvage flat, the ends of selvage per unit space should be slightly larger than or equal to the body. For fabrics with high warp density and high compactness, the warp density of the selvage is the same as that of the body. For fabrics with low warp density

比布身经密可提高30% ~ 50%，甚至100%；对于一般织物，布边经密与布身经密相同，或提高10% ~ 20%。

and low compactness, the warp density of the selvage can be increased by 30% ~ 50% or even 100%. For common fabrics, the warp density of the selvage is the same as that of the body, or increased by 10% ~ 20%.

3. 布边组织与布身结构配合原则 /Principle of Matching the Selvage and Body

（1）为减少生产中的困难，边经纱和边组织应尽可能与布身的经纱和布身的组织相同。

（2）合理的布边结构，最好采用平纹、$\frac{2}{2}$或$\frac{3}{3}$经（纬）重平、方平等同面组织为边组织，边部织物的正反两面，保持纬纱受力状态具有较小的差异性。对于单梭箱的有梭织机来说，采用经重平、方平组织为布边时，要考虑第一纬的投向与左、右边组织的配合，如图4-4-1所示，确保边部经纱能织入。

（3）保证边经纱和布身经纱的织缩率一致。

(1) To reduce weaving difficulties, the weave and the warp threads of the selvage should be as similar as possible to that of the body.

(2) To reduce the difference of the inner stress between the two sides of the fabric at the selvage, the following even-side structures are adopted: plain weaves, $\frac{2}{2}$ or $\frac{3}{3}$ warp and weft ribs, and hopsack weaves. For a single-box shuttle loom, the direction of the first picking should conform with the weaves of the left selvage and the right selvage if warp rib weave or hopsack weave is applied, which is shown in Fig. 4-4-1 to ensure the outest ends in selvage can be woven into the structure.

(3) Measures are taken to ensure similar contraction rate in the selvage and the body fabric.

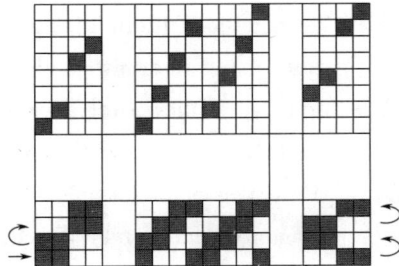

图4-4-1　单梭箱有梭织机第一纬投向与边组织的配合
Correct combination of the direction for the first picking and the selvage weave on a single-box shuttle loom

二、布边织疵及解决方案 /Faults in Selvage and Solutions

1. 布边织疵及产生原因 /Faults in Selvage and Forming Mechanism

织造中，布边可能会产生各种疵点。如果边组织为异面组织，布边织物正、反两面的纬纱受力情况

In weaving, defects may appear at the selvedges. If the selvage is of unbalanced weave, the wefts on the front and back sides of the selvages are stressed differently, and

差异大，往往产生卷边现象。纬纱交织后，纬纱将弯曲横向收缩，导致边部经纱不能平行通过钢筘筘齿，与筘齿摩擦增大，甚至产生边纱断头。

织物内纬纱弹性伸长的恢复（弹性收缩），在织物边部受到的阻力比在织物布身受到的阻力小。因此，边部的经纱密度存在着大于地部经纱密度的内在趋势，导致边部织物的结构相高于地部织物的结构相，边经纱的织缩率与伸长率增大，张力提高，织造中将产生紧边织疵，增加边经纱断头，损害织物的外观和降低织机的生产效率。

如果某些情况下，布边经纱的织缩率小于地经纱的织缩率，则产生木耳边织疵，边道松塌不直。

2. 改善紧边 /Improving Tight Selvage

（1）设法降低边织物的结构相。方法包括减少每筘齿内边经纱的穿入数，采用间歇穿筘（降低边经纱密度）或者增加边综丝内边经纱的穿入根数（相当于边纱变粗或者使用纬重平组织，增加边纱刚度），降低边织物的几何结构相，达到降低经纱织缩率的目的。

（2）提高纬纱织入边部的可能性，提高纬纱可密性，增加纬纱密度和纬纱屈曲波高，降低结构相。方法包括增加边经纱的经向重平组织点，减少边经纱与纬纱的交织次数，提高边部组织的纬向紧度，采用间歇纬的边组织。

3. 改善木耳边 /Improving Wavy Selvage

松弛的木耳边需要采用提高布边结构相的方法来解决。措施有减

curling often occurs. After woven, the weft threads will bend and contract horizontally, so that the warp threads at the selvage cannot pass through the reed wires in parallel, and the friction with the wires increases, and even causes the breakage of threads in the selvedge.

The contraction in width which is caused by the recovery of the elastic elongation of the weft receives less resistance at the selvage than inside the fabric body. The selvage, therefore, is denser than the body, leading to a high geometric structure phase at the selvage, which in turn, resulting in a higher contraction rate in the selvage warp threads. The selvage warp threads are extended more and tend to tight, which might increase warp breakage and damage the fabric appearance and reduce the production efficiency of the loom.

If, in some cases, the contraction rate of the selvage warp is less than that of the body warp, the selvage warps are loose, slack and not straight. The fault is called wavy selvage or stringy selvage.

2. 改善紧边 /Improving Tight Selvage

(1) Try to lower the geometric phase of the selvage. The methods include reducing the number of warps per dent, using intermittent denting plan (to reduce the warp density), or increasing the number of warps in a mail eye of the heddle (equivalent to doubling the warps, using of weft rib weaves or increasing of the warp stiffness) to reduce the geometric phase and reduce the contraction rate of the warp threads.

(2) Improve the possibility of inserting more wefts into the selvedge, improve the capacity of packing the wefts and increase the weft density and weft crimp height, and reduce the geometric phase of the fabric. The method includes using warp rib weaves, reducing the intersections in warp direction, improving the weft tightness, and adopting intermittent weft weaves.

3. 改善木耳边 /Improving Wavy Selvage

To avoid slack wavy selvage, the geometric phase of the selvage should be increased. The methods are as

少边综丝内边经纱的穿入根数，增加每筘齿内边经纱的穿入根数，采用经纱交错次数较多的边组织。

follows: to reduce the number of warp threads in a heddle, to increase the number of warp threads in each reed, and to adopt the weave with more intersections in warp direction.

三、布边设计实例 /Applications

1. 非紧密结构织物 /Untight Fabrics

各种棉平布、中长凡立丁和棉哔叽与部分棉斜纹织物的纬织缩率较大，在织造生产中，边经纱与筘齿存在较大的摩擦。生产这类织物时，若将双根边经纱穿入一根边综丝内，其效果相当于采用较粗的边经纱，具有提高边经纱强力和降低边部结构相的双重作用。部分平纹织物采用$\frac{2}{2}$纬重平组织作为边组织也是同理。

Due to the high contraction rate of the fabrics like cotton plain cloth, valetin of medium length chemical fiber, cotton serge, jean and denim, there is a greater friction between selvage warps and reed wires in weaving. In the production of the fabrics, if two selvage warps are threaded into one heald wire, which is equivalent to using a thicker warp, the dual effect of increasing the warp threads strength and reducing the geometric phase is achieved. The same is true for some plain weave fabrics using a $\frac{2}{2}$ weft rib weave as the selvedge weave.

2. 经向紧密结构织物 /Warp-Wise Tight Fabrics

棉府绸织物、华达呢和卡其类织物等高经密织物的结构相高，经纱织缩率大，纬纱织缩率在2.5%左右，边经纱与筘的摩擦较小，对边经纱无需提出增强的要求，可以每根边综丝内穿入一根边经纱，每筘齿内的边经纱的穿入根数可以小于或等于地经纱的每筘齿内的经纱穿入根数，有时甚至在边经纱空筘，即边部经纱间歇穿筘。通过以上措施，适当地降低织物边部经纱机上密度，以免边部出现过高的结构相。

为了防止斜纹组织的单面卡其织物产生卷边织疵，织造时，可采用如下方法：若经向紧度E_j>100%，采用斜纹人字边组织，每筘齿内地经纱和边经纱均穿入4根；若E_j=90%，采用$\frac{2}{2}$方平组织

In the high warp sett fabrics such as poplin, Gabardine and khaki, etc. the warp contraction rate is high, the weft shrinkage rate is as low as 2.5%, and the friction between the warps and reed is relatively small, there is no need to strengthen the warp thread, and each heald is only threaded into with a single yarn. The number of warp threads per dent at selvage may be less than or equal to the number of warp threads per dent for the body part, and sometimes even empty dent or an intermittent denting plan is applied. Through the above measures, the warp density of the fabric at selvage is reduced properly and an over high geometric phase is prevented.

In order to prevent curling in weaving a drill, several methods are taken. If the covering factor E_j in warp direction exceeds 100%, the herring-bone weave is applied to selvage, and 4 warps are drawn per dent in both selvage and body; while E_j equals to 90%, the $\frac{2}{2}$ hopsack weave is applied to selvage, and 4 warps are drawn per dent in both

为边组织，每筘齿内地经和边经纱均为 4 根穿入；若 E_j=85%，采用 $\dfrac{2}{2}$ 方平组织为边组织，每筘齿内地经纱穿入 4 根，边经纱穿入 6 根；对于 5 枚缎纹组织的棉直贡织物，采用方平为边组织，每根边综丝内穿入一根边经纱，每筘齿内地经纱和边经纱均穿入 3 根或 4 根。

3. 纬向紧密结构织物 /Weft-Wise Tight Fabrics

棉横贡和灯芯绒等织物的纬纱密度大，纬纱织缩率较大，边经纱与筘齿间摩擦剧烈，使边部经密与布身经密差异更大。对边经纱不仅要求强力高，对布边结构更有降相的需要。

在设计横贡织物的边部结构时，往往采用每根边综丝内穿入多根边经，并辅以间歇穿和经向重平组织点。图 4-4-2 为棉横贡织物的边部织造上机图解。地经纱采用每筘齿内穿入 2 根或 8 根经纱，边经纱采用每筘齿内穿入 3 ~ 6 根经纱，并采用间隙穿筘，每根边综丝内的边经纱穿入根数为 2 ~ 4 根。

设计灯芯绒织物的边部结构时，除了以上措施外，甚至还要采用间歇纬的边组织来提高织物边部纳入纬纱的可能性。图 4-4-3 为中条灯芯绒织物的边部织造上机图解。布身每筘齿内穿入 2 根或 3 根经纱，边经纱采用每根边综丝内穿入 2 根边经纱，每筘齿内穿入 4 ~ 6 根边经纱，辅以间歇穿，并利用地组织构成间歇纬的边组织。

selvage and body; while E_j equals to 85%, the $\dfrac{2}{2}$ hopsack weave is applied to selvage, and 6 warps are drawn per dent in the selvage and 4 warps per dent for the body. For the venetian of satin weave (5-shaft satin), only an end is drawn into a single peddle, the hopsack weave is used in selvage and 3 or 4 warps are threaded into a dent across the width of the fabric.

In the high weft density fabrics like sateens and corduroy fabrics, the weft contraction rate is relatively high, and there is a greater friction between the selvage warps and the reed wires, so that the warp density of the selvedge is more different from the fabric body, which requires not only the strong warp threads, but also the requirement of decreasing the geometric phase.

In designing the structure of the selvage in such high weft sett fabrics, more than one end is threaded into the mail eye of a single heddle, combining with the intermittent denting and the warp rib weave at selvages. Fig. 4-4-2 shows the looming plans for a sateen fabric. The warp yarns at body part are threaded with 2 or 8 warp yarns per dent, and the selvage warp yarns are threaded with 3 to 6 warps per dent, with empty dents, and the number of ends threaded in each selvage heald changes from 2 to 4.

In designing the selvage structure of corduroy fabrics, in addition to the above measures, the intermittent weft weave is even used to improve the possibility of inserting more wefts. Fig. 4-4-3 shows the looming plans for a corduroy fabric of medium-width cord. There are 2 or 3 body ends per dent, 2 selvage warps per heddle, and 4 to 6 selvage warps pen dent, supplemented by intermittent denting plan. The weave of the selvage is actually the same as the body except that the ends are doubled by threading two ends into a mail of single heald wire and some dent splits are empty.

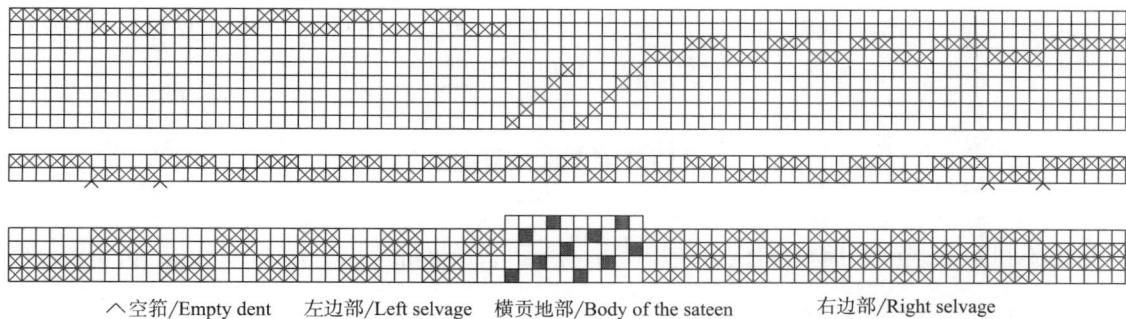

ㅅ空筘/Empty dent　　左边部/Left selvage　　横贡地部/Body of the sateen　　右边部/Right selvage

图 4-4-2　横贡织物的边部设计
The designing of selvage for a sateen

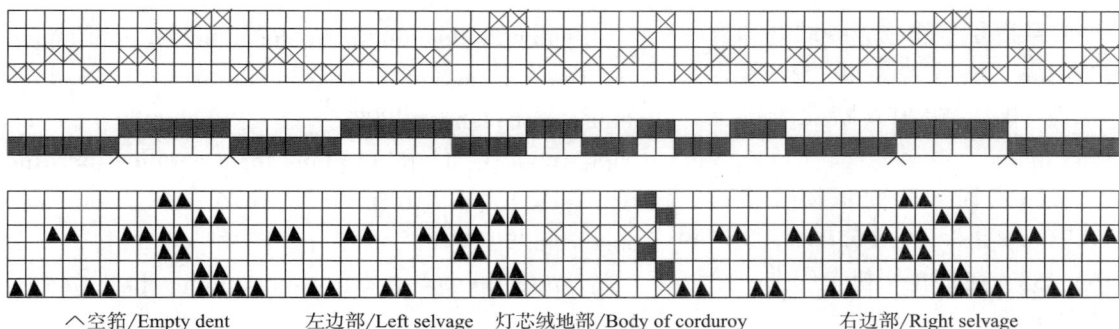

ㅅ空筘/Empty dent　　左边部/Left selvage　　灯芯绒地部/Body of corduroy　　右边部/Right selvage

图 4-4-3　中条灯芯绒织物的边部设计
The designing of selvage for a corduroy fabric of medium-width cord

4. 粗纺毛织物 /Woolen Fabrics

粗纺毛织物坯呢即使边部平直，若边部组织交错次数多，在缩呢整理中边部的经向缩呢率小于地部织物的经向缩呢率，在成品织物中可能显现木耳边织疵。故设计时，要求边部织物的经向紧密度略小于地部织物的经向紧密度，使缩呢过程中织物的边部能随着地部织物一起收缩，缩呢后织物保持边部平直。故平纹组织粗纺呢绒设计中，边组织采用 $\frac{2}{2}$ 方平或 $\frac{2}{2}$ 经重平组织，纬纱可密性增加，降低边部纬向紧密度。在缩呢过程中，边部和地部缩呢率趋于一致。

In the raising processing of the woolen fabric with high intersections in selvage, the warp contraction rate and shrinkage at selvage is less than that of the body. Even if the selvage is flat and straight, the wavy selvage will appear on the finished woolen fabric. Therefore, it is required that the covering factor of selvage is slightly less than that of the body in warp direction, so that the selvage can contract and shrink with the body in the process of raising, and selvage keeps straight after raising. Therefore, in designing the woolen fabric of plain weave, the capacity of packing wefts is increased and the tightness of the selvedge in weft direction is reduced by using the $\frac{2}{2}$ hopsack weave or $\frac{2}{2}$ warp rib weave. Therefore, the contraction rates and shrinkage rates of the selvage and body tend to be the same in felting process.

习题 /Questions

1. 绘制 $\dfrac{2}{3}$ 经重平织物上机图。

2. 绘制麻纱织物上机图。

3. 绘制 $\dfrac{3\quad 1}{1\quad 2}$、$\dfrac{3\quad 1\quad 2\quad 1}{2\quad 1\quad 1\quad 3}$ 变化方平组织的组织图。

4. 某麦尔登织物，纱线线密度为 83.3tex，充实率（相当于紧密率）为 90%。绘制其踏盘织机上机图。

5. 已知棉华达呢织物规格为（16.7tex×2）×（16.7tex×2），476×258，在踏盘织机上织造，绘制其上机图。

6. 分别绘制下列急斜纹组织图。

（1）$\dfrac{5\quad 3\quad 4}{1\quad 1\quad 3}$ 为基础组织，S_j=-2；

（2）$\dfrac{5\quad 1\quad 1}{3\quad 3\quad 3}$ 为基础组织，S_j =-3。

7. 已知某织物的基础组织为 $\dfrac{5\quad 1\quad 1}{2\quad 2\quad 1}\nearrow$，经纬纱密度为 133 根 /10cm × 72 根 /10cm，S_j 为 2，计算该斜纹线的倾斜角，并绘制该织物的组织图。

8. 绘制以 $\dfrac{4\quad 1\quad 1\quad 5}{1\quad 1\quad 1\quad 4}\nearrow$ 为基础组织，织物经纬密度近似，斜纹线的倾斜角为 63° 的急斜纹组织上机图。

9. 绘制以 $\dfrac{4}{3}$ 为基础组织，S_w=-2 的缓斜纹组织图。

1. Draw the looming plans for a $\dfrac{2}{2}$ warp rib weave.

2. Draw the looming plans for a hair cord weave.

3. Draw the looming plans for the fancy basket weaves $\dfrac{3\quad 1}{1\quad 2}$, $\dfrac{3\quad 1\quad 2\quad 1}{2\quad 1\quad 1\quad 3}$ respectively.

4. Draw the looming plans for a melton on a tappet loom. The thickness of the woolen yarn is 83.3tex and the tightness of the fabric is 90%.

5. Draw the looming plans for a cotton gaberdine [（16.7tex×2）×（16.7tex×2），476×258]fabric on a tappet loom.

6. Draw the steep twill weaves under the following conditions:

(1) Based on $\dfrac{5\quad 3\quad 4}{1\quad 1\quad 3}$ with S_j=-2; (2) Based on $\dfrac{5\quad 1\quad 1}{3\quad 3\quad 3}$ with S_j=3.

7. Draw the steep twill weave and calculate the twill diagonal angle of the fabric requiring the following conditions: based on $\dfrac{5\quad 1\quad 1}{2\quad 2\quad 1}\nearrow$, S_j=2, containing 133 ends and 72 picks per 10 cm.

8. Draw the looming plans for the steep twill weave of the fabric under the following conditions: based on $\dfrac{4\quad 1\quad 1\quad 5}{1\quad 1\quad 1\quad 4}\nearrow$, containing equal ends and picks per 10 cm, and the twill diagonal angle is 63°.

9. Draw the recrined twill weave base on $\dfrac{4}{3}$ twill and S_w=-2.

10. Draw the horizontal curved twill base on $\dfrac{3\quad 1}{3\quad 2}$ $\nearrow$ with S_j = 1, 1, 0, 1, 0, 1, 0, 1, 0, 0, -1,

10. 以 $\frac{3\ \ 1}{3\ \ 2}\nearrow$ 为基础组织，按下面经向飞数的变化作曲线斜纹，经向飞数分别为 1，1，0，1，0，1，0，1，0，0，-1，0，-1，0，-1，0，-1，-1。

11. 按照下列已知条件，绘制经山形斜纹组织的上机图。

（1）基础组织为 $\frac{3\ \ 1}{2\ \ 2}$ 斜纹，K_j 分别为 7、8、9；

（2）基础组织为 $\frac{2\ \ 1}{2\ \ 1}$ 斜纹，K_j 分别为 8、9、10。

12. 按照下列已知条件，绘制纬山形斜纹组织的组织图。

（1）基础组织为 $\frac{3\ \ 1}{1\ \ 2}$ 斜纹，K_w =6、7、10；

（2）基础组织为 $\frac{2\ \ 2}{1\ \ 3}$ 斜纹，R_w =12、14、16。

13. 以 $\frac{2\ \ 2\ \ 1}{1\ \ 2\ \ 2}$ 斜纹为基础组织，以 9 根经纱构成右斜、12 根经纱构成左斜，或 9 根经纱构成右斜、6 根经纱构成左斜的规律排列经纱，绘制经山形斜纹组织的组织图。

14. 按照下列已知条件，绘制锯齿斜纹组织的组织图。

（1）基础组织为 $\frac{2\ \ 1}{3\ \ 2}$ 斜纹，K_j=8，S_j =4；

（2）基础组织为 $\frac{3\ \ 2}{2\ \ 2}$ 斜纹，K_j=6，S_j =3。

15. 按照下列已知条件，绘制人字形破斜纹组织的上机图。

0, -1, 0, -1, 0, -1, -1 respectively.

11. Draw the looming plans for the horizontal waved twills under the following conditions:

(1) Based on $\frac{3\ \ 1}{2\ \ 2}$ twill, and K_j is set as 7, 8 and 9 respectively.

(2) Based on $\frac{2\ \ 1}{2\ \ 1}$ twill, and K_j is set as 8, 9 and 10 respectively.

12. Draw the looming plans for the vertical waved twills under the following conditions:

(1) Based on $\frac{3\ \ 1}{1\ \ 2}$ twill, and K_w is set as 6,7 and 10 respectively.

(2) Based on $\frac{2\ \ 2}{1\ \ 3}$ twill, and R_w is set as 12, 14 and 16 respectively.

13. Draw the weave pattern for a horizontal waved weave requiring the following conditions: based on $\frac{2\ \ 2\ \ 1}{1\ \ 2\ \ 2}$ twill weave, 9 ends forming a right−handed diagonal twill line; 12 ends forming a left−handed diagonal twill line; 9 ends forming a right−handed diagonal twill line, and 6 ends forming a left−handed diagonal twill line.

14. Draw the looming plans for the zigzag twills under the following conditions:

(1) Based on $\frac{2\ \ 1}{3\ \ 2}$ twill, K_j =8 and S_j =4;

(2) Based on $\frac{3\ \ 2}{2\ \ 2}$ twill, K_j = 6 and S_j =3;

15. Draw the looming plans for the herringbone twills under the following conditions:

(1) Based on $\frac{1\ \ 3}{3\ \ 1}$ twill, K_j =8;

(2) Based on $\frac{2\ \ 1}{1\ \ 2}$ twill, K_j =6.

16. Draw the looming plans for a broken twill (in narrow sense) based on a $\frac{4}{4}$ twill.

（1）基础组织为 $\dfrac{1}{3}\dfrac{3}{1}$ 斜纹，K_j=8；

（2）基础组织为 $\dfrac{2}{1}\dfrac{1}{2}$ 斜纹，K_j=6。

16. 以 $\dfrac{4}{4}$ 斜纹组织为基础组织，设计一个变位破斜纹组织。

17. 以 $\dfrac{3}{3}$ 斜纹组织为基础组织，设计一个飞断破斜纹组织。

18. 按照下列条件，绘制菱形斜纹组织的上机图。

（1）基础组织为 $\dfrac{3}{1}\dfrac{1}{3}$ 斜纹，K_j=8，K_w=8（要求交界处清晰）；

（2）基础组织为 $\dfrac{3}{1}\dfrac{1}{3}$ 斜纹，K_j=10，K_w=6（要求交界处清晰）；

（3）基础组织为 $\dfrac{2\ 2}{1\ 2}\dfrac{1}{2}$ 斜纹，K_j=5，R_j=10，R_w=10；

（4）基础组织为 $\dfrac{2}{3}\dfrac{1}{2}$ 斜纹，R_j=14，R_w=14；

（5）基础组织为 $\dfrac{2\ 2}{1\ 2}\dfrac{1}{2}$ 斜纹，K_j=12，R_w=8。

19. 以 $\dfrac{2}{2}$ 斜纹为基础组织，作同一方向斜纹线为 3 条的芦席斜纹组织图。

20. 绘制 8 枚变则缎纹的组织图。

21. 设计 3 个 8 枚加强缎纹的组织图。

22. 绘制缎背华达呢织物、斜纹板司呢和海绵组织的组织图。

17. Draw the looming plans for a skip twill based on a $\dfrac{3}{3}$ twill.

18. Draw the looming plans for the diamond twills under the following conditions:

(1) Based on $\dfrac{3}{1}\dfrac{1}{3}$ twill, K_j =8 and K_w =8 (requiring a clear border);

(2) Based on $\dfrac{3}{1}\dfrac{1}{3}$ twill, K_j =10 and K_w =6 (requiring a clear border);

(3) Based on $\dfrac{2\ 2}{1\ 2}\dfrac{1}{2}$ twill, K_j =5, R_j =10 and K_w =10;

(4) Based on $\dfrac{2}{3}\dfrac{1}{2}$ twill, R_j =14 and R_w =14;

(5) Based on $\dfrac{2\ 2}{1\ 2}\dfrac{1}{2}$ twill, K_j =12 and R_w =8.

19. Draw a $\dfrac{2}{2}$ based entwining twill weave with 3 twill lines running on the surface of the pattern.

20. Draw an 8-shaft irregular satin/sateen weave.

21. Draw 3 8-shaft double satin/sateen weaves.

22. Draw the weave patterns of satin-back gaberdine, twill basket and sponge weave respectively.

第五章　联合组织 /Combined Weaves

联合组织是将两种或两种以上的原组织或变化组织，按不同方法组合而成的新组织。构成联合组织的方法多种多样，可能是两种组织的简单并合，也可能是两种组织纱线的交叉排列，或者在某一组织上按另一组织的规律增加或减少组织点数等。由于组合后织物组织中的浮长线长度与分布的变化，新组织的织物外观可能与原来组织相似，也可能明显不同，产生新的外观特征。

应用较广且具有特定外观效应的联合组织有条格组织、绉组织、透孔组织、蜂巢组织、凸条组织、网目组织、浮松组织、凹凸组织（劈组织）、平纹地小提花组织等。

Combined weave is the new weave formed by two or more basic weaves or their derivatives by various methods. The combination can be just putting the two foundation weaves in tandem, or arranging the threads in the weave at regular intervals, or adding marks according to other weaves. Due to the change of the floats and their distribution, the structure and the appearance of the new weave can be similar to or quite different from the original ones.

The combined weaves include: stripe and check weaves, crepe weaves, open gauze weaves, honeycomb weaves, Bedford cord weaves, distorted-effect weaves, huckaback weaves, pique weaves, dobby spot weaves, etc. Combined weaves are usually used to show the color and weave effects.

第一节　条格组织 /Stripe and Check Weaves

两种或两种以上的基础组织并列配置形成条格组织。由于不同基础组织的织物外观不同，织物表面呈现清晰的条带或格形外观。条格组织包括纵条纹组织、横条纹组织、方格组织、格子组织，分别如图 5-1-1（a）（b）（c）（e）所示，广泛应用于服装用织物、被单、手帕等。

Stripe or check weaves are formed by arranging two or more basic weaves or their derivatives at intervals horizontally, or vertically or at both directions. There are distinct stripes or checks appearing on the surface of the fabrics especially when colors are applied. The weaves include longitudinal stripe weave, horizontal stripe weave, check weave, two-way stripe weave, as shown in Fig. 5-1-1(a)(b)(c)(e). These weaves are extensively used in apparels, sheeting, handkerchief, etc.

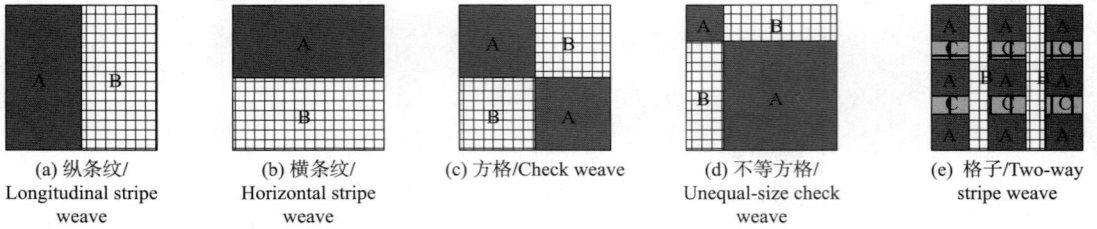

(a) 纵条纹/
Longitudinal stripe
weave

(b) 横条纹/
Horizontal stripe
weave

(c) 方格/Check weave

(d) 不等方格/
Unequal-size check
weave

(e) 格子/Two-way
stripe weave

图5-1-1　各类条格组织
Schematic of various stripe and check weaves

一、纵条纹组织 /Longitudinal Stripe Weaves

当两种或两种以上的组织左右并列时，每个基础组织各自形成纵条纹，这些组织的组合称为纵条纹组织。

If two weaves are arranged in tandem, each foundation weave will form a stripe, and longitudinal stripe weave appears on the surface.

1. 设计原则 /Principles of Designing Longitudinal Stripe Weave

在纵条纹组织两条纹的分界处，要求界线分明。因此在设计纵条纹组织时，在分界处的相邻两根经纱的组织点应尽量配置成"底片翻转法"的关系，即经、纬组织点相反，则相邻两根纱线被分隔开，不易产生相互移位，条纹效应明显，如图5-1-2所示。图5-1-3（a）（b）所示组织的基础组织相同，但图5-1-3（b）所示组织效果好，图5-1-3（a）所示组织设计不当，在期望的交界处出现较多的纬浮长，两根纱线聚拢，导致条纹容易移位，交界不清楚。

若选用的组织不能达到这个要求，则可在交界处添加一根另一组织或另一颜色的纱线。但尽量不要增加织造的复杂性，如图5-1-4所示。

A clear boundary is required at the junction of the two weaves for a longitudinal stripe weave. Therefore, when designing the structure, the interlacing points of the two adjacent ends at the boundary should be set by "herringbone reversing method" as far as possible. In other words, the neighboring interlacing points should be opposite, so the two adjacent ends are separated and not easy to turn over each other and blur the stripe effect, as shown in Fig. 5-1-2. In Fig. 5-1-3 where two weaves are combined by the 2 same base weaves. The weave as shown in Fig. 5-1-3(b) has a good stripe effect while the weave as shown in Fig. 5-1-3(a) has an inappropriate design. There are so many weft floats forming at the expected border in the weave as shown in Fig. 5-1-3(a), so the two ends at the borders will aggregate and blur the stripe effect.

Sometimes, two extra ends of different colors may be introduced at the border when it is difficult to use herringbone reversing at the cost of increasing the complexity in weaving, which is shown in Fig. 5-1-4.

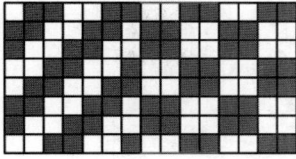

图5-1-2　边界严格底片翻转法的纵条纹组织

A clear stripe weave by reversing the border

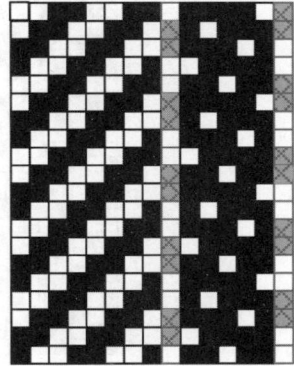

(a)　　　　　　　　　　(b)

图5-1-3　边界清晰/不清晰的纵条纹组织的边界配置

Clear and unclear borders of longitudinal stripe weaves

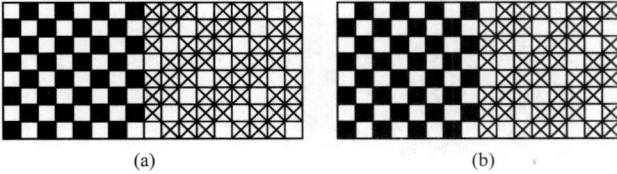

图5-1-4　边界添加其他色纱的纵条纹组织

A longitudinal stripe weave with colored border

2. 纵条内经纱根数修正 /Correction the Number of the Ends in Stripes

若组成纵条纹的两个基础组织的经纬纱循环根数分别为 R_{j1}、R_{j2}、R_{w1} 与 R_{w2}，则纵条纹组织纬纱循环根数为 $R_w=R_{w1}$ 与 R_{w2} 的最小公倍数，纵条纹组织循环经纱数 R_j 是各纵条纹中经纱数之和。

为了尽可能形成破界，达到条纹清晰的目标。根据条纹宽度和密度，将每一纵条的经纱修正为该纵条每箅齿穿入数的整倍数，同时为该纵条组织经纱循环根数的整数倍。

若每个条子的经纱根数较多，在组织图下方用"基础循环根数 × 循环数"表示此处经纱的根数。图 5-1-5 是经过修正经纱根数后的由四个经纱条子组成的纵条纹组织。

If the repeat number in warp direction and weft direction are R_{j1}, R_{j2}, R_{w1} and R_{w2} respectively, the ultimate repeat number R_w in weft direction will be the lowest common multiple of R_{w1} and R_{w2}, and the ultimate repeat number R_j in warp direction is the sum of the ends in all stripes.

In order to form a broken and clear border, the number of ends n each stripe will be corrected as the multiple of the number of ends in a split of reed based on the width and the warp density of the stripe, and the number should be corrected as the multiple of the repeat number of the weave in the stripe as could as possible.

If a numerous of ends are used in stripes, the number of ends can be represented by the manner like "the number of weave units × warp repeat number". Fig. 5-1-15 shows the number of ends in a longitudinal stripe weave with 4 stripes.

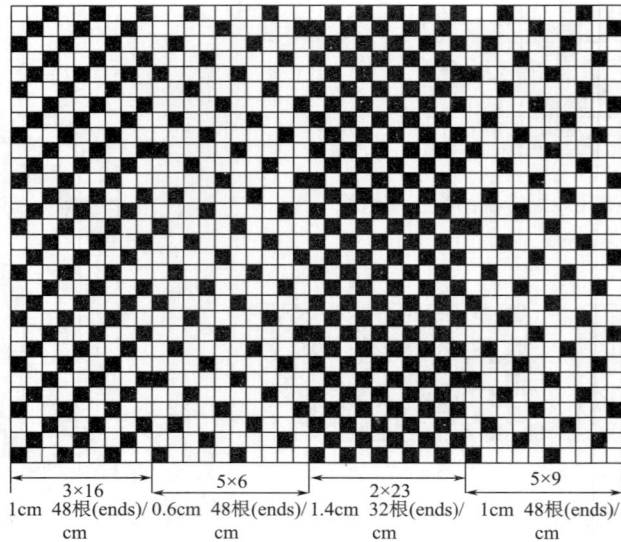

3×16	5×6	2×23	5×9
1cm 48根(ends)/ cm	0.6cm 48根(ends)/ cm	1.4cm 32根(ends)/ cm	1cm 48根(ends)/ cm

图5-1-5　修正经纱根数的纵条纹组织
Correcting the number of ends in a longitudinal stripe weave

3. 纵条纹组织上机 /Looming for Longitudinal Stripe Weaves

设计纵条纹组织时，采用各种组织的交错次数不要差异太大；否则，将造成经纱织缩显著不同，使织造产生困难。上机时，穿综宜采用间断穿法，如图 5-1-6 所示。经纱根纱应该为筘齿穿入数的倍数，处于不同条带的相邻经纱应该穿入不同的筘齿中。如果必须把交错次数相差大的组织配置在一起，采取的措施有：①采用双织轴；②加大交错次数少的组织的经密；③在准备各工序加大交错次数少的组织的张力。

横条纹组织织物使用较少，其设计思路、经纬组织循环根数的计算类似纵条纹组织，采用顺穿法穿综，如图 5-1-7 所示。

In a longitudinal stripe weave, the numbers of intersections of base weaves are expected to be similar to avoid the tight ends or slack ends in weaving. Grouped drafting as shown in Fig. 5–1–6 is suitable to stripe weaves. The number of the ends should be the multiple of the threads pen dent, and the two neighboring ends of different stripe should be threaded into different dents respectively. If intersections of the base weaves differ very much, several measures might be taken: ① two weaving beams are employed; ② increase the warp density of the weave of less intersection; ③ increase the tension of the ends of the weave of less intersection during the warping process.

Horizontal stripe weaves, though not common, can be designed based on the principles for the longitudinal stripe weaves. Usually the weave is straight drafted, as shown in Fig. 5–1–7.

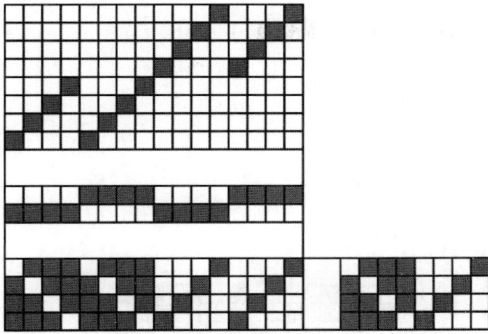

图5-1-6 纵条纹组织上机图
The looming plans for a longitudinal stripe weave

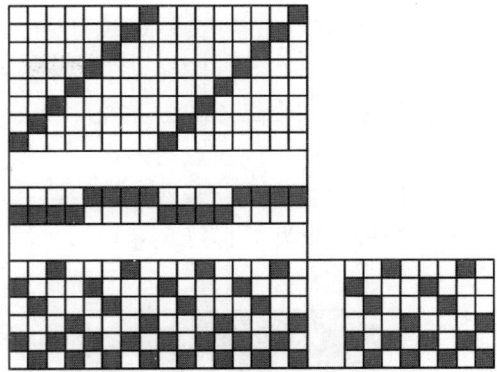

图5-1-7 横条纹组织上机图
The looming plans for a horizontal stripe weave

二、方格组织 /Check Weaves

1. 外观特点 /Description

方格组织织物外观呈现方格效应，所表现方格大小可相等（图5-1-8），也可不等（图5-1-9）。其经纱循环根数与纬纱循环根数相等，即 $R_j=R_w$。选用经面或纬面组织为基础组织，沿着经向和纬向间隔配置而获得格形。注意尽可能确保在同一纬纱上，这两种组织的交织次数一样，避免纬纱出现波浪型弯曲。因此，通常采用同一组织，经过底片翻转法，得到方格组织。此时处于对角位置的两部分应配置相同的组织。

The façade of the fabrics takes on the check effect, the four portions of the check can be equal or unequal in size as shown in Fig. 5–1–8 and Fig. 5–1–9. The sizes of the repeat in two directions are equal, viz., $R_j=R_w$. The check effect is achieved by alternately placing a warp–faced weave and a weft–faced weave along both the vertical and horizontal directions. Care should be taken to ensure the same number of intersections of the two weaves on the same weft as much as possible to avoid wavy effect of the weft. Therefore, the warp–faced (or weft–faced) weave is usually set as base weave at first, and the weft–faced (or warp–faced) weave is obtained by herringbone reversing. The weaves in a diagonal position of a check are actually the same.

2. 设计要点 /Key Points to Designing

（1）应注意分界处界线分明，即分界处相邻两根纱线上的经纬组织点必须相反。

（2）基础组织起始点选择。在绘作方格组织时，要求位于对角位置的两部分的相同组织的组织点

(1) A clear border is required, which means the two adjacent interweaving points on the border are herringbone reversing.

(2) The base weave should commence at a right thread to ensure that the pattern line which is formed by the interweaving points continues evenly in the diagonal weaves.

可以连续，其织物外观整齐美观。图 5-1-10 所示的方格组织外观效应不整齐。

（3）不要在四个格子的分界处形成平纹组织（中心组织点与周边 4 个相邻组织点均不同），导致此处织物薄、凹陷，如图 5-1-11 所示。

Fig. 5-1-10 shows an imperfect check weave.

(3) Try to avoid forming a plain weave in the center of the check as it will lead to a low-lying effect at the center as shown in Fig. 5-1-11.

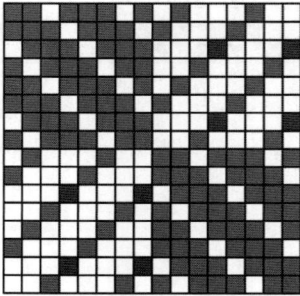

图5-1-8　等大小方格组织
A square check weave

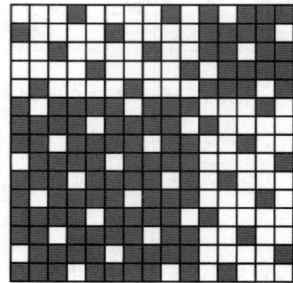

图5-1-9　不等大小方格组织
A unequal check weave

图5-1-10　纹路不连续不美观的方格组织
A imperfect check weave

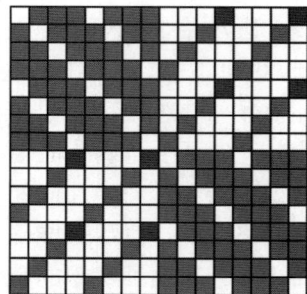

图5-1-11　中心凹陷不平整的方格组织
A center-sagged check weave

3. 组织起始点调整 /Adjusting the Commencement of the Base Weave

为了作图方便，一般左下角采用纬面组织。欲使位于对角位置的相同组织的组织点连续，则可按下列两种方案之一来决定绘制组织的起始点位置。

（1）沿经向调整。观察基础组织的经纱，从中找出左右两根相邻

For the convenience of drawing, a weft-sided weave is generally adopted at the lower left corner. There are two methods to adjust the commencement of the weave to ensure the continuity of the pattern line in the diagonal direction.

(1) Regulating along the warp direction. Observe the warps of the base weave, and find out the two neighboring ends with the following characteristics: the distance between

经纱的单独组织点（一般用纬面组织求作）与上下边缘距离相等的两根纱线。将右侧经纱作为组织的第一根经纱，左侧经纱作为组织循环最右（最后）一根经纱配置。

（2）沿纬向调整。观察基础组织的纬纱，从中找出上下两根相邻纬纱的单独组织点（一般用纬面组织求作）与左右边缘距离相等的两根纱线。将上侧经纱作为组织的第一根纬纱，下侧纬纱作为组织循环最上（最后）一根纬纱配置。

图5-1-12是某八枚缎纹组织的纱线顺序调整原理。图5-1-12（a）所示为采用沿经向调整的思路，图5-1-12（b）所示为采用沿纬向调整的思路，调整后的结果相同。图5-1-13是根据调整顺序后的基础组织设计的2种方格组织。

the single mark on the left end and the top edge of the weft-faced base weave equals to that of the single mark on the right end and the bottom edge. The right side of the end is arranged as the first end of the new weave and the left side as the rightest (last) end of the new weave.

(2) Regulating along the weft direction. Observe the wefts of the base weave, and find out the two neighboring picks with the characteristics: the distance between the single mark on the upper pick and the left edge of the weft-faced base weave equals to that of the mark on the lower pick and the right edge. The upper side of the pick is arranged as the first pick of the new weave and the lower pick as the top (last) pick of the new weave.

Fig. 5-1-12 illustrates how to adjust the commencement of an 8-shaft sateen. Fig. 5-1-12(a) shows how to regulate along warp direction while Fig. 5-1-12(b) shows how to regulate along weft direction. The final results are the same. Fig. 5-1-13 shows the 2 possible proper check weaves based on the regulated base weaves.

沿经向,左右相邻纱线找与上下两侧等距离的组织点/Regulate along warp direction

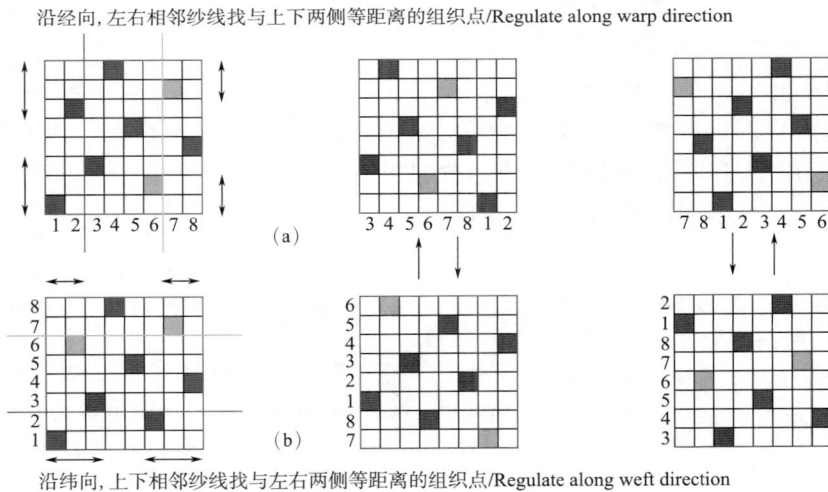

沿纬向,上下相邻纱线找与左右两侧等距离的组织点/Regulate along weft direction

图5-1-12　调整组织起始点位置的两种方案
Two methods to regulate the start yarn of the check weave

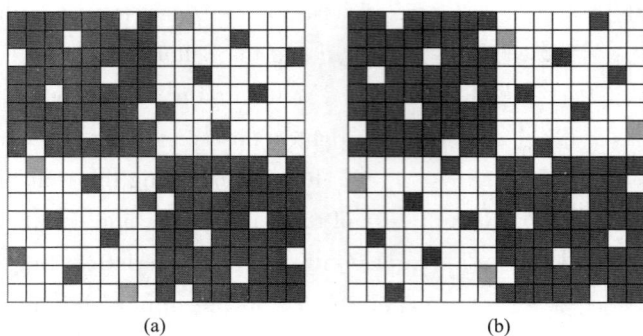

(a)   (b)

图5-1-13　根据调整顺序后的基础组织设计的方格组织
Two check weaves after regulating the start thread

三、格子组织 /Two-Way Stripe Weaves

由纵条纹组织和横条纹组织联合构成的大格型花纹，作图原则和方法与纵条纹组织类似。多用于手帕。

图 5-1-14 是格子组织结构示意图。设计时，纵条 a 需要的综页数 A 是 1cm（地组织经纱循环数，横向四枚缎条组织的经纱循环根数）；纵条 b 需要的综页数 B 是纵向四枚缎条组织的经纱循环根数，故综页总数是 $A+B$。采用间断穿综法即可织造。织造时，嵌织的缎纹提花组织均需增大纱线密度，以突出缎条效应，保持织物平整。对于经向缎条，加大筘穿入数，增加经纱密度；对于纬向缎条，采用停送停卷装置，增加纬纱密度。

The two-way stripes are the combination of the longitudinal stripes and the horizontal stripes. The drawing and looming are similar to the horizontal stripes. This type of the weave is usually of large pattern and usually used for the handkerchief.

Fig. 5-1-14 shows the pattern of a two-side check weave comprising of plain weave, $\frac{3}{1}$ satinette and $\frac{1}{3}$ satinette. The number of the heads (A) used for the stripe a is the lowest common factor of 2 (R_j of the plain weave) and 4 (R_j of the $\frac{1}{3}$ satinette), and the number of the heads (B) used for the stripe b is 4 (R_j of the $\frac{3}{1}$ satinette). Thus, the total number of heads is $A+B$. The weave is grouped drafted. To accentuate the satin effect and keep a smooth fabric, the stripe b is high-sett by placing more ends in a split of reed. For the stripe part formed by $\frac{1}{3}$ satinette, the weft density is also increased by intermittent stop taking-up motion.

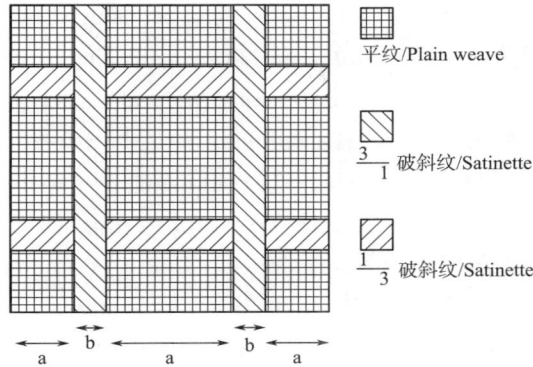

图5-1-14 格子组织结构示意图
Schematic diagram of the two-way stripe weave

第二节 绉组织 /Crepe Weaves

绉组织 Crepe Weaves

一、绉效应 /Crepe Effects

织物组织中不同长度的经纬浮点，在纵、横方向错综排列，使织物表面具有分散的、规律不明显的、微微凹凸的细小颗粒，呈现绉效应，这类组织称为绉组织。绉组织织物包括表面粗犷型的摩力呢、正面呈现无规则凹凸条形的旦斯绉、采用双层组织的厚型丝绸呢类的重绉等，绉组织织物较平纹织物手感柔软、丰厚、弹性好、表面反光柔和。织物正反面效果相同。起绉的织物，其外观呈现不同形态的绉纹，织物手感富有弹性，抗皱性能好，表面光泽柔和。

The crepe weave is the weave having a random distribution of floats so as to produce an all-over rough, pebbly effect in the fabric to disguise the repeat. An irregular distribution of the floats and intersections of the warp and weft threads in such a manner as will prevent the occurrence of stripes and lines in any direction. The examples of the crepe fabrics include Marocain of rough-surface, Moss crepe of uneven surface and double-layer silk Crepon of heavy weight. The weaves contain little or no twilled or other prominent effect, and give the cloth the appearance of being covered by minute spots or seeds. Compared with the plain weave fabric, the crepe fabric feels soft, rich, resilient and soft reflective. Both sides are exactly alike and therefore reversible.

二、起绉方法 /Methods of Making Crepe Weaves

得到绉效应的方法有：①利用强捻纱以平纹或其他简单组织交织

Crepe fabrics also include any type or weight of fabrics made with crepe yarns either twisted or chemically crinkled.

163

后经印染使织物表面起绉；②利用不同张力或不同收缩性能的纱线间隔排列而获得起绉外观；③利用轧纹工艺获得起绉外观；④利用绉组织获得起绉外观。

最常见的方法是利用强捻纬纱和普通经纱织造，在后整理时纬纱产生大量收缩形成绉效应。需要在纬向握持织物，避免过度收缩，可以在张力下烘干暂时定型得到。此类绉织物在使用前必须先测试，确保遇湿热不再收缩。

由绉组织形成的绉效应尺寸稳定，故本节主要介绍由绉组织形成的绉效应。

Crepe effect can be obtained by the use one of the following manners: ① hard-twist filling yarns (or arranging S-twisted yarns and Z-twisted yarns alternately); ② chemical or thermal treatment to provide different shrinkage in the finished fabric; ③ embossing, by passing the fabric between heated engraved-embossing rollers, either into a softened thermoplastic fabric or in combination with resin; and ④ employing crepe weaves.

The most common form of crepe fabrics is produced by using hard-twist weft and ordinary warp and considerable width shrinkage takes place during finishing. It may be necessary to hold the fabric out in width to prevent in contracting too much, and it can be temporarily set in this position by drying it while under tension. Crepe fabrics of this type are always tested for sensitivity before use so that heat and moisture treatments can be controlled or even entirely eliminated.

Crepe effect is quite stable by employing crepe weaves. This section only deals with the crepe effects by crepe weaves.

三、起绉原理与设计要点 /Principles and Key Points of Designing Crepe Weaves

在一个绉组织循环内，经、纬纱的浮长长短不一，又沿不同方向交错配置。浮线较长的组织点，经、纬纱之间结构较松；而浮线短的组织点，经、纬纱之间结构较紧。结构较松的长浮线分布在结构较紧的短浮线之间，较松的组织点就在较紧的组织点间微微凸起，形成细小的颗粒状。细小的颗粒均匀分布在织物表面形成绉效应。由于绉组织织物表面均匀分布了细小的颗粒状组织点，对光线形成漫反射，所以光泽较柔和。因为组织中有不规则分布的松弛长浮线，所以织物手感松软、厚实、有弹性。

设计绉组织时，要注意：①经、

In a crepe weave, the lengths of the warp and weft float vary. The threads with longer float are comparatively slack and tend to form a tiny hump while the threads with short float are tight and tend to lie. The tiny humps distribute randomly to form pebbly spots and scatter and diffuse the light. Therefore, the fabric is of soft luster. Crepe fabric feels soft, thick, full and resilient due to the long, slack floats irregularly distributed among the interweaving points.

In designing a crepe weave, the principles listed should be followed: ① the length of the warp and weft floats should not exceed 3 points (about 2mm), otherwise it will damage the appearance of fine particles or spots; ② the floats of various lengths are evenly distributed in all directions, and vertical, horizontal and oblique lines should be avoided; ③ the larger the weave repeat, the smaller the possibility of grain formation, but too many heads may be employed, which causes looming

纬浮线不能过长，连续浮长线不超过3个组织点（2mm），否则会破坏细小颗粒状外观；②不同浮长的组织点沿各个方向均匀分布，切忌出现纵、横、斜向的纹路；③组织循环越大，形成纹路的可能性越小，但综片过多会造成上机困难；④不能有大群相同的经（纬）组织点聚集在一起，否则会使织物表面产生光亮和暗淡区域；⑤组织循环内各根经纱的交织次数不宜相差太大，否则会影响开口清晰度，造成织造困难，布面不平整。

difficulties; ④ there should not be a large group of the marks or blanks gathered together, otherwise the surface of the fabric will produce bright and dark areas; ⑤ the interweaving frequency of each warp thread in a weave repeat unit should not differ too much to ensure the same amount of tensile strain upon all these threads uniformly, otherwise it will affect the clarity of the shedding and cause weaving difficulties and tend to create more or less conspicuous blemishes in the cloth produced, in consequence of the taut and the slack threads pulling unequally.

四、绉组织设计 /Designing of Crepe Weaves

绉组织设计的方法主要有以下几种。

Several methods are often used to design the crepe weaves.

1. 增点法（重叠法）/Overlapping the Interlacing Points

以原组织或变化组织为基础，按另一组织的规律增加组织点构成绉组织，如图5-2-1（a）所示；或者将两种组织重叠得到绉组织，如图5-2-1（b）所示。图5-2-1（b）所示的组织是将经组织点按照缎纹组织分布的方式，添加在平纹组织上形成的。

The weaves are constructed by overlapping of one weave over another as shown in Fig. 5-2-1(a), or by adding marks or blanks in certain orders to a base weave as shown in Fig. 5-2-1(b). The crepe weave as shown in Fig. 5-2-1(b) is formed by adding marks in sateen orders to a plain weave.

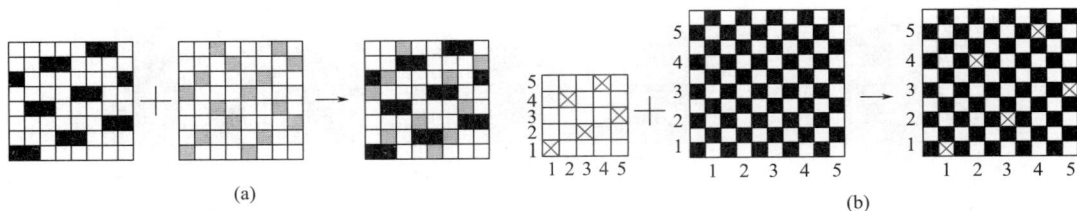

(a)　　　　　　　　　　(b)

图5-2-1　增点法构成绉组织
Over lapping the interlacing points

2. 间隔排列纱线法 /Insertion of One Weave over Another

该方法将两个不同的组织间隔

This method of constructing crepe weaves consists

排列形成。将一个组织的经（纬）纱，按一定的比例（1:1、1:2）移绘到另一个组织的经（纬）纱线间，形成绉组织，如图5-2-2所示。

of inserting two different weaves one over the other. It is noteworthy that at least one weave is an irregular weave to produce an irregular effect. Usually, the ends (picks) of two base weaves are alternately arranged in the ratio of 1:1 or 2:2, then a new crepe weave is formed.

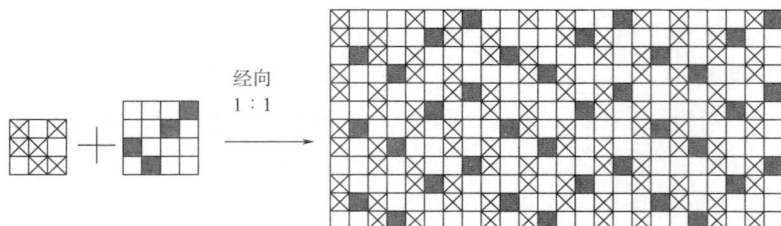

图5-2-2　间隔排列纱线法构成绉组织
Alternately arranging the ends of different weaves

3. 调序法 /By Adjusting the Order of Threads

绉组织通过调整同一种组织的纱线次序构成。用这种方法时，一般是以变化组织为基础组织，然后变更基础组织的经（纬）纱排列次序而成。图5-2-3所示绉组织以一个复合斜纹为基础，首先经向调整顺序，然后纬向再次调整顺序得到。

The method of constructing crepe weaves consists of selecting a derivative of the basic weave and changing the orders of the threads. The crepe weave in Fig. 5-2-3 is based on a composed twill weave, and formed by changing the order of the ends then by rearranging the order of picks.

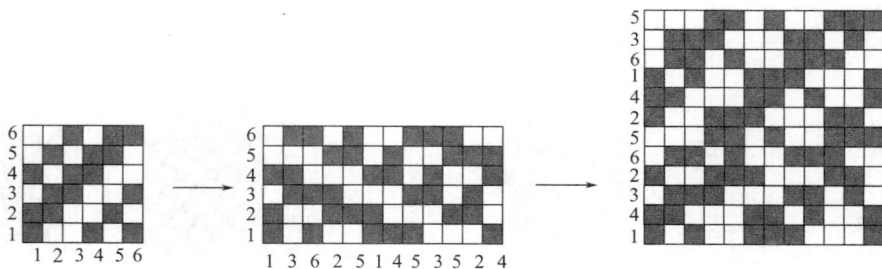

图5-2-3　调整纱线排序法构成绉组织
Adjusting the order of threads in both directions

4. 旋转法 /By Rotating

以一个同面组织（或者近似同面组织）为基础单元，使其顺时针

This method of constructing crepe weaves can be described as following: select a base weave as a unit, rotate it

或逆时针旋转、组合构成绉组织。步骤如下：首先旋转90°，得到一个组织单元，再旋转两次，又得到两个组织单元。将这四个组织单元按图5-2-4所示的方法组合，得到绉组织，其组织循环是基础单元循环的2倍。

for 90° and obtain a new weave as another unit, then employ the method for another two times, therefore, 3 new units are formed. The new crepe weave is obtained by combining the 4 units together as shown in Fig. 5-2-4. The repeat of the generated crêpe is 2 times of the base weave.

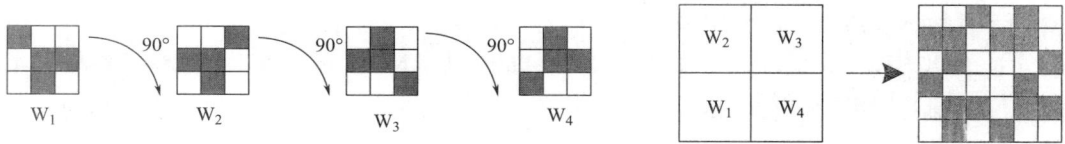

图5-2-4　旋转组织法构成绉组织
Rotating and combining the weave components

5. 省综设计法 /By Saving the Number of Healds

以上几种设计方法，因为组织循环较小，得到的绉组织效应一般并不理想。若无规律的组织循环大，一般需要的综页多，对织机有要求。为了实现在综页少的织机上织造组织循环大的绉织物，在实际生产中，一般先确定综页数，通过扩大纹板的长度，增加经纱上不同的交织规律，增大 R_w；借鉴调整纱线次序的思路，调整穿综顺序，不断组合，在不增加综页数的条件下，增大 R_j，可以得到较好绉效应的绉组织。这种方法称为省综设计法。步骤如下：

（1）确定采用的综框数 n，一般使用4～8页综框。本例选用6页综框。

（2）确定组织循环图的范围。一般要求 R_j 和 R_w 大致相同。本例中，R_j=60，R_w=40。

（3）确定每片综的提升规律，

The crepe effect obtained by the previous methods is not good due to the small size of the weave repeat. In order to get a better crepe effect, the weave repeat number has to be enlarged, which requires more healds. It is required to design a more complicated interlacement for each end upon a base weave with small R_j. In practice, the number of healds is usually determined first, and R_w is increased by enlarging the number of the pattern lags and increasing the different interlacements on the warp threads. Similar to the idea of adjusting the order of threads to obtain a crepe weave, by adjusting the order of drafting, R_j can also be increased without adding the number of healds. Finally, a satisfactory crepe weave can be obtained by using less healds. The procedures of designing a crepe weave by saving the number of healds can be described as following:

(1) Determine the number n of the healds used, usually between 4~8 healds. In the example, 6 healds are used.

(2) Define the weave range, and usually R_j is similar to R_w. eg . R_j=60, R_w=40.

(3) Draw all the possible interlacements for each pick as shown in Fig. 5-2-5. $n/2$ marks will be lifted for each

画出纹板图的全体可能状态，如图 5-2-5 所示。一般每纬提起 $n/2$ 个组织点，故可能的规律有 $C_n^{n/2}$ 种。因此，有 6!/[3! × (6-3)!]=20 种规律。实际使用中，要打乱这个顺序，经纬浮长都不要超过 3 个组织点。最好相邻两块纹板有且仅有一个连续的经组织点、一个连续的纬组织点。图 5-2-6 是按照该方法确定的实际纹板图。若 $R_w > C_n^{n/2}$，有些纹板可以重复利用，但必须相隔至少 3 根纬纱。

（4）画穿综图和组织图。首先，把经纱循环数分成若干组，每一组的经纱数等于综页数，修改穿综顺序即可。穿同一综框的相邻两根经纱之间也至少间隔 3 根经纱。图 5-2-7 是采用省综设计法得到的某绉组织的上机图。省综设计法也可以认为是将纹板图作为组织图，通过不断调整穿综顺序得到不同的组织，然后将所有的组织组合得到的。显然，省综法设计的绉组织穿综采用照图穿法，每筘齿穿入数为 2 ~ 4 根。若使用 8 页综框，一般不宜采用连续 4 个相同组织点作为可能的纹板，因此，可以采用的纹板数量是 C_8^4-8。

省综法需要的纹板还可以用两种不同的复合斜纹组织用 1:1 或者 2:2 间隔排列得到。在不同的纹板上，复合斜纹的起点和方向随机变化。图 5-2-8 是用该法设计的绉组织，纹板图采用 $\dfrac{3\quad 1}{3\quad 1}$ 斜纹与 $\dfrac{2\quad 2}{3\quad 1}$ 斜纹的纬纱 2:2 间隔排列。纹板图中，第一纬从第 3 经开始向

pick, therefore, there are $C_n^{n/2}$ possible interlacements. For a 6-shaft lifting plan, there are 6!/[3! × (6-3)!] =20 possibilities on a pick. In practice, the interlacing order is random to ensure the length of all floats is not more than 3 points, and there should have and only have a continuous mark and blank of the two neighboring pattern lags simultaneously as shown in Fig. 5-2-6. If $R_w > C_n^{n/2}$, some of the pattern lags can be reused for different picks, but at least 3 picks apart.

(4) Draw the drafts and weave design. First, divide all the ends into several groups in the drafting plan. The number of the ends in a group is equal to the number of the healds. By changing the order of drafting in each group, a crepe weave is obtained. It should be noted that the two ends drawn in the same healds have to be placed 3 ends apart. Fig. 5-2-7 shows the looming plans for a crepe weave based on the method. The core of the method is to calculate the weave pattern based on lifting plan and drafting plan. Obviously, the drafts are based on the design, and 2~4 ends are placed in a split of reed. For an 8-shaft crepe weave, it is not suitable to use 4 consecutive same interweave points on a pick, thus, the number of the potential pattern lags is C_8^4-8.

Another example shows that two base weaves can be used in this method. Two composed twills, $\dfrac{3\quad 1}{3\quad 1}$ twill and $\dfrac{2\quad 2}{3\quad 1}$ twill, are alternately placed along weft direction at ratio 2:2 as shown in Fig. 5-2-8. By changing the starting point, direction of the movement and the arrangement ratio, the pegging plan is designed. Filling No. 1 starts from end No. 3 to right; filling No. 2 starts from end No. 6 to left; filling No. 3 starts from end No. 5 to right; filling No. 4 starts from end No. 7 to left .... Finally, the pegging plan is finished and a crepe weave is obtained.

右排列；第二纬从第5经开始向左
排列；第三纬从第5经开始向右排
列；第四纬从第7经开始向左排
列……再设计穿综图，得到绉组织。

图5-2-5 所有可能的纹板（20块）

All the potential 20 pattern lags

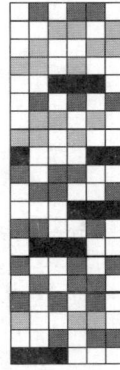

图5-2-6 符合省综设计法要求的某纹板图

Pattern lags designed by saving hearlds

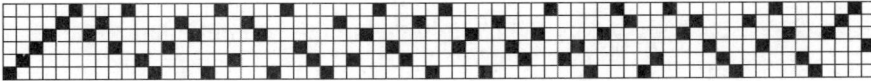

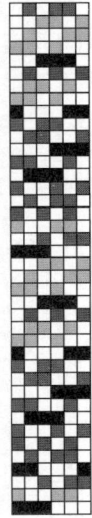

图5-2-7 某绉组织的上机图

The looming plans for a crepe fabric

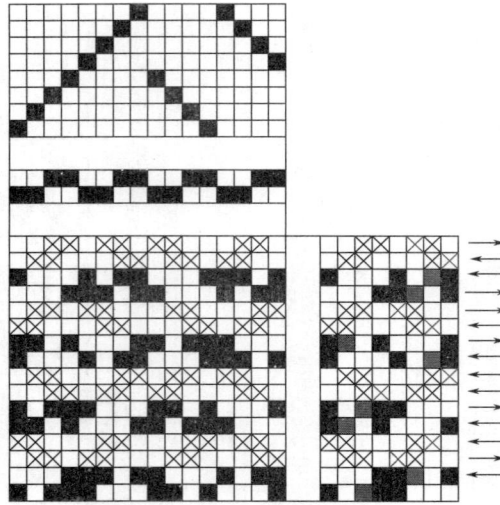

图5-2-8　由2个复合斜纹构成的纹板图生成的绉组织
The pegging plan combined by 2 composed twills

图 5-2-9 是用省综法设计的树皮绉织物组织。与一般的绉组织不同，树皮绉织物要求表面有自然、逼真的树皮状纹路，要求组织图上有凹凸不平、长短不一、大致垂直或略有倾斜的浮长线。纹板图由 $\frac{5}{1}$ 斜纹组织点与 $\frac{1}{1}$ 平纹组织点经向组合而成。为了防止背面纬浮长线移位，每隔两根纬纱安排一根平纹点纬纱。这样，经向浮长最长 5 个组织点，背面纬向浮长最长 7 个组织点。

Fig. 5-2-9 shows the looming plan of a bark crepe weave which is different in that natural bark texture is required to appear in the surface of the fabric. Therefore, there are several long floats which form roughly vertical or slightly inclined lines in the weave. In designing, the pegging plan is drawn by combining $\frac{5}{1}$ twill weave and $\frac{1}{1}$ plain weave in vertical direction. In order to void an over long float, a plain interlacing horizontal space is inserted every two picks. Thus, the warp covers as many as 5 picks in face side and the weft covers as many as 7 ends in back side.

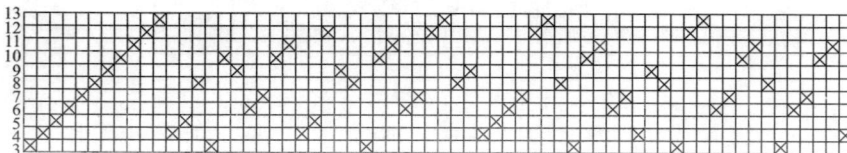

3 4 5 6 7 8 9 10 11 12 13

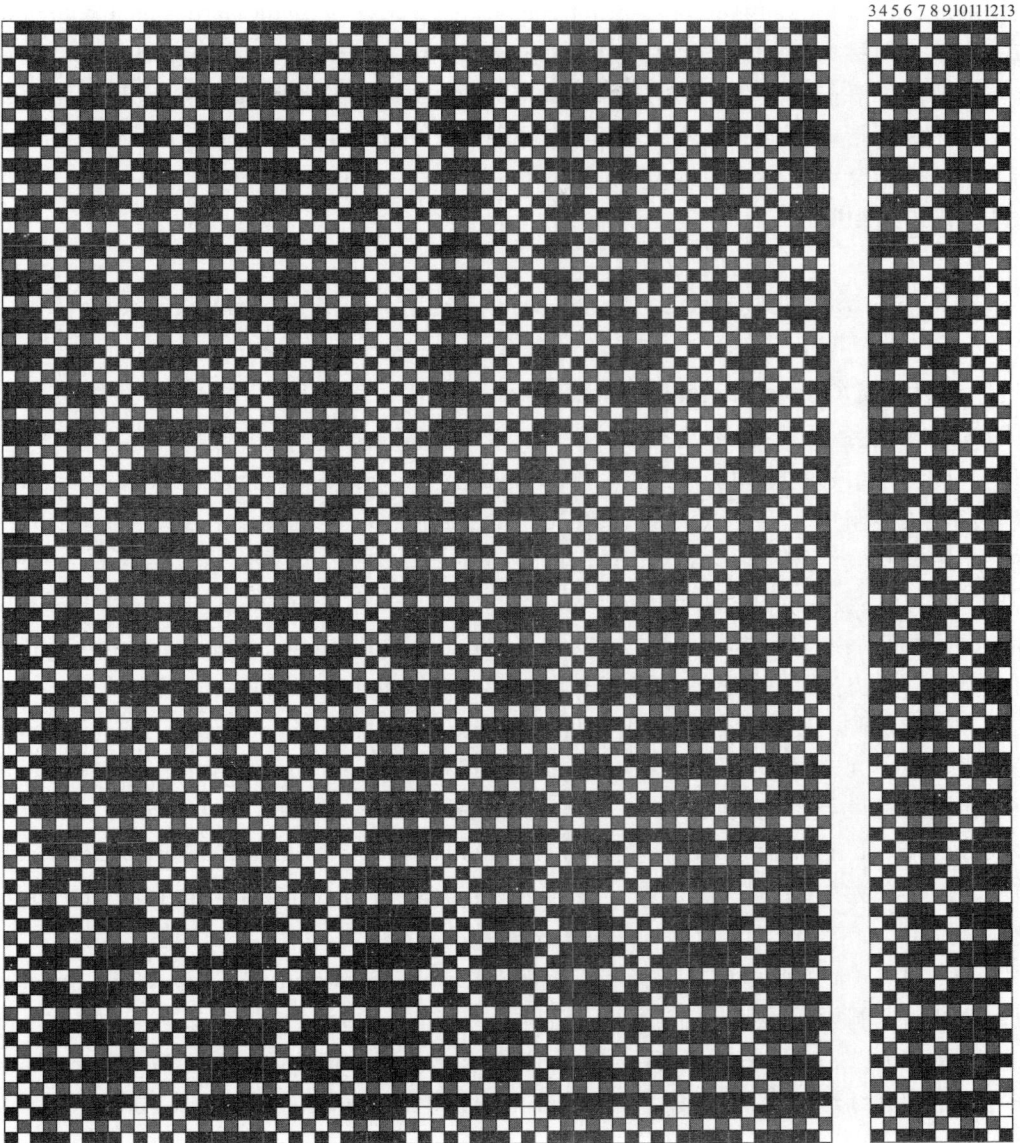

图5-2-9 树皮绉织物组织
The looming plans for a bark crepe fabric

第三节 透孔组织 /Open Gauze Weaves

透孔组织

一、外观特点 /Description

透孔组织由平纹组织与重平组

Open gauze weave is a combination of weaves having

织联合形成，在织物表面具有均匀
分布的小孔，故称为透孔组织。由
于这类织物的外观与复杂组织中由
经纱相互扭绞而形成孔隙的纱罗织
物相类似，因此又称之为"假纱组
织"或"模纱组织"。

interlacing that tends to form the warp ends into groups (with empty spaces intervening) in the cloth, thereby giving an imitation of the open structure that is characteristic of leno fabrics formed by twisting the ends. It is one kind of two mock lenos.

二、组织特点与透孔的形成原理 /Construction and Formation of the Structure

1. 常规解释 /Conventional Explanation

图 5-3-1 所示是典型的透孔
组织。可以看出，透孔组织由平纹
组织和长浮长线组成，也可以认为
由重平组织与平纹组织间隔排列而
成，甚至可以认为由一个小单元经
过多次底片翻转后组合而成。故组
织可分成上下左右 4 个等分单元，
在单元分界线两侧，组织点相反而
相互分离。在单元之间，长浮长线
的长度相同、跨度起始点相同，共
同形成分离作用，有助于形成纵向
和横向缝隙。单元内部，长浮长线
拉拢纱线聚集。

以图 5-3-1（a）所示的透孔
组织为例说明孔洞形成原理。第3、
4 根经纱及第 6、7（1）根经纱都
是按平纹组织和纬纱相交织，其经
纬组织点相反，因此第 3、4 根经
纱及第 6、7（1）根经纱就不易互

Typical open gauze weaves are shown in Fig. 5-3-1. These weaves are composed of plain weaves and the long floats. They can also be regarded as the warp rib weaves are alternately distributed in the plain weave, or they are composed of 4 weave units being counterchanged. The weave is divided into 4 units, and each unit is opposite to its neighbor, which causes the sections are forced apart. The floats are of the equal length, same commencement and ending position relative to another series of threads, and the common separation forces help the formation of the vertical or horizontal gaps between units. However, the order of the interlacings permits the threads within the unit to readily approach each other.

The weave as shown in Fig. 5-3-1(a) is taken as an example to explain how the gauze is formed. Both ends No. 3 and No. 4 are interwoven with the filling yarn alternately but just opposed thereby tend to be forced apart to form a gap between them. At same time, the long float on weft thread No. 2 and No. 5 aggregate the ends No. 1, No. 2 and No. 3 to form a

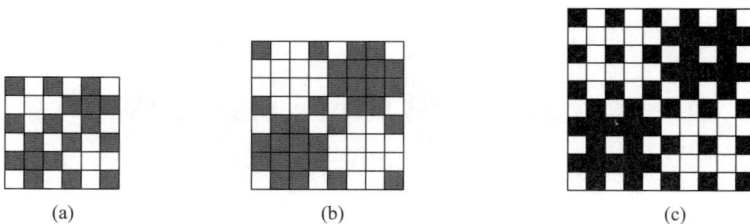

(a) (b) (c)

图5-3-1　透孔组织
Open gauze weaves

相靠拢。第2、5根纬纱浮长线使第1、2、3根经纱向一起靠拢，第4、5、6根经纱也向一起靠拢，故第3、4根经纱之间及第6、1根经纱间形成纵向缝隙。

同理，第3、4根纬纱上下两侧组织点底片相反，不易靠近；第6、1根纬纱上下两侧组织点底片相反，不易靠近。在经浮长线的拉拢聚集下，第1、2、3根纬纱靠拢，第4、5、6根纬纱也集聚靠拢，在第3、4根纬纱之间及第6、1根纬纱之间形成横向缝隙。这样就使织物表面出现了孔眼，如图5-3-2所示。图5-3-3是织物实物图。

group, ends No. 4, No. 5 and No. 6 to form another group. So do the ends No. 6 and No. 7 (1). Therefore, vertical gaps appear on the surface.

Similarly, the interweaving points along the wefts No. 3 and No. 4 are just opposite, so, the two wefts are separated each other, so do the wefts No. 6 and No. 1. Under the gathering forces of the long warp floats, the weft threads No. 1, No. 2 and No. 3 are grouped to get closer, and the weft threads No. 4, No. 5 and No. 6 get closer as well. Therefore, 2 horizontal gaps appear between the weft threads No. 3 and No. 4, No. 6 and No. 1 respectively. The holes are formed on the surface duo to the gaps of two-way as shown in Fig. 5-3-2. Fig. 5-3-3 shows a photo of the real fabric of the weave.

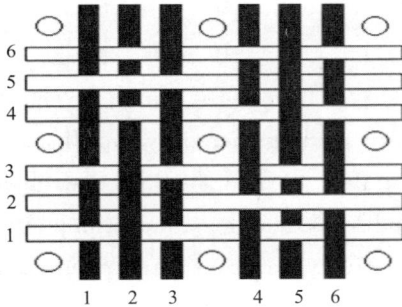

图5-3-2 透孔孔隙分布示意图
Diagram of splits in a open gauze weave

图5-3-3 透孔组织织物实物图
The image of the fabric of open gauze weave

2. 深入分析 /Deep Analysis

如图5-3-2所示，第1纬上，第3、4经因为组织点相反交错分开，隔离成缝隙。但第1、2、3经之间组织点也是相互相反的（交错），为何这些经纱能够集聚到一起？此外，第4、5、6经也是如此。第3、4、6纬的情况与第1纬相同。如果横向看，第1、3、4、6经也能够将第1、3、4、6根纬集

But another problem arouses. On the weft No. 1, ends No. 3 and No. 4 are separated due to the intersections of the neighboring interweaving points (Fig.5-3-2). There are intersections between No. 1, No. 2 and No. 3, then how can ends No. 1, No. 2 and No. 3 be grouped together? The same situation happens for ends No. 4, No. 5 and No. 6 on the weft No. 1. It is the same for wefts No. 3, No. 4 and No. 6, or for ends No. 1, No. 3, No. 4 and No. 6. The principles can only be explained based on the overlapping of the floats.

聚在一起。其原理需要用浮长线的重叠效应分析。

（1）纵向缝隙，经纱被纬浮长线拉拢分离情况分析。根据该组织各组织点对应经向浮长线分布图5-3-4，因为在第2（5）经上连续3个组织点的经浮长超过（或低于）左右两侧，可以认为经纱之间实际已经发生了局部重叠。

(1) Vertical Gap, Along the Warp Direction. According to the diagram of lengths of warp floats in Fig. 5-3-4, the warp float lengths of the 3 consecutive interlacing points exceed or less than their neighbors of left side or right side, which indicates local overlapping actually happens between the warp threads.

1	-3	1	-1	3	1
-1	-3	-1	1	3	-1
1	-3	1	-1	3	1
-1	3	-1	1	-3	1
1	3	1	-1	-3	-1
-1	3	-1	1	-3	1

图5-3-4　经浮长分布图
Lengths of warp floats

在左下角第2纬处，第1、2、3经容易被纬浮长线拉拢聚集。因为第2经浮长长，高度特高，导致其与第1、3纬纱交织弯曲程度小甚至实际可能不存在交错情况，截面变化大致变成图5-3-5下方的情况。故第2纬的浮长比较容易拉拢相邻的第1、3纬处的第1、3经，使其易于靠近。因此，第1、2、3经在左下角容易靠拢。

在右下角第2纬处，第4、5、6经容易被纬浮长线拉拢聚集。因为第5经特低，导致第1、3纬弯曲程度小，截面变化大致变成图5-3-6下方的情况，故第2纬浮长比较容易拉拢相邻的第1、3纬处的第4、6经，使其易于靠近。因此，第4、5、6经在右下角容易靠拢。

右上角的情况与左下角类似，

On the pick No. 2 at bottom-left corner, the ends No. 1, No. 2 and No. 3 are easily gathered by weft floats. Since there is a long warp float, the end No. 2 is relatively higher than ends No. 1 and No. 3, and the crimps with picks No. 1 and No. 3 are relatively decreased and even vanished thereby forming the cross-section of the fabric similar to the under part of Fig. 5-3-5. Consequently, the ends No. 1 and No. 3 have the tendency to approach even at wefts No. 1 and No. 3. As a result, the ends No. 1, No. 2 and No. 3 get closed at the bottom-left corner of the weave.

On the pick No. 2 at bottom-right corner, the ends No. 4, No. 5 and No. 6 are easily gathered by weft floats. Since there is a long warp float on the reverse side, the end No. 2 is relatively lower than ends No. 1 and No. 3, and the crimps with picks No. 1 and No. 3 are relatively decreased and even vanished thereby forming the cross-section of the thread similar to the under part of Fig. 5-3-6. Consequently, the ends No. 4 and No. 6 have the tendency to approach even at wefts No. 1 and No. 3. Thus, the ends No. 4, No. 5 and No. 6 are easily aggregated at

左上角与右下角类似。故在整个组织范围内，第1、2、3经容易聚集，第4、5、6经容易靠拢，在第3经与第4经之间、第6经与第1（7）经之间形成纵向缝隙。

the bottom-right corner of the weave.

The weave unit at top right corner is just the same as that at the bottom left corner, and the top left corner is corresponding to bottom right corner. Therefore, ends No. 1, No. 2 and No. 3 are easy to aggregate, ends No. 4, No. 5 and No. 6 are easy to approach, and longitudinal gaps are formed between ends No. 3 and No. 4, No. 6 and No. 1 (7).

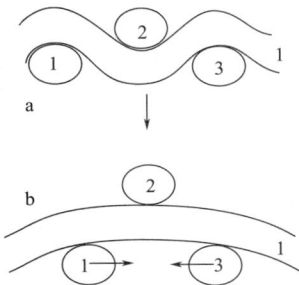

图5-3-5 第1纬与第1、2、3经交织变化过程示意图
Geometric changing in interlacing pick 1 and ends No. 1, No. 2 and No. 3 at lower left quarter

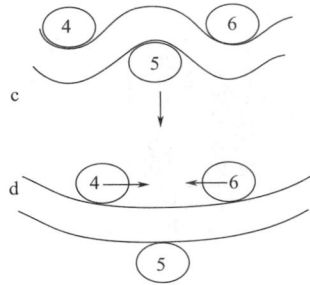

图5-3-6 第1纬与第4、5、6经交织变化示意图
Geometric changing in interlacing pick 1 and ends No. 4, No. 5 and No. 6 at lower right quarter

（2）横向缝隙，纬纱被经浮长线拉拢分离情况分析。图5-3-1组织中各组织点对应的纬浮长线长度如图5-3-7所示。因为在第2（5）纬连续3个组织点的纬浮长超过（或低于）上下两侧，可以认为纬纱之间实际已经发生了局部重叠。

在左上角第2经处，第4、5、6纬容易被经浮长线拉拢聚集。因为第5纬特高，导致第4、6纬弯曲程度小，截面变化大致变成图5-3-8右图的情况，故第4、6纬处，第2经的经浮长比较容易拉拢第4、6纬靠近。因此，第4、5、6纬在左上角容易靠拢。

在右上角第5经处，第4、5、6纬容易被经浮长线拉拢聚集。因为第5纬特低，导致第4、6纬

(2) Horizontal Gap, Along the Weft Direction. According to the diagram of lengths of weft floats in Fig. 5-3-7, the weft float lengths of the 3 consecutive interlacing points exceed or less than their neighbors of upper side and lower side, which indicates local overlapping actually happens between the weft threads.

On the end No. 2 at upper left corner, the wefts No. 4, No. 5 and No. 6 are easily gathered by warp floats. Since there is a long weft float on the face side, the end No. 5 is relatively higher than ends No. 4 and No. 6, and the crimps with picks No. 4 and No. 6 are relatively decreased and even vanished and thereby forming the cross-section of the thread similar to the right part of Fig. 5-3-8. Consequently, the ends No. 4 and No. 6 have the tendency to approach even at wefts No. 4 and No. 6. Thus, the wefts No. 4, No. 5 and No. 6 are easily aggregated at the top-left corner of the weave.

On the pick No. 5 at upper right corner, the ends No. 4, No. 5 and No. 6 are easily gathered by weft floats. Since there

弯曲程度小，截面变化大致变成图 5-3-9 右图的情况，故第 4、6 纬处，第 5 经的经浮长比较容易拉拢第 4、6 纬靠近。因此，第 4、5、6 纬在右上角容易靠拢。

is a long weft float at the reverse side, the end No. 5 is relatively lower than wefts No. 4 and No. 5, and the crimps with picks No. 4 and No. 6 are relatively decreased and even vanished thereby forming the cross-section of the thread similar to the right part of Fig. 5-3-9. Consequently, the ends No. 4 and No. 6 have the tendency to approach even at wefts No. 4 and No. 6. Therefore, the wefts No. 4, No. 5 and No. 6 are easily aggregated at the top-right corner of the weave.

Keys to Form Open-Gauze Effects

-1	1	-1	1	-1	1
3	3	3	-3	-3	-3
-1	1	-1	1	-1	1
1	-1	1	-1	1	-1
-3	-3	-3	3	3	3
1	-1	1	-1	1	-1

图5-3-7　纬浮长分布图
Lengths of weft floats

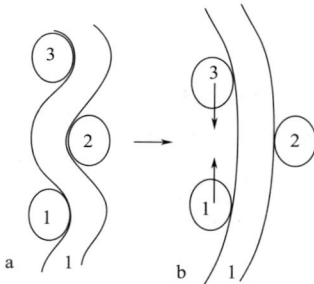

图5-3-8　第1经与第1、2、3纬交织示意图
Geometric changing in interlacing end 1 and picks No. 1, No. 2 and No. 3 at top left quarter

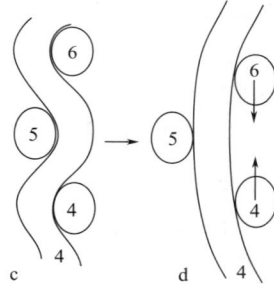

图5-3-9　第4经与第4、5、6纬交织示意图
Geometric changing in interlacing end 1 and picks No. 4, No. 5 and No. 6 at bottom right quarter

右上角的情况与左下角类似，左上角与右下角类似。故在整个组织范围内，第 1、2、3 纬容易聚集，第 4、5、6 纬容易靠拢，在第 3 纬与第 4 纬之间、第 6 纬与第 1（7）纬形成横向缝隙。

The weave unit at top right corner is just the same as that at the bottom left corner, and the top left corner is corresponding to bottom right corner. Therefore, wefts No. 1, No. 2 and No. 3 are easy to aggregate, wefts No. 4, No. 5 and No. 6 are easy to approach, and transverse gaps are formed between wefts No. 3 and No. 4, No. 6 and No. 1 (7).

3. 形成透孔效应的关键 /Keys to Form Open-Gauze Effects

综上，透孔形成的关键是：①浮长线的共同拉拢作用，要求浮长长

In summary, the keys to form open-work effects are listed as following: ① the threads are aggregated to group

度相同、浮长位置一致；②孔洞或者缝隙处具有破界；③被拉拢的相邻纱线具有高度差，使得拉拢效果突出、形成的孔隙大；④有适当的穿筘方式，成组聚集的经纱穿在同一筘齿中，相邻组之间空筘；⑤采用间隙卷取方式，孔洞效应更加明显。

together at the different units by the long floats of the warp and weft simultaneously. The long floats of one series of threads have the same commencement and terminal in another series; ② there are broken borders for 4 units to separate the threads; ③ the grouped adjacent threads have the tendency to overlap due to different height, which makes the adjacent threads aggregate without difficulty; ④ proper denting helps the threads within the weave unit grouped. The ends that are gathered in a group should be threaded into the same dent, and the empty dent between the neighboring group gives the salient open-work effects; ⑤ the open-work effects may be more obvious with the aid of an interrupted take-up device, and such a device is confined to an all-over open cloth.

三、简单透孔组织的组织图绘制 /Designing of Simple Open Gauze Weaves

简单透孔组织组织图的绘制步骤如下：

（1）确定组织循环纱线数，一般 $R_j=R_w$，且取一个奇数的 2 倍，如 6、10、14，但也可以是 8。

（2）在意匠纸上画出组织图的范围，并将组织分成上下左右 4 等份。

（3）在组织图的左下角填绘基础组织。画出"十""井""卅"字型的浮长线。

$R_j=2\times(2\times n+1)$，n 是纵向长浮长个数，对于"卅"型透孔组织，$n=2$。

$R_w=2\times(2\times m+1)$，m 是横向长浮长个数，对于"卅"型透孔组织，$m=1$。

（4）按照底片翻转的关系画出其余的组织点。

The following describes the procedures of drawing a simple open gauze weave:

(1) Determine the proper even number as the size of the weave repeat R_j and R_w, say, 6, 8, 10, or 14.

(2) Draw the square range on the point paper and divide the weave into 4 parts.

(3) Draw the weave unit at the bottom-left of the design like the symbols "十""井""卅".

$R_j=2\times(2\times n+1)$, n: number of fine lines in vertical spaces. For symbol "卅", $n=2$.

$R_w=2\times(2\times m+1)$, m: number of fine lines in horizontal spaces. For symbol '卅', $m=1$.

(4) Draw the other 3 parts by reversing or counterchanging the interweaving points.

四、透孔织物上机要点 /Looming for Open Gauze Weaves

（1）透孔织物的密度不宜过大，否则透孔效应不明显，失去

(1) The threads in of open gauze fabric should not be too overcrowded, otherwise the pen-work effect is not obvious

薄、轻、松、爽等特性。

（2）浮长越长，孔眼越大，但一般服用织物透孔组织的浮长线不超过 5 根。否则，织物过于松软，影响透孔效应。

（3）穿综采用照图穿法或间断穿法，采用 4 片综即可织造。

（4）穿筘时将成束的经纱穿入同一个筘齿内，或每组经纱之间空一筘。图 5-3-10 为透孔组织的上机图。

and the fabric will lose its characteristics of thin, light, loose and crisp.

(2) The longer the float length is, the bigger the pore will be. However, if the length of float is too long, the fabric is too soft and, in turn, the open-work effect is damaged. Therefore, the float length in an open gauze weave commonly used in clothing fabric is rarely greater than 5.

(3) The fabric is threaded by grouped drafting or design drafting, and can be woven with 4 healds.

(4) The ends that are aggregated in a group should be threaded into the same dent, and empty dent may improve the open-work effects. Fig. 5-3-10 shows the looming plans of an open gauze weave with different drafting.

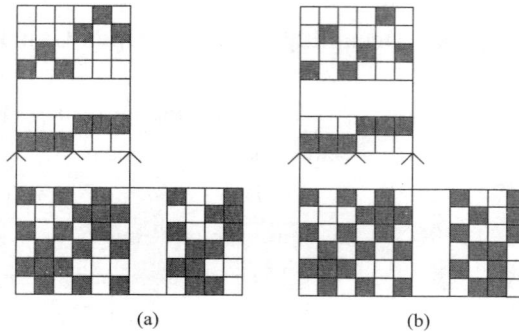

图5-3-10　透孔组织上机图

The looming plans for an open gauze weave

五、变化透孔组织 /Variations of Open Gauze Weaves

还有一些其他形式的透孔组织，如图 5-3-11 和图 5-3-12 所示。花式透孔组织是在平纹组织作为地部的基础上，添加一些透孔组织形成。一般组成透孔组织的 4 个单元至少有 3 个单元可以确保较好的透孔效果。若只有对角 2 个单元，拉拢力量不足，只能形成浮松组织的效应。

The units of the open gauze weave can be arranged as the form of a patten, and combined with the plain ground to produce a fancy effect as in Fig. 5-3-11 and Fig. 5-3-12. Usually, 3 units will ensure better open-work effects. Two units which are arranged diagonal cannot provide the ability to gather or separate the threads, and only the huckaback effects appear on the surface of fabric.

(a)

(b)

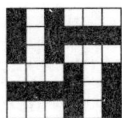

图5-3-11 变化的透孔组织

A varied open gauze weave

图5-3-12 花式透孔组织

Fancy mock leno weaves

六、透孔组织的应用 /Applications of Open Gauze Weaves

透孔组织织物因多孔、轻薄、凉爽，易于散热、透气等特点，一般用作稀薄的夏季服装用织物和装饰织物，如各种网眼布和花式透孔织物等。透孔组织广泛用于棉、麻、丝织物，毛织物中应用较少。涤纶等合成纤维织物中采用透孔组织既增添了花纹，又改善了合成纤维透气性差的缺点。透孔组织有时单独使用，如用于廉价的窗帘。若用于轻薄的女式衣服，如罩衫、围裙等，多与其他组织联用。最常见的用途是形成经向透孔条纹效应和隔开的小提花花纹效果。

Due to its characteristics of porous, light, cool and easy to dissipate heat and ventilate, open gauze weaves are generally used for thin summer clothing and decorative fabrics, such as various kinds of mesh fabrics and fancy perforated fabrics. Open gauze weaves are widely used in cotton, linen and silk fabrics, but not in wool fabrics. The porous structure used in polyester and other synthetic fabrics not only increases the pattern, but also improves the poor permeability of synthetic fabrics. Open gauze weaves are sometimes used alone, as in canvas cloths, and in cheap fabrics for window curtains; but for light dress fabrics, blouses, aprons, etc. they are, to a large extent, employed in combination with other weaves. However, as the most common occurrence of the perforated effect is in the form of stripe or isolated dobby spot figure.

第四节 蜂巢组织 /Honeycomb Weaves

蜂巢组织

一、外观特点 /Description

织物表面呈现四周高、中间低的凹凸四方形，且形如蜂巢，故称

The honeycomb weave is a weave that produces a textured surface fabric with a pattern of squares or diamond

蜂巢组织。在织物表面，经纱和纬纱形成了凸起和凹处，形成了明显的蜂窝状效果。该织物正反面都有经纬长浮长线，手感松软、丰厚，缩水率较大，如图5-4-1所示。有两种蜂巢组织：①普通蜂巢组织，蜂巢效果各处一样；②勃拉东蜂巢组织，正面效应突出，且蜂巢大小不一。

shapes similar in appearance to a honeycomb. On the fabrics, cellular or waffle appearance is prominent on the surface. The warp and weft threads form ridges and recesses (hollows), which give a cell-like appearance. Warp and weft threads floats on both sides make the cloth soft and moisture absorbent(Fig. 5-4-1). There are two types of weaves which produce this effect: an ordinary honeycomb weave, which gives a marked cellular effect on the face and back of the cloth; a Brighton honeycomb weave, which develops the effect more prominent on the face but in a less regular manner and with large and small cells.

图5-4-1　蜂巢织物外观形态
The image of the fabric of a honeycomb weave

二、组织特点与蜂巢的形成原理 /Construction and Formation of the Structure

蜂巢组织是在单个组织点的菱形斜纹基础上延伸浮长线而成。图5-4-2所示组织为一个简单的蜂巢组织的若干循环。蜂巢组织被单个组织点的2根菱形斜纹线分成4个部分。一组对角为经组织点，另一组对角为纬组织点。蜂巢组织的四周分别是经浮长线和纬浮长线，浮长线逐渐向中间过渡缩短直至成平纹组织点。组织的四周浮长长、高度高、厚度大、较为松弛；中间平纹部分结构紧密，凹下，交错次数多，面积扩张，形成洼地。由

A honeycomb weave is formed by extending the floats of a diamond base of single diagonal line. In a typical honeycomb weave as shown in Fig. 5-4-2, the weave is divided into 4 parts by 2 rhombus lines. One pair of the opposed sections in one direction are filled with marks while the other pair of the opposed sections in another direction are kept with blanks. Ridges occur where the long floats of warp and weft are formed at the border of the cells; and the lengths of the floats decrease gradually to form the hollows at the center of the fabric where the threads interweave in plain interlacing. The loose structure (long floats) and the tight structure (a plain weave) are alternately arranged, which makes the gradual transition of the fabric

于松紧组织相间配置，织物表面由低逐步过渡到高，由高逐步过渡到低，形成四方形凹凸形状，而不是组织上纹路显示的菱形形状。在该组织上，有两处平纹点，分别是 A 点、B 点。A 点虽然浮长值不大，但处于织物正面经浮长线和正面纬浮长线的交汇处，经纬纱都有向正面比较高的地方运动的趋势，以减少纱线的应力。B 点则处于织物反面经浮长线和反面纬浮长线的交汇处，在织物反面是比较高的地方，故该点有向反面高处运动的趋势，以减少纱线的应力，因此在织物正面处于低洼的位置。如果将织物组织底面翻转，可以发现反面与正面组织完全相同，也是四周高、中间低的浮长配置，但是经纬向都刚好相差半个组织循环。这意味着，织物正面高处是反面的凹下处。

　　蜂巢组织形成的凹凸蜂窝格型是否为方形取决于以下几个因素：经纬最长浮长线长度的比例关系（有时是 R_j 与 R_w 的对比）、织物的经纬密

surface from thin to thick, from thick to thin, and forming a cellular shape. The plain weave tightens the threads and causes a depression to be formed. Although the weaves are constructed on a diamond basis, the cellular formation makes the patterns appear rectangular in the cloth. There are two places where plain interlacement occurs, i.e. A and B. Although the length of float at interlacing point A is only 1, it crosses the long warp float and long weft float, therefore, point A is still at high place to avoid excess inner stress in thread. However, interweaving point B crosses the long warp float and long weft float at the reverse of the fabric, so, interweaving point B locates the hollow of the cell. If the honeycomb fabric is reversed, the reversed weave is just the same as the obverse weave but with difference of a half-repeat, which means the highest place on the face is just the lowest place of the back whilst the lowest point on the face is the highest point of the back.

The shapes of the cells are dependent on the following factors: the ratio of the length of the longest warp float and the length of the longest weft float (sometimes, ratio of R_j and R_w); ratio of the warp thread density and the weft thread density; ratio of the warp count and the weft count. A prominent effect of the fabric is obtained by the coarse thread, high thread density and long floats. The waffle

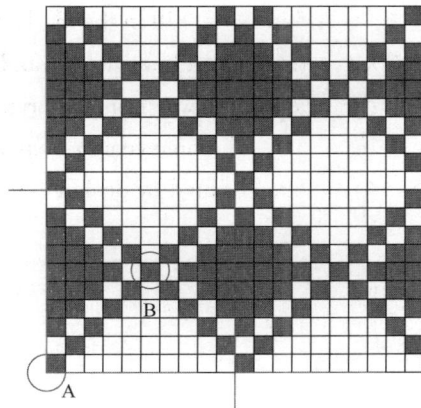

图5-4-2　蜂巢组织
A honeycomb weave

度对比、经纱与纬纱的粗细程度比例关系。经纬纱线越粗、浮长线越长、织物经纬纱的密度越大，效果越显著。疏松织物的蜂巢效应不明显。

effects are obscure with an open structure.

三、简单蜂巢组织的组织图绘制 /Designing of Ordinary Honeycomb Weaves

既然蜂巢组织是在单个组织点的菱形斜纹基础上形成的，因此其组织循环根数的计算类似菱形组织。蜂巢组织组织图的绘制步骤如下：

（1）选定基础组织。常用$\frac{1}{4}$、$\frac{1}{5}$、$\frac{1}{6}$纬面斜纹为基础组织。假设要设计的蜂巢组织是基于$\frac{1}{4}$斜纹组织，则 R_b= 1+4=5。

（2）确定组织循环。若 K 为其基础斜纹组织的循环根数，则 $R_b=K_j=K_w$。

$R_j =R_w =2\,K_j–2 = 2 \times 5–2 = 8$（或 $2K_w – 2$)。

（3）填绘单个组织点的菱形斜纹。

（4）菱形斜纹的斜纹线把整个组织分成四个部分，然后在其相对的两个三角形内（上和下两部分或左和右两部分）填绘经组织点。填绘时与原来的菱形斜纹之间空一个纬组织点。

图 5–4–3 是以$\frac{1}{4}$斜纹为基础的蜂巢组织。穿综采用照图穿法，节约综框数量。

The ordinary honeycomb weaves are designed based on the diamonds with a single twill line. Therefore, R_j and R_w of the honeycomb weave are calculated as the diamond weave. The procedures of designing honeycomb weaves are described as following:

(1) Determine the base twill and calculate the base repeat R_b. The diamond weaves are usually based on $\frac{1}{4}$, $\frac{1}{5}$ and $\frac{1}{6}$ twill. Supposing the honeycomb weave is based on a $\frac{1}{4}$ twill, then, R_b=1+4=5.

(2) Calculate the R_j and R_w according to the way for diamond weave. If K is the repeat of the base twill weave, $R_b=K_j=K_w$.

$R_j =R_w = 2\,K_j - 2 = 2 \times 5 - 2 = 8$ (or $2K_w - 2$)

(3) Fill the marks of the single twill line on the diamond weave.

(4) The rhomboid line divides the whole weave into four parts. Fill in the marks within the two opposite triangles (upper and lower or left and right), keeping the filled marks a square away from the original rhomboid twill line.

The honeycomb weave is shown in Fig. 5–4–3, and is often drafted by design to reduce the healds used.

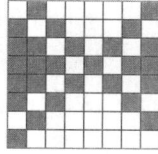

图5-4-3 以 $\dfrac{1}{4}$ 斜纹为基础的蜂巢组织

The honeycomb weave based on a $\dfrac{1}{4}$ twill

四、变化蜂巢组织 /Variations of Honeycomb Weaves

1. 基础斜纹线错位 /Dislocation of the Base Marks

组织循环大小与简单的蜂巢组织相同，把对角斜纹线错开一格为基础，再在此基础上填绘经长浮点而形成蜂巢组织。图 5-4-4 是以 $\dfrac{1}{5}$ 斜纹为基础，$K_j=K_w=6$ 的变化蜂巢组织的组织图绘制的过程中基础错位斜纹线的两种绘制方法。R_j 和 R_w 仍然保持不变（10），首尾 2 根经纱和首尾 2 根纬纱形成高处。由于只能用顺穿法，故使用综框数量较多。

The base marks are not forming a regular rhomboid but a misplaced rhomoid. The arrangement usually makes a rectangle cellular rather than a square cellular appearance. Fig. 5-4-4 shows the drawing procedures of such a variation based on $\dfrac{1}{5}$ twill and $K_j=K_w=6$. Therefore, R_j and R_w keep unchanged; 2 ends (first and last) and 2 fillings (first and last) form the ridges. Straight drafting can only be used, so more drafts are required.

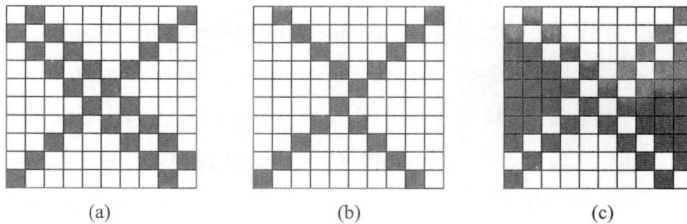

(a)　　　　　(b)　　　　　(c)

图5-4-4 斜纹线错位的变化蜂巢组织图绘制过程

The procedures of designing a honeycomb weave based on the dislocated base mark

2. 基础斜纹隔空一行 /Separating Opposite Vertex of the Base Marks

将单个组织点菱形斜纹线变成顶点相对且隔一纬的上、下两个山形斜纹。然后在左、右两侧对

The base marks are divided into two isolated or separated parts by moving opposite vertex of the base marks as shown in Fig. 5-4-5. The other marks are filled

角区域内填绘经组织点。该组织 $R_j=2K_j-2$，$R_w=2K_w$。图 5-4-5 是以 $\dfrac{1}{4}$ 斜纹为基础，$K_j=K_w=5$ 的变化蜂巢组织图。如果横向隔开，基础斜纹线如图 5-4-5（a）所示，则 $R_j=2K_j$，$R_w=2K_w-2$，得到图 5-4-5（b）所示的组织。这种组织中最长的经纬浮长长度相同，一般采用山形穿综。如果基础斜纹线被经纱隔开，如图 5-4-5（c）所示，则得到图 5-4-5（d）所示的组织，此时需要增加两页综框。

in the way similar to the ordinary honeycomb weave. If the two parts are separated by a filling thread as shown in Fig. 5-4-5(a), $R_j=2K_j-2$, $R_w=2K_w$, the honeycomb weave as shown in Fig. 5-4-5(b) is obtained; otherwise, the two parts are separated by an end as shown in Fig. 5-4-5(c), $R_j=2K_j$, $R_w=2K_w-2$, and the weave as shown in Fig. 5-4-5 (d) is obtained. From the diagrams, the weave has the same longest warp float and weft float. The weave is pointed drafted. If the bask mark is separated by an end, 2 more heals are needed.

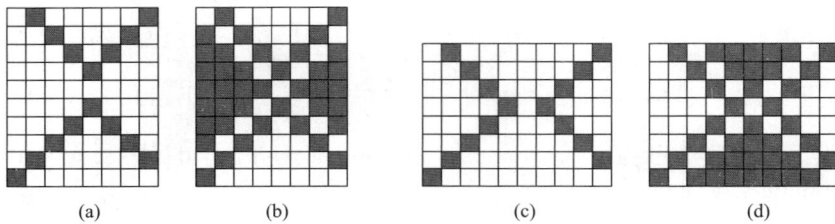

(a) (b) (c) (d)

图5-4-5　对顶点移动的变化蜂巢组织的绘制流程
The procedures of designing a honeycomb weave based on moving opposite vertex

3. 两条基础斜纹线 /Double-Stitch Base Marks

随着蜂巢组织的循环增加，纱线交织次数明显下降，织物牢度下降。组织循环大时，建议在单个组织点菱形斜纹线的下方，隔一个纬组织点，再作一条平行的斜纹线。通过增加交织次数，在不影响蜂巢织物特点的情况下，增强牢度。图 5-4-6 是以 $\dfrac{1}{8}$ 斜纹为基础，$K_j=K_w=9$ 的变化蜂巢组织图。

As the honeycomb weaves increase in size, the threads are proportionately less frequently interlaced, thereby producing a weaker texture. It is advisable therefore, to construct the larger weaves on the basis of what is termed a double-stitch diamond. By increasing the degree of interlacement of threads, a fabric of firmer texture is produced without destroying the salient features of the honeycomb weave. The double-stitched honeycomb weave in Fig. 5-4-6 is based on a $\dfrac{1}{8}$ twill with $K_j=K_w=9$.

4. 勃拉东蜂巢组织 /Brighton Honeycomb Weaves

勃拉东蜂巢组织形成大小不同的两个蜂巢，其组织循环数必须

Brighton honeycomb weaves produce alternate small and large cells, both longitudinally and transversely.

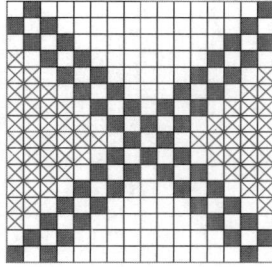

图5-4-6 双菱形斜纹线变化蜂巢组织
A double–stitched honeycomb weave

是 4 的倍数。在单个组织点菱形斜纹的左斜纹线下方，隔一个纬组织点，再作一条平行的左斜纹线。然后在左、右两侧对角区域内填绘经组织点，各形成一个菱形区域。其经、纬最长的浮长线等于（$R/2$）–1。以 $\frac{1}{6}$ 斜纹为基础，$K_j=K_w=7$ 的勃拉东蜂巢组织绘制过程如图 5-4-7 所示。该组织的长浮长线形成的蜂窝高处分布如图 5-4-8 所示。由于大小蜂巢间隔分布，最终形成的织物实际效果如图 5-4-9 所示。

The cellular appearance is similar to that of honeycomb weaves, but it is not reversible. The size of the repeat of a Brighton weave must be a multiple of 4. Draw another left–handed base line one square away from the original left–handed twill base line, and the whole weave is divided into 4 sections. In each section, draw a diamond with the longest float of ($R/2$)–1. The Brighton honeycomb weave in Fig. 5–4–7 is based on $\frac{1}{6}$ twill with $K_j=K_w=7$. The distribution of the ridges formed by the long floats is shown in Fig. 5–4–8, from which the two cells of different size are placed at interval at two directions, therefore, the appearance of the weave is shown in Fig. 5–4–9.

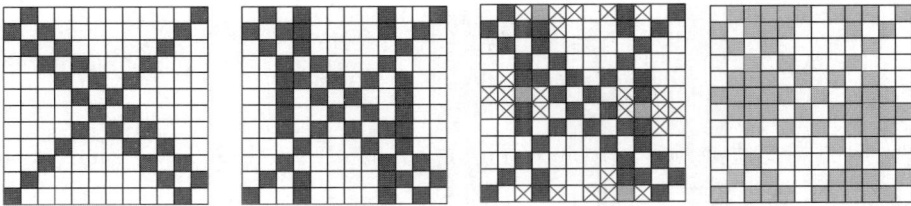

图5-4-7 勃拉东蜂巢组织绘制过程
The procedures of designing a Brighton honeycomb weave

五、蜂巢组织上机与应用 /Looming and Applications of Honeycomb Weaves

与菱形斜纹组织一样，简单的蜂巢组织采用山形穿法或照图穿法。变化蜂巢组织织物可根据情

Like diamond twill weaves, the ordinary honeycomb weaves are pointed drafting or drafted by design. However, the variations of the honeycomb weaves will be drafted by

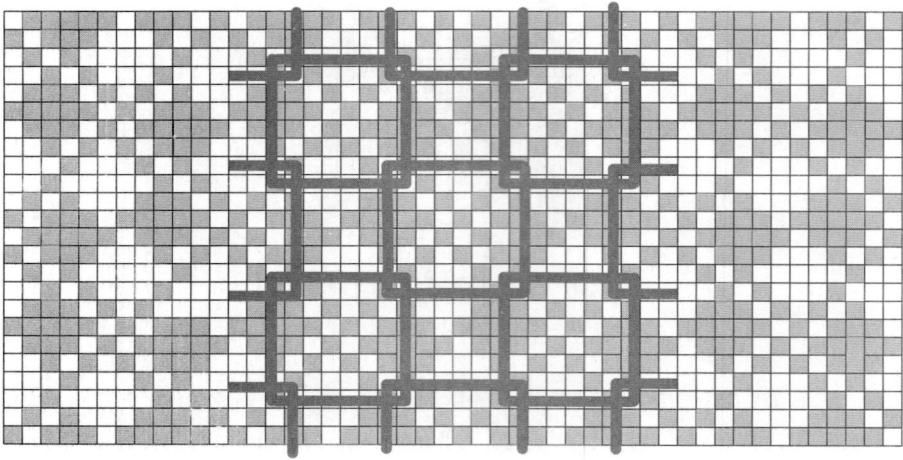

图5-4-8　勃拉东蜂巢组织的蜂窝高处分布

The distribution of the ridges of a Brighton honeycomb weave

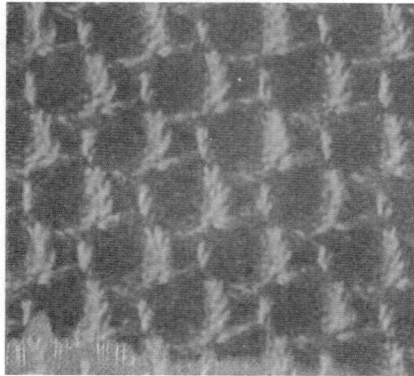

图5-4-9　勃拉东蜂巢织物外观

The appearance of a Brighton honeycomb weave

况，穿法采用照图穿法或顺穿法。较细的蜂巢组织织物常用于女装和童装，较粗纱线或股线的蜂巢织物则用于装饰用的帷幔、沙发套、窗套、抹布等。

design or straight drafted. Fine yarn honeycomb weave fabrics are used for women's and children's wear; heavier and plied yarn fabrics for home furnishings such as draperies, slipcovers, and bedspreads.

第五节 凸条组织 /Bedford Cord Weaves

凸条组织

一、外观特点 /Description

在织物正面产生纵向、横向或斜向的凸条纹，反面有浮长线浮起的组织，称为凸条组织。织物外观呈现纵向、横向或斜向的灯芯条，反面呈纬浮线或经浮线的效应。根据条子的宽度，通常在凸条的背面有 2 根或多根芯线。

Bedford cord weaves are a variety of fabrics characterized by a series of more or less pronounced plain or twilled ribs or cords, lying in the same direction as warp threads, with weft floating somewhat freely at the back of the ribs, and usually with one, two or more wadding threads (according to the width of ribs) lying loosely between.

二、组织特点与凸条的形成原理 /Construction and Formation of the Structure

凸条组织由浮长较长的重平组织和平纹或斜纹组织按照一定比例组合而成，如图 5-5-1 所示。平纹或斜纹等简单组织起固结浮长线的作用，并隆起在织物正面，故称为"固结组织"。若固结纬重平的纬浮线则得到纵凸条组织，若固结经重平的经浮线则得到横凸条组织，若把纵向凸条排列改变成斜向则得到斜向凸条。图 5-5-1 ～图 5-5-5 是不同的凸条组织。

Bedford cord weaves are proportionally combined by the weft rib or warp rib with long floats and the binder weave with more interlacing such as plain, twills. The binder weave floats and bulges on the face. If with the weft rib, the longitudinal cord weaves are formed; if the warp rib is combined, the transverse cord weaves are formed. The fancy cord weaves such as twill cord weaves can also be made by changing the direction of the rib. Various types of Bedford cord weaves are shown from Fig. 5-5-1 to Fig. 5-5-5.

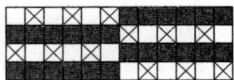

图5-5-1　纵凸条组织1（1∶1正面）
The Bedford weave 1 (1∶1, face)

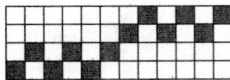

图5-5-2　纵凸条组织2（2∶2反面）
The Bedford weave 2 (2∶2, back)

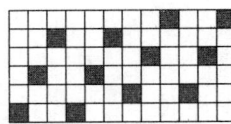

图5-5-3　纵凸条组织3
The Bedford weave 3

凸条组织中浮长线和固结部分交替分布。图 5-5-6 所示纵凸条组织的纬浮长线分布如图 5-5-7 所示。从第 1 经到第 6 经，连续 6 个

The float rib and the binder area are distributed alternately. The distribution diagram of the lengths of weft floats of the Bedford cord weave (Fig. 5-5-6) is shown in Fig. 5-5-7. From the ends No. 1 to No. 6, the filling threads

187

浮长都是第2、4纬高，第1、3纬低，在纬纱密度大的情况下，形成局部重叠，正面只能看见第2、4纬（即固结组织），而第1、3纬在下层，只能在背面看见（即浮长线）；同理，从第7经到第12经，正面只能看见第1、3纬（固结组织），第2、4纬则被覆盖，只能在背面可见。

No. 2 and No. 4 are always higher than their neighbor filling threads No. 1 and No. 3. With high weft sett, the filling threads tend to overlap, thereby only the filling threads No. 2 and No. 4 (binder weave) are visible on the face, and the long floats of filling threads No. 1 and No. 3 sink at the back of the fabric and become invisible on the face. Similarly, from ends No. 7 to No. 12, only the filling threads No. 1 and No. 3 (binder) are visible from the face, and the long wefts float on the other side of the fabric.

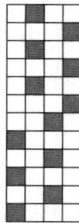

图5-5-4 横凸条组织
The horizontal Bedford weave

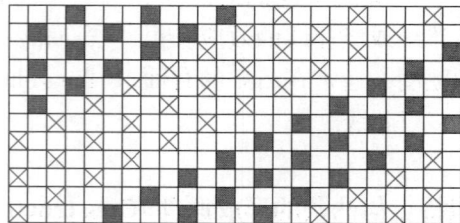

图5-5-5 斜凸条组织
The Bedford weave of twill effect

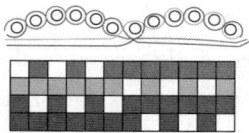

图5-5-6 纵凸条组织截面结构示意图
Cross-section of the Bedford fabric

1	-1	1	-1	1	-7	-7	-7	-7	-7	-7	-7
-7	-7	-7	-7	-7	-7	1	-1	1	-1	1	-7
-7	1	-1	1	-1	1	-7	-7	-7	-7	-7	-7
-7	-7	-7	-7	-7	-7	-7	1	-1	1	-1	1

图5-5-7 纵凸条组织纬浮长线分布图
The lengths of floats in warp direction

在纬纱重叠的同时，纬浮长线拉拢另一系统（经纱）的纱线，占据的空间小，应力小，浮长线松弛；同时，固结部分交错多，且纱线伸长大，应力大，因此，只能向两侧扩张侵占浮长线区域，以减小纱线内在应力。这样，同一根纬纱在固结部分占据的纱线长度空间大，在浮长线区域占据的纱线长度空间小。除了浮长线中间的第3、4、9、10根经纱外，其他经纱每间隔一根纬纱，都有交替向左、向

Since there are more intersections in the binder area, the slack floats are liable to retract straightly while the tight binder part expands to reduce the inner stress. Therefore, the float part occupies less space while the binder part takes more space. The taut ends except for No. 3, No. 4, No. 9 and No. 10 will change the trend of motion every other pick as in Fig. 5-5-8. However, the taut ends are unable to change their positions frequently but almost keep the straight status to reduce the inner stress. To compromise the movements, the over expanded binder part protrudes to form humps or pronounced ribs while the floats sink and shrink at the back to reduce the inner stress. Therefore, the filling threads and

右运动的趋势，如图 5-5-8 所示。但这些经纱受到较大的上机张力，且长度大致一定，无法频繁左右移动，而是基本保持直线状态，以减少内应力。因此，占据较大空间的纬纱固结部分只能向空间发展隆起形成凸条，使得经纱和纬纱均处于较小应力的稳定状态。图 5-5-9、图 5-5-10 是典型凸条组织织物的正反面外观图。

the warp threads are all placed at stable shapes. Fig. 5-5-9 and Fig. 5-5-10 show the face and back side of a typical fabric of Bedford cord weave.

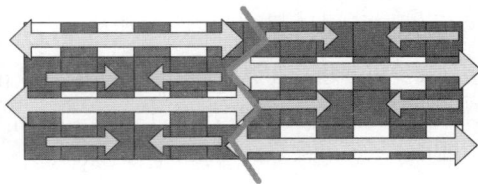

图5-5-8 经纱移动方向
Horizontal movement tendency of warp yarns

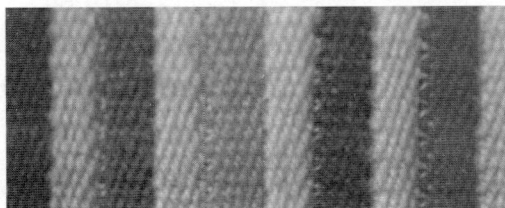

图5-5-9 凸条组织的正面外观形态
Obverse side of the Bedford cord weave

图5-5-10 凸条组织的反面外观形态
Reverse side of the Bedford cord weave

三、简单凸条组织的组织图绘制 /Designing of Simple Bedford Cord Weaves

（1）重平组织（基础组织）和固接组织的选择。基础组织一般选用 $\frac{4}{4}$ 重平、$\frac{6}{6}$ 重平，固结组织一般选用平纹、$\frac{2}{1}$ 斜纹、$\frac{1}{2}$ 斜纹或

(1) Choose the base rib and the binder weave. $\frac{4}{4}$ rib and $\frac{6}{6}$ rib are usually chosen as the base rib. The normal binder weaves are plain weave, $\frac{2}{1}$ twill, $\frac{1}{2}$ twill or $\frac{2}{2}$

$\dfrac{2}{2}$斜纹。浮长线一般不小于 4 个组织点，且应为固结组织纱线循环的整倍数。浮线过短、过长，凸条效应皆不理想。

（2）确定基础重平组织与固结组织的排列比 $m:n$。一般常用 1:1 或 2:2，排列比太大容易在织物正面暴露浮长线的痕迹。

（3）计算 R_j、R_w。

①纵凸条组织

R_j = 基础纬重平组织经纱循环根数 = R_{j1} =2× 纬浮长的长度

R_w =lcm［lcm（基础重平组织的纬循环数，m）/m，lcm（固结组织纬循环数，n）/n）］×（m+n）

②横凸条组织

R_j = lcm［lcm（基础重平组织的经循环数，m）/m，lcm（固结组织经循环数，n）/n）］×（m+n）

R_w = 基础经重平组织纬纱循环根数

这里下标 1 是固结组织，下标 2 是重平组织。

（4）一个组织循环的范围内，绘制重平组织。先按排列比填绘基础组织，再在重平组织的浮长线上填绘固结组织。固结组织填绘在基础组织的浮长线上。

例：如果要绘制图 5-5-6 所示的凸条组织，其浮长线长度为 6，采用平纹固结组织，纬纱方向的组织与条子的排列比为 1:1，则有：

twill. The length of floats is usually not less than 4, and should be the multiple of the repeat of the binder weave. The satisfactory effects can only be achieved by a suitable rib.

(2) Determine the arrangement ratio $m:n$ of the base rib and the binder weave. The ratio is usually 1:1 or 2:2, and too large the ratio will cause the long floats to be exposed on the face of the fabric.

(3) Calculate R_j and R_w

① For longitudinal cords:

$R_j = R_{j1} = 2 \times$ length of the long weft float

R_w = lcm [lcm（R_{w1}, m）/m, lcm（R_{w2}, n）/n] ×（m+n）

② For transverse ribs:

$R_w = R_{w1} = 2 \times$ length of the long warp float

R_j = lcm [lcm（R_{j1}, m）/m, lcm（R_{j2}, n）/n] ×（m+n）

Where, subscript 1 and 2 denote the rib weave and binder weave respectively.

(4) Determine the range of the weave on point paper. Draw the base rib according to the arrangement ratio, then fill the marks on the floats according to the binder weave.

Design a Bedford cord weave based on $\dfrac{6}{6}$ weft rib and a plain weave binder (Fig. 5-5-6). The arrangement ratio of wefts on the base rib and the binder weave is 1:1.

$$R_j = R_b = 6+6=12$$
$$R_w = \text{lcm} [\text{lcm}（R_{w1}, m）/m, \text{lcm}（R_{w2}, n）/n] \times（m+n）$$
$$= \text{lcm} [\text{lcm} (2,1)/1, \text{lcm} (2,1)/1] \times（1+1）=4$$

四、增加凸条隆起效果的方法 /Measures to Enhance the Cords

为了增加凸条的隆起程度，可以采取下列方法：

（1）一定程度下，组织反面浮长线的长度越长，条纹凸起的程度越显著。

（2）浮长线的收缩力越大，条纹凸起得越大。纱线有一定的捻度（丝要加一定的捻度），纱线有弹性，都可增加收缩力，增加条纹的隆起程度。

（3）同时加大织物经纱与纬纱的密度，凸条更清晰。加大浮长线所在纱线系统密度，有助于纱线重叠凸起；加大凸条方向的纱线系统密度，促使浮长线与其交织频率加大，浮长线张力加大，利于条子凸起。

（4）各凸条间加入平纹组织作为分隔组织。分隔组织凹下，凸条与对比后，隆起更显著。

（5）加入芯线。芯线使凸条丰满，且有增重、增强的作用。芯线与纬纱实际并不交织，总是处于上层纬纱之下，下层纬纱之上。图 5-5-11 是有芯线的纵凸条组织的部分上机图，图 5-5-12 则是该组织的纬浮长分布图。可以看出，第 4、5 根经纱与纬纱交织时，第 1、2 纬（下层）浮在纬纱之上，而第 3、4 纬（上层）则沉在纬纱之下，故实际该经纱在两层纬纱之间，不与纬纱交织，是芯线。同理，第 14、15 根经纱与纬纱交织时，第 1、2 纬（此时处于上层）沉在纬纱之下，而第 3、4 纬（此时处于下层）则浮在纬纱之上，故

The following measures can be used to have a better effect for the fabrics of Bedford cord weaves.

(1) To a certain extent, the longer the length of the floats on the opposite side of the fabric, the more prominent the degree of cords rising.

(2) The higher the contraction force of the float is, the higher the ridge is raised. Stretch yarns, yarns of hard twist, yarns of high tension increase the force of contraction and increase the degree of uplift of cords.

(3) Increase the fabric setts in both warp and weft direction. With the increment of the thread density along the cord or rib, the intersection along the float increases and the tension of the float is improved, which is in favor of forming a distinct cord. If the thread density of the float series increases, the float and the binding weave tend to overlap and promote to form the cord or rib.

(4) Insert plain weaves as cutting weaves between cords. The cutting threads interweave on the plain principle with all picks of weft, thereby forming a furrow or cutting, which sharply divides the cords and makes the rib more prominent.

(5) Add wadding threads. For the purpose of giving the ribs or cords greater prominence, and also to increase the weight, bulk and strength of the fabric, one, two or more extra warp threads are sometimes introduced in each cord to serve as wadding. The thick wadded ends always float over the lower wefts and sink under the upper wefts. The wadded ends can be judged by the floats. The wadding ends actually don't interweave with the wefts. Fig. 5-5-11 exemplifies the looming plans of a Bedford cord weave with wadding ends while Fig. 5-5-12 shows the diagram of the lengths of the weft floats. It will be seen that wadding threads are always raised along with all face threads of the same cords when it is required to place weft at the back; but they remain down when weft interweaves with face threads, to form the ridge of a cord, whereby they lie between the face

实际在该两层纬纱之间，不与纬纱交织，也是芯线。

（6）改变纱线排列比，将固结组织与基础组织的排列比由2:2改为1:1。这两种排列方式区别不大，但是1:1的排列效果略优，因为凸条可以分布得更均匀，凸条组织质地更加紧密，每个条子分隔更清晰。

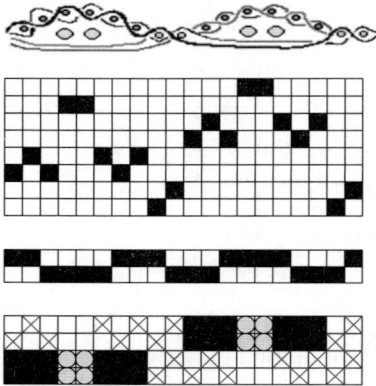

图5-5-11　有芯线的纵凸条组织
The looming plans for a Bedford weave with core ends

of a cord and the floating weft. The ends No. 4 and No. 5 always float over the filings No. 1 and No. 2(both are at the back layer) and beneath the fillings No. 3 and No. 4 (both are upper weft). Therefore, ends No. 4 and No. 5 lie between the rib face cloth and the weft floats on the underside and should be judged as wadding threads. Similarly, ends No. 14 and No. 15 are also wadding threads because they always float beneath the upper wefts No. 1 and No. 2, and over the back wefts No. 3 and No. 4.

(6) Change the thread arrangement ratio from 2:2 to 1:1. There is little difference between the two systems, but slightly superior results obtain with an alternate arrangement of picks, as these are more perfectly distributed in cloth. It is also capable of producing a closer texture, and forms a clearer cutting between the cords, which appears more distinct.

图5-5-12　有芯线的纵凸条组织的纬浮长分布
The lengths of the weft floats of a Bedford weave with core ends

Looming for Bedford Cord Weaves

五、凸条织物上机要点 /Looming for Bedford Cord Weaves

纵凸条组织织物穿综采用间断穿法。交织次数较多的平纹等分隔组织宜穿入前综。芯线不可见，可用较粗的原料，因为其织缩为0，芯线必须单独穿在另一个织轴上，且加大张力。芯线一般穿入最后面的综页中，或者紧挨在分隔组织之后，有时也可穿在最前的综页上，相当随意。纵凸条组织中经组织点

Grouped drafting applies to the longitudinal Bedford cord weaves. Ends of the cutting weave should be drawn on the front healds due to their frequent interlacement. Wadding threads sink under the cord and float on the long floats, and are invisible. Therefore, coarse yarns can be used as wadding threads. The rate of contraction being nil, the wadding threads are wound on a separate beam with high tension. Wadding threads are drawn through two healds placed immediately in front of those governing cutting

较多，也可采用反织法。

　　为了使凹下部分更加明显，在穿筘时，平纹部分的固结经纱分穿在不同的筘齿中。但有时，平纹部分固结经纱穿在同一筘齿中。根据织物的精细程度，平纹分隔组织每筘齿穿入数可为2入、3入，或者更多，凸条部分每筘齿3入或4入。通常，凸条部分的每筘穿经数略大于固结部分。每个条子中经纱根数也是筘齿穿入数的影响因素。

　　横凸条组织穿综一般采用顺穿法。当经密大时，可将综片扩大一倍，但必须顺穿。

and face threads respectively, in accordance with usual practice. Sometimes the healds governing cutting threads are placed in front, followed by those governing wadding and face threads respectively; but this is quite optional. The face downward weaving can be applied to the Bedford cord weaves if the warp dominates the weave.

In order to develop fully the sunken lines, the plain cutting ends should be separated by the splits of the reed. In some cases, however, the pairs of plain ends are dented together. Two, three, or more ends are passed through each split according to the fineness of the cloth; and sometimes the plain ends are woven two per split, and the cord ends three or four per split. The number of ends in the width of a cord has some influence upon the order of denting.

For transverse Bedford cords, straight drafting applies. When there is a large warp density, the number of the healds should be doubled.

六、凸条组织的应用 /Applications of Bedford Cord Weaves

　　凸条组织立体感强，质地松厚，富有弹性，在各类织物中均有应用。根据用途，织物可以轻薄，也可以厚重。轻薄和中等重量的凸条织物主要用于女士礼服面料、夏装和节日服装，厚重型织物主要用于特点鲜明的男士服装，如花式背心、马裤、军服、运动服和骑士装。该组织还可以通过改变条子宽度和方向（图5-5-13），或者使用色纱进一步变化。在提花织物中既可作地组织，也可作花组织或点缀组织。

The Bedford cord weaves have strong three-dimensional bold feeling, loose and thick texture, and are full of elasticity, used in all kinds of fabrics. Bedford cord weaves are produced from light to relatively heavy cloths, according to the particular use for which they are intended. The lighter and medium fabrics are chiefly used as ladies' dress materials, ladies' light summer and holiday clothing; whilst the heavier and coarser fabrics are generally made up into men's clothing of a special character, as fancy vests, breeches, military, sporting and riding suits. By means of variegated cords as shown in Fig. 5-5-13, and colored threads of warp, Bedford cord weaves have more variations. In Jacquard weaving, Bedford cord weaves can be employed as ground, or small detached sprigs or simple geometrical forms evenly distributed in the fabric.

图 5-5-13　斜向凸条组织

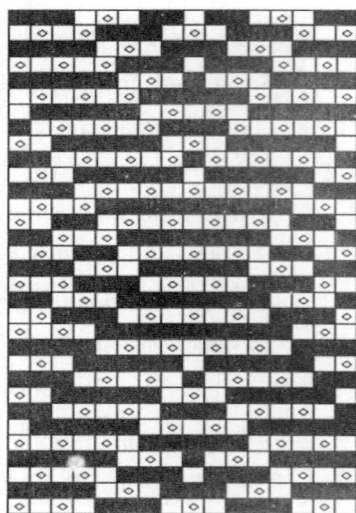

The Bedford weave with diagonal twill effects

第六节　网目组织 /Spider Weaves

一、外观特点 /Description

在织物上分布着曲折的长浮线形成的蜘蛛网络状的组织，称为网目组织，如图 5-6-1 所示。网络状曲折长浮线是经纱的，形成经网目组织，也称网目经，如图 5-6-2 所示，织物效果如图 5-6-3 所示。网络状曲折长浮线是纬纱的，形成纬网目组织，也称网目纬。网目纱在织物表面偏离了原来的直线，形成正弦线或者波状曲折线，表面如同需要半综才能生产的纱罗织物。因此，网目组织也用来模拟纱罗织物的外观，以纬网目更加常见。

网目织物中网目纱可能是平行的曲折条子，也可能是对称相向的

A spider weave is a net like or distorted effect on the face of fabric by floating and deflecting either the ends or the picks. The effects of the fabric are shown schematically in Fig. 5-6-1. The threads may be distorted in warp direction or weft direction. Two weaves which are distorted in warp direction are shown in Fig. 5-6-2 and the effect of the fabric is shown in Fig. 5-6-3. Some threads, usually of weft, are pulled in opposite directions at different points, thereby causing them to deviate or deflect from their original straight line, and to assume sinuous lines of a more or less wavy or zigzag character, not unlike that of a net "leno" effect, produced by means of a "doup" or "leno" harness. Therefore, distorted effect provides another way to imitate the lenos.

The threads may be waved in the same direction uniformly to produce a series of parallel waves, or they

钻石形、菱形、S形或者其他线形
网状效果。

may be waved in opposite directions to produce diamond, lozenge, ogee, and other simple linear effects.

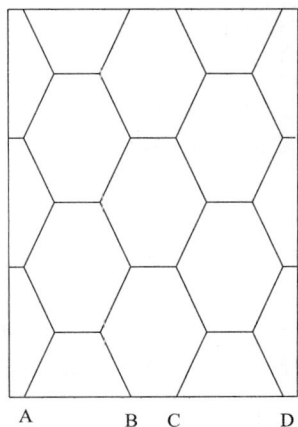

(a) 经网目组织/Distorted effect in warp direction

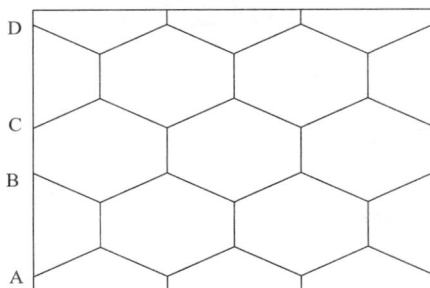

(b) 纬网目组织/Distorted effect in weft direction

图5-6-1 网目组织外观示意图

Schematic diagram of spider weave fabrics

图5-6-2 经网目组织

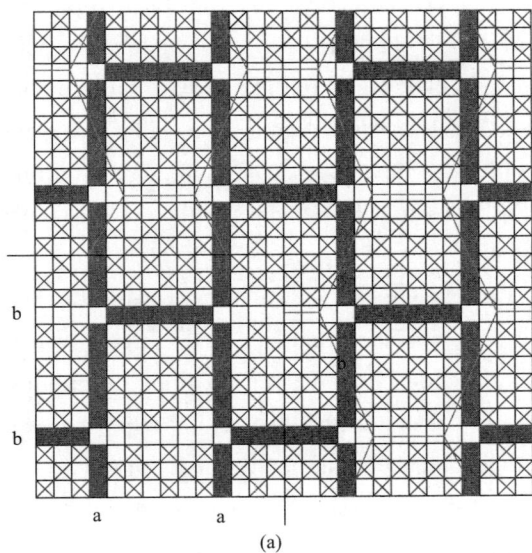

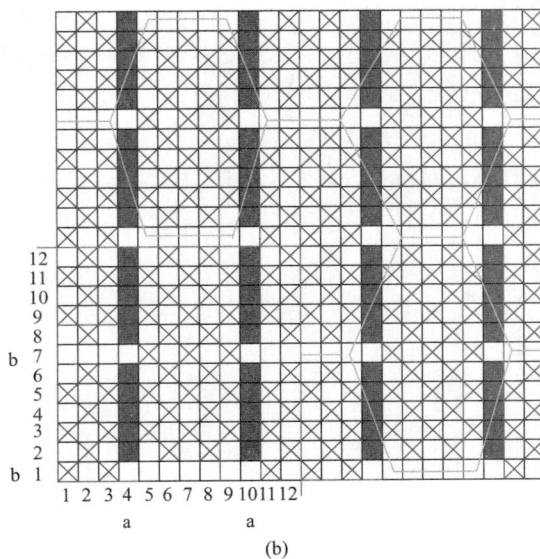

A spider weave in warp direction

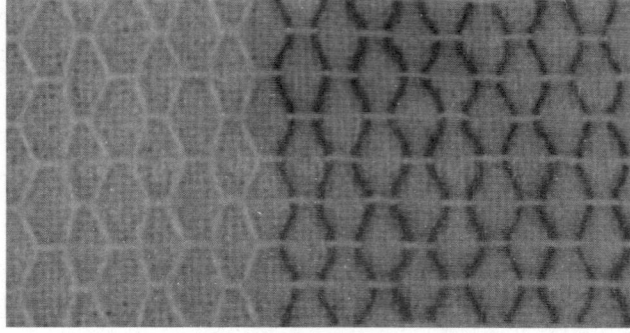

图5-6-3　经网目组织织物

The image of the fabric of distorted effect

二、组织特点与网目的形成原理 /Construction and Formation of the Structure

网目组织一般由平纹组织或者斜纹组织形成的地组织与间隔分布的经纬浮长线构成。在一个系统方向上，只有一种长浮长线，偶尔沉在另一系统纱线之下，这个系统纱线是网目纱［图5-6-2（a）的a纱］。由于在另外一个系统一定宽度上，某些纱线在织物正反面交替出现长浮长［图5-6-2（a）的b纱］，将网目纱拉拢形成蛛网状的曲折效应，这些纱线称为牵引纱。为了避免拉拢纱过于松弛，牵引纱的部分反面浮长被取消，增加交织点。这样，在牵引纱上长浮长线与地组织交替出现，如图5-6-2（b）中的b纱所示。

图5-6-2（b）的经网目组织中，第4、10根经纱由于浮长线长，高度高，容易浮在织物其他纱线之上，形成局部重叠。第1、7根纬纱在织物表面存在纬浮长线，也同样容易浮在织物表面。纬浮长同时对第4、10两根经纱向不同的方向拉拢，因经纱位置远远高于其相邻经纱，运动阻碍小，导致经纱曲折

A spider weave usually comprises of the alternately displaced floats and ground weave made of plain or twill. In one direction, the long float in the distorted thread a in Fig. 5-6-2(a) occasionally passes under the other system of threads. The long float will be drawn to zig-zag effect. But in other direction, the warp float and weft float in pulling thread b as shown in Fig. 5-6-2(a) appear alternately. In order to avoid an over slack pulling thread, parts of the floats in the back face of the fabric are cancelled and more intersections increase the tension. Therefore, on the pulling thread b in Fig. 5-6-2(b), long floats and the ground weave appear alternately.

In the spider weave as shown in Fig. 5-6-2(b), the long floats in the ends No. 4 and No. 10 are higher, which makes it easy to float and overlap on their neighbors. Similarly, the weft floats on the wefts No. 1 and No. 7 are higher than their neighbors and tend to float on the fabric. Meanwhile, the weft floats gather the ends No. 4 and No. 10 at different directions. Since at a higher position, the ends get less obstacle from its neighbors and are easily distorted to form a zigzag effect. In the weave, the long weft floats are the pulling wefts while the two ends No. 4 and No. 10 are the distorted ends.

In a warp distorted effect repeat, one or several

扭曲成蛛网状。这两根纬纱称为牵引纬；而被拉拢的两根经纱称为网目经。

在经网目组织循环中，每隔一定根数的地经，配置有单根或若干根网目经。网目经的组织由经长浮线与单个（或两个甚至三个）纬（经）组织点组成，通常为$\dfrac{5}{1}$、$\dfrac{7}{1}$，两条网目经之间的地经根数的多少决定着网目的大小。每隔一定根数的纬纱配置一条纬浮长线。每两条纬浮长线之间相隔的纬纱根数等于网目经的连续经浮点数。相邻两条纬浮长线必须交叉配置。

纬网目组织的构成原理与经网目组织类似。网目组织中，若网目纱是经（纬）纱，则拉拢牵引纱必定为纬（经）纱。图5-6-4 所示为双纱加强的纬网目组织。

distorted ends are arranged at intervals. The interlacement of the distorted end comprises of the long float and single (double or triple) sinkers, usually set as $\dfrac{5}{1}$ or $\dfrac{7}{1}$. The number of the ground ends between the two distorted ends determines the size of the distorted effect. A long weft float is arranged at certain intervals and the number of the weft threads between the two weft floats equals to the number of the warp float of the distorted end. It should be noted that the two neighboring weft floats are arranged in intervals.

The construction and the principle of weft distorted weave are similar to the warp distorted effect weave. In a spider weave, the distorted ends are pulled by long weft floats while the distorted wefts are pulled by long warp floats. Fig. 5-6-4 shows a weft distorted weave with doubling distorted wefts and doubling pulling ends to enhance the effects.

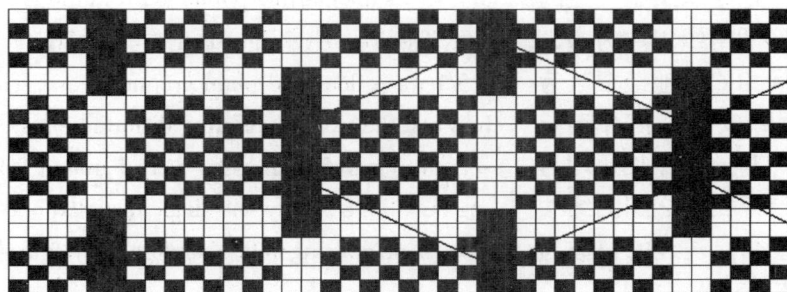

图5-6-4　双纱加强的纬网目组织
An enhanced weft-distorted effect weave

三、简单网目组织的组织图绘制 /Designing of Simple Spider Weaves

以经网目为例，介绍简单网目组织的组织图绘制方法，其步骤如下：

A warp distorted effect is taken as an example to elaborate the procedures of designing a spider weave. The processes are explained as following.

（1）确定地组织。一般采用平纹，或者原组织斜纹组织。

（2）配置网目经与牵引纬。

（3）确定组织循环大小。

R_j =（两条网目经之间的地经根数 ＋ 每条网目经的根数）×2

R_w =（两条纬浮长线之间的纬纱根数 ＋ 每条纬浮长线的根数）×2

（4）填绘组织图。在网目经上增加经组织点，形成网目经上的经浮长线；同时，在牵引纬上去掉部分经组织点，形成纬浮长线。具体方法是：在网目经上按其沉浮规律填绘组织点；在两网目经浮长端点的纬纱之间空出纬浮长线，形成拉拢纬并使相邻两条纬浮长线呈交叉配置；在与纬浮长线两端点相邻的组织点处填入经浮点，并以此为起点填绘地组织。

如图 5-6-2（b）所示的网目组织，以平纹为地组织绘制一经网目组织。网目经的组织规律为 5/1，两根网目经之间相隔的地经根数为 5，每隔 5 根地纬安排一根纬浮长线。每条网目经与地纬浮长线均为单根。即两条纬浮长线之间的纬纱根数为 5，每条纬浮长线的根数为 1，因此，$R_j=R_w=(5+1) × 2=12$。

(1) Draw the ground weave in the repeat. Usually, a plain weave or a regular twill is used as the ground weave.

(2) Determine the positions of the distorted ends and the pulling weft floats.

(3) Calculate the size of the weave repeat.

R_j =（number of the ground ends between two zigzag lines ＋ number of the distorted ends in a zigzag line）×2

R_w =（number of the weft threads between two drawing weft floats ＋ number of the weft threads in a pulling weft float）×2

(4) Fill in the marks into the weave. Add the marks on the distorted ends to form the warp floats of given length. Remove part of the marks on the pulling wefts to form the weft floats and ensure the long weft floats opposite each other between the two neighboring pulling wefts. The procedures can be described as the followings: draw the marks on the distorted ends by interweaving mode, the parts between the two ends points of the two distorted ends on the wefts are kept as blanks to form the pulling wefts, and ensure the pulling wefts crossing each other, and the other ends points next to the distorted ends are filled with marks and set as the starting points to fill in the ground weave.

The weave as show in Fig. 5-6-2(b) is combined with a 5/1 warp float and plain ground weave. There are 5 ends between the two neighboring distorted ends, and a weft float with the length 5 is arranged every 5 ground wefts. To each zigzag line, there is only one single distorted end which is drawn by a single drawing weft. Therefore, $R_j=R_w=(5+1) × 2=12$.

四、网目织物上机要点 /Looming for Spider Weaves

网目组织上机时采用照图穿法。制织网目组织时，两网目经纬纱间至少要间隔 5 根地部纱线。网目经的交织次数少，浮长线长，另用一轴，且送经量大，使之更易浮起重叠，被拉弯曲。生产中，也有

The spider weave is drawn by design. In weaving the fabric, the two neighboring distorted threads should be placed at least 5 threads apart. The distorted ends are usually placed on a separate beam and are given in more rapidly than the ground ends. Occasionally, the distorted ends and the ground ends are drawn on the same beam.

网目经与地经共用一轴的工艺，但并不多见。穿筘时要将网目经夹在地经之间穿入同一筘齿。若与条格组织联用，则含有网目的筘齿经穿入数要高于地组织。图 5-6-5 是某经网目组织的穿综穿筘图。

When used in stripe form the ends which form the zig-zag effect should be somewhat crowded in the reed and put into the middle of the dent split surrounded by its neighboring ground ends as shown in Fig. 5-6-5.

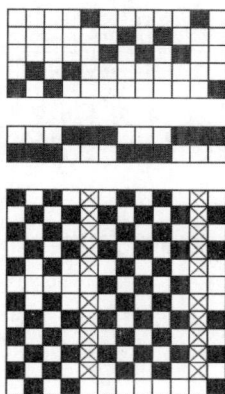

图5-6-5　经网目组织穿综穿筘
The looming plans for the spider weave in warp direction

五、增加网目效果的方法 /Measures to Enhance Distorted Effects

为了加强网目效果，可用较粗、双纱或彩色纱作为网目纱、牵引纱，如图 5-6-4 所示的纬网目组织；或者在被拉拢经（纬）纱的牵引纬（经）取消部分经、纬纱的交织点，如图 5-6-6（a）所示。图 5-6-6（b）中，将松弛的网目纬纱织紧，使中间的纬纱挤压到织物表面突出，在牵引经纱上增加一些平纹点，使经纱张力增加，促使松弛的网目纬纱分开。网目纬纱的扭曲程度取决于松弛的网目纬和交织频繁的地纬纱之间的缩率差异大小。地组织纬向收缩越大，网目纬的效果越好。

In order to develop more fully the distorted effect, coarse thread, doubling thread or colored thread can be used as distorted threads or pulling threads to make distinct effects as shown in Fig. 5-6-4. Another way is that part of the interweaving points of the threads near the pulling thread are removed as shown in Fig. 5-6-6 (a). In Fig. 5-6-6(b), the loosely woven picks are beaten up close together so that those in the center are forced prominently to the surface, and are in a proper position for being drawn together, and then the plain interweaving of the floating ends produces the most suitable conditions for forcing the distorted picks apart. The degree of distortion varies according to the difference in the shrinking of the distorted picks, which float loosely, and the ground picks, which interweave frequently; hence the best results are obtained when a ground texture is formed that shrinks considerably in width.

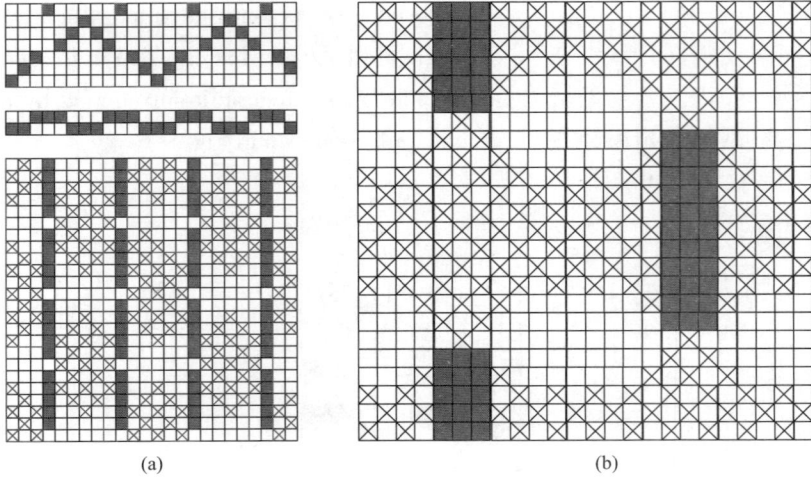

(a)　　　　　　　　　　(b)

图5-6-6　取消部分组织点的网目组织
Enhancing the distorted effects by removing interweaving points

六、变化网目组织 /Variations of Spider Weaves

网目组织还可以有各种变化，如各段网目经长度不一，如图 5-6-7 所示；或者网目经相互平行，一个朝向，如图 5-6-8 所示。网目组织常与其他组织联合用于服装及装饰用织物。图 5-6-9 所示为变化的网目组织。

The distorted effect can be varied by changing the length of the float as shown in Fig. 5-6-7, or the arrangement of the pulling threads to control the direction of the zig-zag line as show in Fig. 5-6-8. More variations are shown in Fig. 5-6-9.

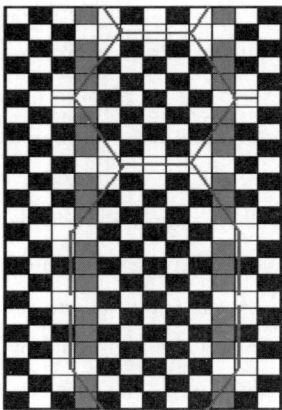

图5-6-7　长短不一的经网目组织
Distorted effect of varied size

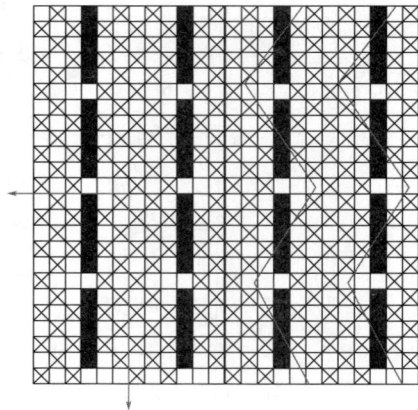

图5-6-8　单向经网目组织
Distorted effect of single direction

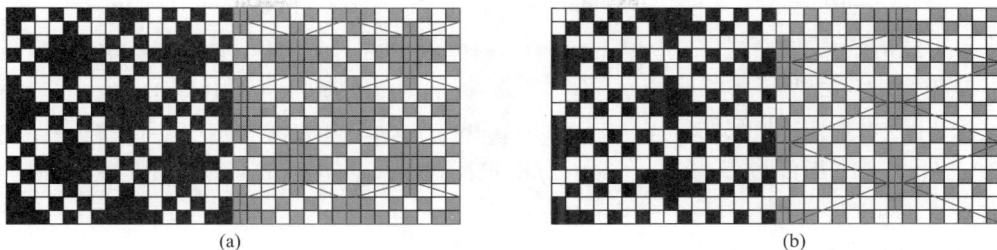

图5-6-9　变化网目组织
Variations of distorted effect weaves

第七节　小提花组织 /Dobby Spot Weaves

小提花组织

在织物表面运用两种或两种以上的组织经变化而形成花纹、且能在多臂织机上织造的组织称为小提花组织。通过精心设计，小提花组织也可以在织物上形成精美的图案。

Dobby spots or dobby figure weave is obtained by combining two or more weaves or their variations to form patten and woven on a dobby loom. With careful design, beautiful spots of figure appear on the fabric.

一、小提花组织的构成 /Construction of Dobby Spot Weaves

小提花织物由两种不同结构的部分组成：起花组织和地组织。起花部分通常由经向、纬向或者两个方向上的浮长线组成。花纹布局可以按照某种规律，也可以是散点。采用对称图形花纹，可以节约综框。起花部分也可以由蜂巢或透孔等其他组织构成。浮长散点部分可以均匀地分布在表面上，也可以像平纹或缎纹组织一样，按一定的规律排列成一定的花纹。

由浮长线形成的图案规律性排布是为了与地组织部分形成对比。如果经纬纱颜色、材质差异显著，对比更加强烈。即使经纬纱完全相

Dobby spot weaves comprise of two parts of different constructs: the figure part and ground part. The figure part or spot part is generally the pattern composed of floats, which can be lengthwise or transverse, or both. The figures layout can follow a certain patten or just detached spots. The pattern is usually symmetrical to save the healds used. The figure part can also be formed by a honeycomb weave, a open gauze weave or others. The detached spots can be distributed evenly on the surface or arranged in a certain interlacement like a plain weave or satin weave.

Dobby spots are formed by floating the ordinary weft or warp threads on the surface of the cloth in an order that is in contrast with the interlacing in the ground. The figures show most prominent when the warp and weft threads are in different colors or materials; but if the two series of threads are alike the

同，由于浮长光照的差异，织物表面的差异仍然清晰可见。

小提花织物应以循环小的组织为地组织，平纹地小提花织物是主要类型之一。平纹组织因其紧密，外观细洁，不粗糙，对比纹样也不过分突出，更适合作为地组织。不过，地组织也可采用斜纹或者其他组织。小提花织物多数是色织物，经纬纱全部或部分采用异色纱，也可配一些花式线间隔地在织物表面形成强烈对比，或者采用花经或花纬在织物表面形成图案。

二、小提花织物设计原则 /Principles of Designing Dobby Spot Weaves

在设计小提花织物时，要注意遵循以下原则：

（1）花型主要起装饰作用，且精细而均匀分散地分布在表面。起花组织与地组织之间的轮廓线应该保证清晰，使花纹轮廓清晰、准确且不变形。

（2）起花部分的浮长线不应该太长。棉毛织物经纱浮长线的长度通常不超过3～5个组织点，丝织物不超过7～9个组织点。

（3）在起花部分，经纱浮长与地组织的交织次数不应相差太多，以避免在花纹部分出现松散结构和交织结构的差异。一般来说，经纱平均浮长应控制在1～1.3个组织点，以保证采用单轴织造，减少工艺的复杂性。

（4）每次开口提综数尽可能均匀，因此花型配置应相对分散。

（5）因为起花部分只起点缀的作用，所以织物的密度一般可与地

difference in the reflection of the light from the different weave surfaces is sufficient to render the figures clearly visible.

Dobby spot fabrics should take the weave of small repeat as the ground part, and are mostly based on plain ground. The plain weave is preferred for the ground part since it is firm, fine clean, not coarse, and the pattern is not too prominent. However, the twill or other weaves can also be used for ground part. By employing fancy threads in which spots of contrasting color occur at intervals, and by introducing extra warp or extra weft threads, the spots are formed on the surface of the fabric.

Principles of Designing Dobby Spot Weaves

In designing a dobby spot weave, the principles below should be followed:

(1) Figures play an ornament role, and are fine and distributed evenly in the surface. The contour distinguishing the figures and the ground should be clear, so that the outline of the pattern is clear, accurate and not deformed.

(2) The floats of the figure part should not be too long, and the float length on ends should not exceed 3 ~ 5 weaving points for cotton and wool fabrics, or no more than 7~9 for silk fabrics.

(3) The intersections between the warp floats and the plain weave in the figured part should not be too different, so as to avoid the difference of loose structure and interweaving contraction in the figured part. In general, the average length of the warp floats should be controlled within 1 ~ 1.3 to ensure weaving with single-beam.

(4) The number of lifted heddles is as even as possible at each shedding, so the pattern spots should be relatively evenly distributed.

(5) As the figured part only acts as an ornament, the density of the fabric is generally the same as that of the ground weave. Therefore, normal denting applies to the

组织相同。采用平筘穿法，不用花筘。

（6）设计花型时用综页数不能超过织机的最大容量，上机的总综框数（包括边纱）不能超过16页综。

（7）经纬浮长起花效果差异明显，纬纱浮长线因为捻度小、蓬松、明亮，纬向一般织缩大，纬起花浮长更加显著。

dobby design.

(6) The total shafts used shall not exceed the maximum capacity of the dobby loom. The total harnesses of the loom (including selvage) shall not exceed 16 shafts.

(7) Even if the two series of threads are alike the difference in the reflection of the light from the different weave surfaces is sufficient to render the figures clearly visible. Other things being equal, the weft usually forms brighter and clearer spots than the warp: because it is more lustrous and bulkier on account of containing less twist; and because cloths generally contract more in width than in length, the weft thus being brought more prominently to the surface than the warp.

三、小提花织物设计步骤 /Procedures of Designing Dobby Spot Weaves

小提花织物设计步骤如下：

（1）在方格纸上设计起花部分的花型纹样，根据花型轮廓，在方格纸上填入浮长组织点。其余部分用地组织填充。

（2）根据经纬纱密度和花纹大小确定组织循环纱线数。

The followings are the procedures of designing a dobby spots weave.

(1) Draft the spot figures of the pattern. Simple spot figures are designed directly upon point paper, and the squares are then filled in along the outline. The other part is filled with the ground weave.

(2) Determine the size of weave repeat unit according to pattern size, warp density and weft density.

四、小提花组织设计实例 /Examples of Designing Dobby Spot Weaves

图 5-7-1 所示为一个小提花图案的绘制与填入方格纸的过程。在这个设计中，由经浮长或纬浮长或经纬浮长线联合形成花型。图 5-7-2 则是将花纹图案以平纹形式排列在织物中。图 5-7-3 中，将图案设计成对称图形，以减少综框使用数量。小提花组织常用照图穿法、间断穿法。

Fig. 5-7-1 demonstrates how a figure is drafted and transposed onto a design paper. In this design, the figures

Examples of Designing Dobby Spot Weaves

are formed by long warp floats, weft floats or the combination of them. In Fig. 5-7-2, the patterns are arranged in plain form, and the pattern as shown in Fig. 5-7-3 is designed symmetrically to reduce the number of harnesses used. Dobby spots weaves are design drafted or grouped drafted.

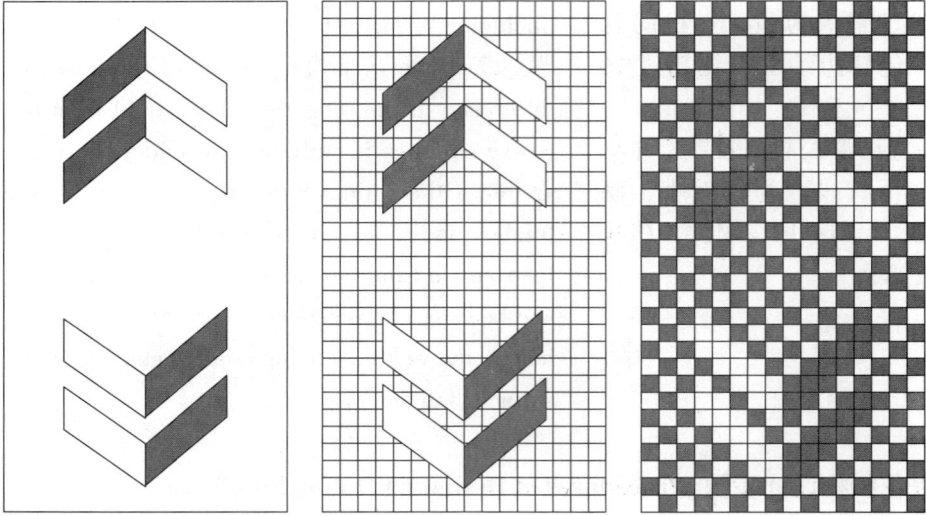

图5-7-1　小提花图案设计过程

The procedures of designing a dobby figure

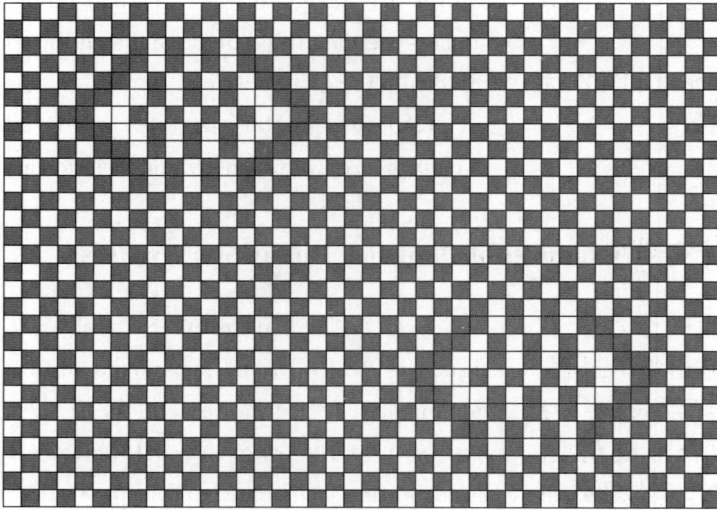

图5-7-2　以平纹形式排列图案的小提花组织

A dobby design arranged in plain form

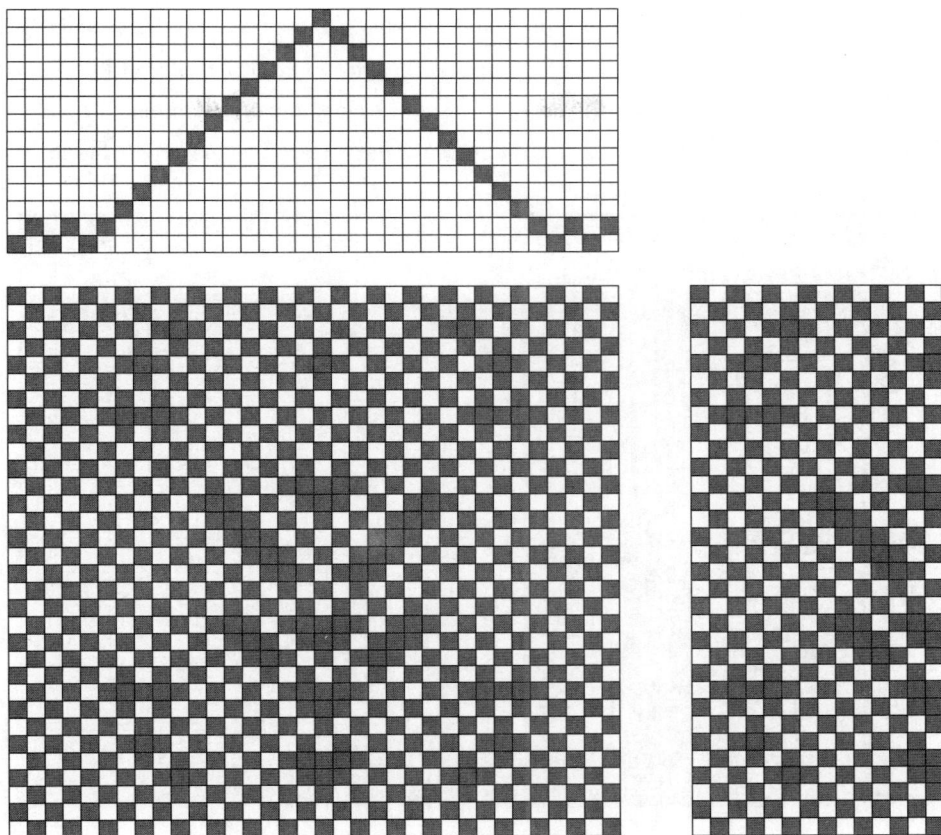

图5-7-3 对称小提花图案上机图

The looming plans for a symmetrical dobby design

第八节 配色模纹效果 /Color and Weave Effects

配色模纹效果 Color and Weave Effects

平纹组织与色彩结合可以产生条带效应。利用不同颜色的纱线与其他织物组织配合，在织物表面能构成各种不同的花形图案，反而使色纱排列被忽略。图 5-8-1 所示织物分别采用 $\frac{3}{1}$ 破斜纹、破菱形格子组织，$\frac{2}{2}$↗斜纹、$\frac{2}{1}$↗斜纹、$\frac{2}{2}$↗斜纹和 $\frac{2}{2}$ 方平组织，说明

Combining with colored threads, plain fabrics show stripes or cross-over effects. If a small-group color patterning of warp and/or weft is applied to other weaves, various distinct patterns and effects develop, and often the color order of the threads is not apparent. Figs. 5-8-1 (a) ~ (f) shows six woven structures with $\frac{3}{1}$ satinette, diaper check, $\frac{2}{2}$↗ twill, $\frac{2}{1}$↗ twill, $\frac{2}{2}$↗ twill and $\frac{2}{2}$ matt respectively, which indicates that the pattern and

织物的外观不仅与组织结构有关，而且与经纬纱颜色的配合有关。图 5-8-1（c）织物采用同一种组织，因配色不同，故织物外观不同。

appearance is related to the weave, order of warping and order of wefting. Although the fabrics as shown in Fig. 5-8-1(c)(e) apply the same weave, the patterns formed are different due to the difference of the color orders in warp and weft.

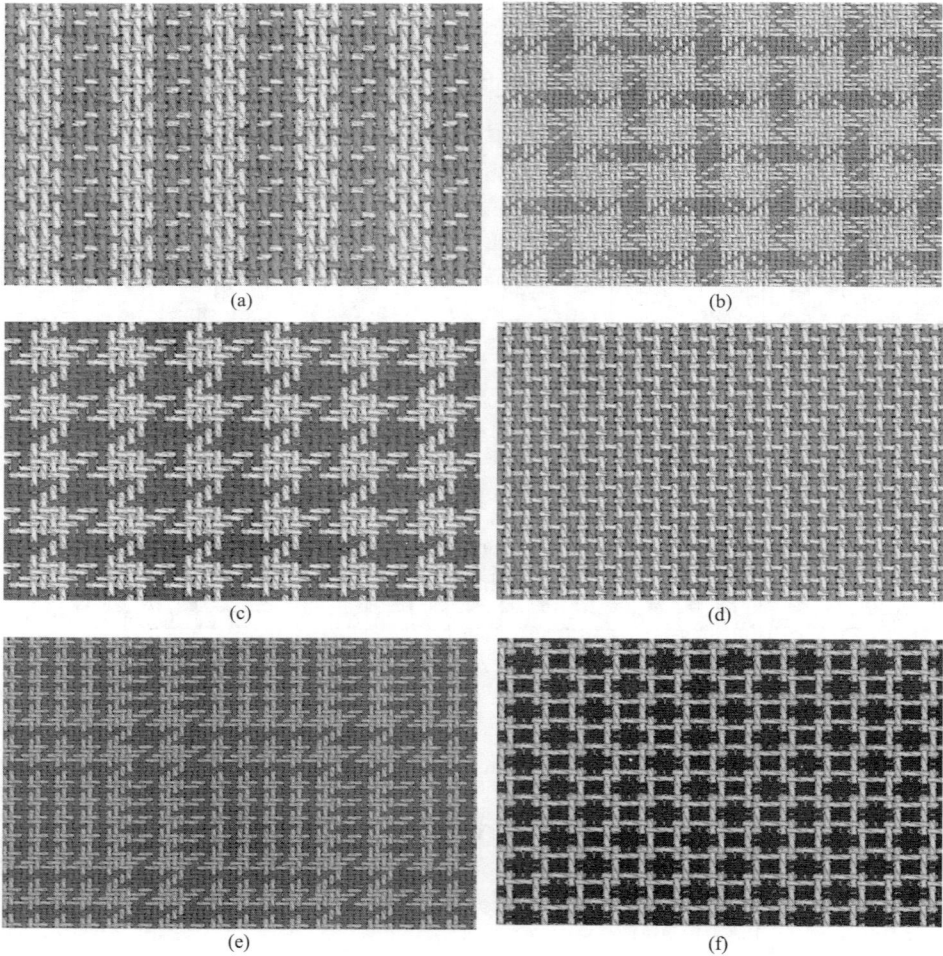

(a)

(b)

(c)

(d)

(e)

(f)

图5-8-1 不同的色织物
Various colored fabrics

一、配色模纹图 /Color and Weave Effect Diagram

配色模纹图是色纱与组织配合产生的 2 种及以上的颜色形成的花纹图案。根据该图，设计者可以预知组织的不同配色效果。图 5-8-1

Color and weave effect diagram is the pattern in two or more colors showing the combination of the colors and the weave. It enables the designer to see the effect that any color plan will produce with a given weave. The effects of

（a）（c）所示组织的配色模纹图效果如图 5-8-2（a）和图 5-8-3 所示，如果将代表经组织点的圆点去除［图 5-8-2（b）］，能很好地反映织物的外观配色效果。要绘制配色模纹图，必须已知织物组织、经纱色排和纬纱色排这三个条件。

the fabrics as shown in Fig. 5-8-1(a)(c) are displayed in Fig. 5-8-2(a) and Fig. 5-8-3 respectively. If the dots indicating the marks are removed, the effects as shown in Fig. 5-8-2(b) will better represent the appearance and pattern of the fabric. To draw a color and weave effect diagram, three factors are required—i.e., the order of warping, the order of wefting, and the weave.

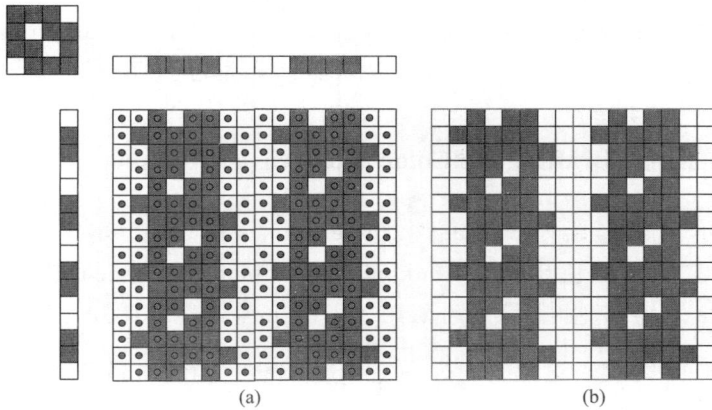

图5-8-2　图5-8-1（a）所示组织的配色模纹图
Color and weave effect for the fabric as shown in Fig. 5-8-1(a)

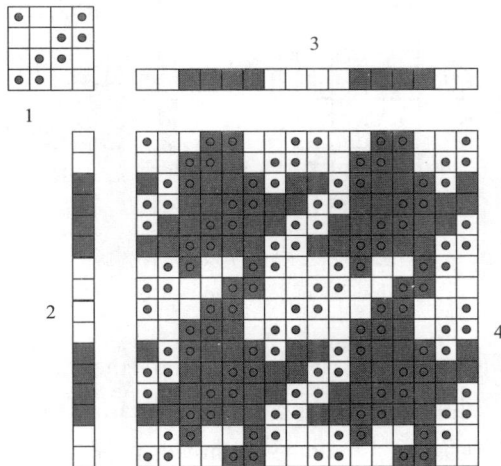

图5-8-3　图5-8-1（c）所示组织的配色模纹图
Color and weave effect for the fabric as shown in Fig. 5-8-1(c)

图 5-8-3 中，配色模纹图被分为 4 个区域。1 区表示组织图；2 区表示各色纬纱的排列顺序，自下

In Fig. 5-8-3, the diagram of color and weave effect can be described in 4 areas. Area 1 is for the weave design; area 2 is for the order of wefting, the arrangement of the

而上填入纬纱颜色；3 区表示各色经纱的排列顺序，在最上一行排列颜色；4 区表示所形成的织物外观，即配色模纹图。

与上机图一样，配色模纹图也在意匠纸上绘制。在配色模纹图中，用浮在上方纱线的颜色填绘方格。经组织点处填绘相应的经纱颜色，纬组织点处填绘相应的纬纱颜色。

picks as to color is indicated up the side of the reserved space; area 3 is the order of warping, the arrangement of the ends as to color is indicated along the top; and area 4 is for the color and weave effect.

Like looming plans, color and weave effects may be readily indicated upon pointpaper. In a color and weave effect diagram, squares are filled with the color of the yarn on top. The risers are filled with the color of the corresponding warp while the sinkers are filled with the color of the corresponding filling.

二、配色模纹图的绘制 /Drafting the Color and Weave Effects

图 5-8-4 所示为如何绘制配色模纹图的步骤。经纱的色排循环是 1A2B2A2C1A，纬纱色排是 2D3B3A，组织是 $\frac{1}{3}$ 斜纹。这里，A、B、C 和 D 是颜色代号。

The examples in Fig. 5-8-4 illustrate in stages the working out of an effect in which the threads are arranged 1A2B2A2C1A in warp, 2D3B3A in weft, while the weave is $\frac{1}{3}$ twill. Here, A,B,C and D represent different color respectively.

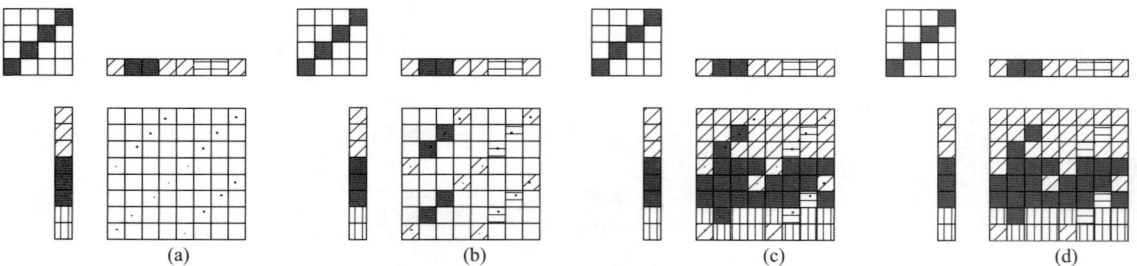

图5-8-4　配色模纹图的绘制
The drawing of color and weave effects

（1）计算配色模纹循环。配色模纹循环数等于色纱循环和组织循环的最小公倍数。组织图的经纱 / 纬纱循环数都为 4，经纱色排和纬纱色排的循环都为 8，故配色模纹循环数在经向与纬向都是 8。

（2）在配色模纹区域，用一个小圆点代替经组织点绘制组织图，如图 5-8-4（a）所示。在色排区

(1) Calculate the size of the effect repeat which is obtained by calculating the lowest common factor of the number of threads in one repeat of the color plan, and in one repeat of the weave. $R_j=R_w=4$, and both the sizes of warping repeats and wefting repeats are 8, therefore, the size of the color and weave effect repeats 8 in both warp and weft direction.

(2) In Fig. 5-8-4(a), the point paper for the effect diagram is inserted in the form of dots according to the

域，根据经、纬纱颜色排列填入相应的颜色，不同的颜色用不同的符号表示。

（3）沿着经纱方向，每个经组织点都用对应的经纱颜色填绘，如图5-8-4（b）所示，直至所有经纱完成。

（4）沿着纬纱方向，每个纬组织点都用对应的纬纱颜色填绘，如图5-8-4（c）所示，直至所有纬纱完成。

（5）将配色模纹图上的小圆点去除，如图5-8-4（d）所示。

为了更好地观察配色模纹效果，建议在经纬向扩大几个循环绘图。

weave, and the marks indicate risers. Fill in the squares of the color plans based on the order of warping and wefting. It should be noted that the different colors should be represented by different marks.

(3) In Fig. 5-8-4(b), the ends are followed vertically in successive order, and where there are weave marks—that is, where the warp is floated on the surface—the squares are filled in the mark corresponding to the end, which is indicated in the column of warping. The process continues until all the vertical spaces are finished.

(4) In Fig. 5-8-4(c), the picks are followed horizontally in successive order, and where there are blanks in the weave—that is, where the weft is floated on the surface—the squares are filled in the mark corresponding to the pick, which is indicated in the row of wefting. The process continues until all the horizontal spaces are finished.

(5) In Fig. 5-8-4(d), the dots are removed, and the complete effect appears.

It is usually preferable for the sketch to be extended over two or more repeats in each direction to better show several repeats of the completed design.

三、根据配色模纹图反求色纱排列与组织 /Determine the Weave and the Orders of the Coloring by the Effects

有时需要基于配色模纹确定色纱排列和织物组织。图5-8-5描述了根据配色模纹求色排循环和组织图的全过程。假设已知配色模纹图如图5-8-5（a）所示。步骤如下：

（1）确定纬纱色排。以每根纬纱上相同颜色的组织点数占优势的颜色定为该根纬纱的颜色，据此，图5-8-5（b）确定了纬纱色排，在纬纱色排图上填入相应的符号。

（2）绘制组织图范围。按照模纹图的大小，确定组织在方格纸的范围，如图5-8-5（c）所示。

Sometimes, it is necessary to determine the color arrangement and the weave based on the color effects. Supposing the effect is shown in Fig. 5-8-5(a), all the stages to solve the problem are illustrated in Fig. 5-8-5.

(1) Determine the order of wefting. In general, the filling color should be determined by the maximum number of the colors used at this picking. Therefore, the order of wefting is determined and the corresponding marks are filled in the plan as shown in Fig.5-8-5(b).

(2) Determine the range of the weave in the point paper according to the size of the effect, which is shown in Fig.5-8-5(c).

(3) Determine the certain risers. Along the horizontal

（3）确定必然经组织点。在图5-8-5（a）所示模纹图上沿纬纱的横向方向，找到与纬纱符号不同的位置，确定此点必为经组织点，在组织图上标记为交叉符号，如图5-8-5（d）所示。

（4）确定色经排列顺序。根据5-8-5（d）所示组织图的必然组织点对应在图5-8-5（a）中对应的符号，确定色经排列顺序，并在色排图中填入相应的符号，如图5-8-5（e）所示。

（5）确定必然纬组织点。在图5-8-5（a）所示模纹图上沿经纱的纵向方向，找到与经纱颜色不同的位置，确定此点为必然纬组织点，在组织图上标记为长方形符号，如图5-8-5（f）所示。

（6）标记组织图上其他位置为圆点，表示可经可纬的组织点。

（7）保留所有的交叉符号，去除组织图上所有的长方形标志，根据需要决定是否擦除圆点，得到待求组织，尽可能选用常见的规则组织，如图5-8-5（h₁）～（h₄）所示。

space in the effect as shown in Fig.5-8-5(a), find the squares which are different in marks with the corresponding filling, and determine them as the certain risers, then draw the marks of cross at the corresponding positions in weave pattern as shown in Fig.5-8-5(d).

(4) Determine the order of warping by the known certain risers, and fill the marks for the color of the risers to the corresponding squares in the plan for the order of warping, as shown in Fig.5-8-5(e).

(5) Determine the certain sinkers. In Fig.5-8-5(f), along the vertical space in the effect as shown in Fig.5-8-5(a), find the squares which are different in marks with the corresponding end, and determine them as the certain sinkers, then draw the marks of rectangles at the corresponding positions in weave pattern.

(6) Mark the other squares with spots indicating these are uncertain interlacing points.

(7) Reserve all the marks of cross, and remove all the marks of rectangle. The final weave is obtained when some of the dots are removed. Common weaves are the best choice as shown in Figs. 5-8-5(h₁) ~ (h₄).

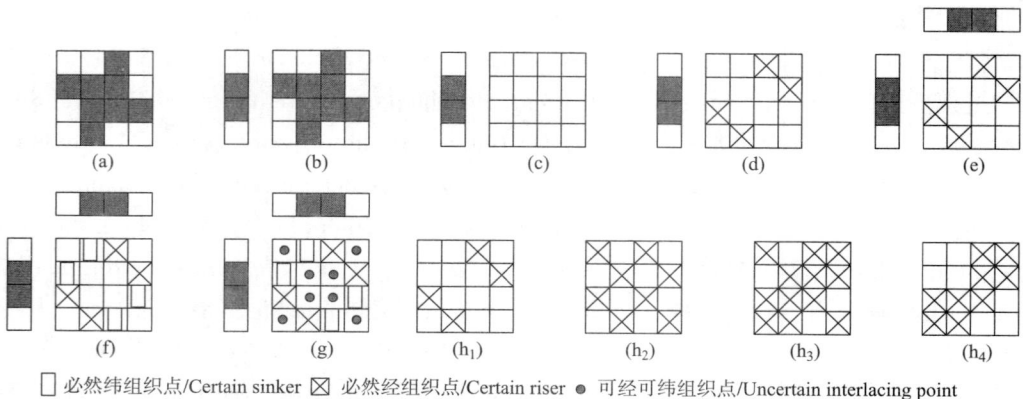

(a)　　　(b)　　　(c)　　　(d)　　　(e)

(f)　　　(g)　　　(h₁)　　　(h₂)　　　(h₃)　　　(h₄)

☐ 必然纬组织点/Certain sinker　☒ 必然经组织点/Certain riser　● 可经可纬组织点/Uncertain interlacing point

图5-8-5　根据配色模纹求色排和组织图
Determine the possible weaves, warping and wefting based on color and weave effects

四、配色模纹类型 /Variations of Color and Weave Effects

简单组织通过配以不同的颜色排列，可得到条带型（图 5-8-2）、犬牙型（图 5-8-3）、完全直线型［图 5-8-6（a）］、鸟眼点状型［图 5-8-6(b)］、台阶型［图 5-8-6（c）］、方格格栅型、回字型和全幅大花型等模纹效果。

The effects produced by applying weaves to simple orders of colorings comprise continuous line effects (Fig. 5–8–2), hound's tooth patterns (Fig. 5–8–3), hairlines[Fig. 5–8–6(a)], bird's–eye and spot effects [Fig. 5–8–6(b)], step patterns [Fig. 5–8–6(c)], check and grid, hollow square, and all–over patterns.

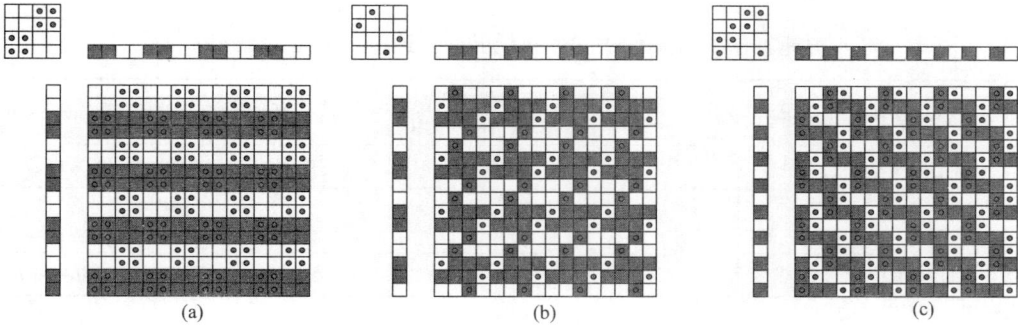

(a)   (b)   (c)

图5-8-6　各类配色模纹

Various color and weave effects

习题 /Questions

1. 以 $\dfrac{2}{2}\nearrow$、$\dfrac{2}{2}$ 纬重平和 $\dfrac{2}{2}\nwarrow$ 为基础组织，设计一个横条纹组织。

2. 某纵条组织的织物，P_j=260 根 /10cm。第一条宽 2.5cm，采用 $\dfrac{2}{2}\nearrow$ 组织，第二条宽 2cm，采用 $\dfrac{2}{2}$ 方平组织，试画出该织物的上机图。

3. 以下列组织为基础，分别作方格组织图，方格组织的 R_j、R_w

1. Design a horizontal stripe weave based on $\dfrac{2}{2}\nearrow$ twill, $\dfrac{2}{2}$ weft rib and $\dfrac{2}{2}\nwarrow$ twill.

2. Draw the looming plans for a vertical stripe weave under the following conditions: containing 260 ends per 10 cm, the first stripe is 2.5cm in width with $\dfrac{2}{2}\nearrow$ weave; the second stripe is 3 cm in width by $\dfrac{2}{2}$ basket weave.

3. Draw the check weaves based on the following weaves. R_j and R_w are chosen by the designers.

自选。

（1）以 $\dfrac{4}{1}$ 斜纹为基础组织；

（2）以 五 枚 缎 纹 为 基础 组织；

（3）以八枚缎纹为基础组织。

4. 按照下面的格子组织纹样（习题图 5-1），设计一个格子花型织物。试求：（1）R_j 和 R_w；（2）穿综说明和纹版图（已知条件见习题表 5-1）。

(1) $\dfrac{4}{1}$ twill; (2) 5-shaft satin; (3) 8-shaft sateen.

4. Design a two-way stripe weave based on the pattern as shown in exercise figure 1. The conditions are listed in the table. Since both R_j and R_w are big numbers, just explain the denting order and the lifting plan of the design.

习题表 5-1　织物规格 /Specifications of the two-way stripe

条纹宽度/ Width of stripes (cm)	条纹密度（$P_j = P_w$）(根/10cm)/ Fabric density: Threads/10cm	条纹组织/ Weaves for the stripes
$a=8$	340	平纹组织/Plain weave
$b=2$	680	6枚不规则缎纹/ 6-shaft irregular sateen
$c=16$	340	平纹组织/Plain weave

5. 自行选择基础组织，分别用增点法、间隔排列法设计绉组织。

5. Design the crepe weaves respectively by overlapping the weave and inserting the weave into another. The base weaves are determined by the designers.

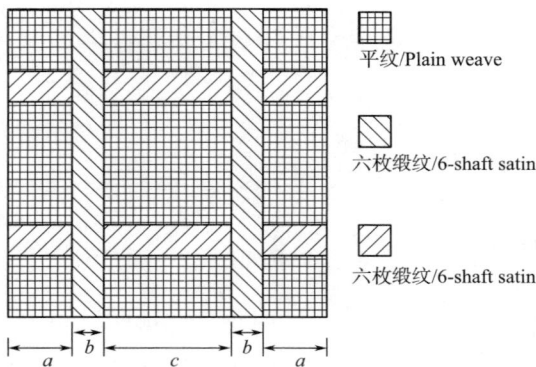

平纹/Plain weave

六枚缎纹/6-shaft satin

六枚缎纹/6-shaft satin

习题图5-1

6. 以 $\dfrac{3\ 1\ 2}{2\ 2\ 1}\nearrow$ 为基础组织，采用调整经纱次序的方法构成一个绉组织。

7. 用旋转法构作一个绉组织，

6. Design a new crepe weave by adjusting the order of the ends with the base $\dfrac{3\ 1\ 2}{2\ 2\ 1}\nearrow$.

7. Design a new crepe weave by rotating a base weave that is chosen by the designer.

基础组织自选。

8. 在平纹组织的基础上设计一个花式透孔组织，花形纹样为 $\dfrac{5}{3}$ 纬面缎纹，每小格代表 6 根经纬纱。

9. 如习题图 5-2 所示组织，从结构上判断，是否具有透孔效应？

8. Design a fancy mock leno weave with a plain weave. The pattern of the perforated effect is arranged at the form of $\dfrac{5}{3}$ sateen, and each weave point in the sateen represents 6 threads.

9. Do the following weaves have the perforated effect just by their constructions?

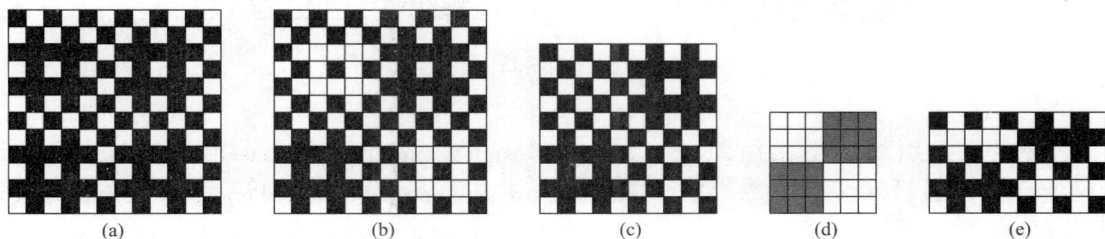

(a)　　　　(b)　　　　(c)　　(d)　　(e)

习题图5-2

10. 以 $\dfrac{1}{5}$ 斜纹为基础组织，绘制蜂巢组织的上机图。

11. 如习题图 5-3 所示组织图的 A 点和 B 点，在织物背面是处于高点还是低点？

10. Draw the looming plans of the honeycomb weave based on a $\dfrac{1}{5}$ twill base.

11. There are points A and B in the weave. Which one is at the ridge of the cell at the back surface? Which one is at the recesses of the cell at the back surface ?

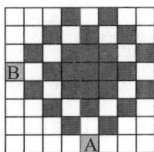

习题图5-3

12. 设 R_j=24，R_w=8，以 $\dfrac{2}{2}\nearrow$ 为固结组织，试绘制凸条组织的上机图（要求消耗动力较少）。

13. 设计以 $\dfrac{4}{4}$ 纬重平为基础组织、固结组织为平纹的纵凸条组织，要求凸条效应明显。

12. Draw the looming plans for a Bedford weave requiring the following conditions: less energy consumption, R_j=24 , R_w=8, with a $\dfrac{2}{2}\nearrow$ binding weave.

13. Draw the looming plans for a Bedford weave requiring the following conditions: prominent cord effects, based on a $\dfrac{4}{4}$ weft rib and plain binder.

14. 纵凸条占据的平纹与浮长的宽度是 7 根经纱，请设计该纵凸条组织。

15. 判断习题图 5-4 所示的纵凸条组织中芯线的所在位置，并绘制出完整的上机图。

14. Draw the looming plans for a Bedford weave requiring the following conditions: the length of the float is 7 and the binder is a plain weave.

15. Judge the locations of the wadding threads and draw the complete looming plans for the weave.

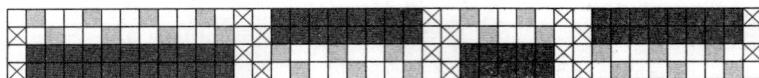

习题图5-4

16. 以平纹组织为基础组织，R_j=12，R_w=16 设计网目经组织，要求含有两根网目经对称。

17. 试设计一个平纹地经纬起花小提花组织的组织图，花型效果如习题图 5-5 所示。

18. 习题图 5-6 所示组织图中，色经排列为 4A4B，色纬排列为 4A4B，求配色模纹图。

16. Design a spider weave based on a plain ground, requiring two symmetrical distorted ends, R_j=12 and R_w=16.

17. Design a dobby spot weave of the pattern as shown in the following design with floats and plain ground.

18. The following weaves are all ordered of 4A4B in both warp and weft directions. Try to draw the color and weave effects.

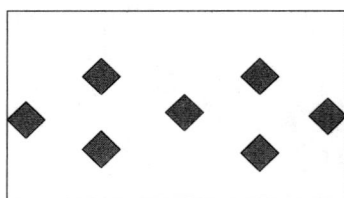

习题图5-5

习题图5-6

19. 试作以 $\dfrac{2}{2}$ 为基础组织，K_j=4 的人字型破斜纹，色经色纬排列均为 4A4B4C4D 的配色模版图。

20. 已知配色模纹如习题图 5-7 所示，求作色纱排列和组织图（要求至少画四个不同的组织），并能在踏盘织机上织造。

19. Draw a herringbone weave based on a $\dfrac{2}{2}$ twill and K_j=4. If both the warp threads and the weft threads are arranged in the color sequence 4A4B4C4D, try to draw the color and effect diagram of the weave.

20. The color and weave effect is shown as following diagram. Try to determine 4 possible weaves, the order of warping and wefting. The weave can be woven on a tappet loom.

21. 习题图 5-8 所示的组织属于浮松组织，试分析其外观结构特点。

21. The weave drawn below is a huckaback weave. Try to describe the construction and the appearance of the fabric of the weave.

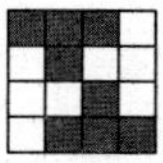

习题图5-7

习题图5-8

第六章　二重及双层组织 /Backed Weaves and Multi−Layer Weaves

Backed Weaves and Multi−layer Weaves

简单组织由一个系统的经纱和一个系统的纬纱交织而成，而复杂组织中经纱或者纬纱至少有一种是由两个或两个以上在织物中起不同作用或者不同功能的纱线组成。复杂组织织物中，有些纱线的作用是作为织物身骨的地布，其他纱线是仅起到装饰作用的起毛纱或者处于织物表层，这些纱线可能相互之间并不平行，如有些起毛纱线竖立，与织物平面垂直。复杂组织包括重组织、双层及多层组织、起毛组织、毛巾组织等。

Simple weaves are composed of one series of warp and weft threads, but in compound weaves, there are more than one series of ends or picks, and some of which may be responsible for the "body" of the fabric, such as ground yarns, whilst others may be employed entirely for ornamental purposes such as "figuring" or "face" yarns. Some threads may be found not to be in parallel formation one to another in either plane. For example, there are pile surface constructions in which some threads may project out at right angles to the general plane of the fabric. Compound weaves include backed weaves, double−layer and multi−layer weaves, fleecy weaves, and terry weaves.

第一节　经二重组织 /Warp−Backed Weaves

经二重组织

重组织由两组或两组以上的经纱与一组纬纱交织，或由两组或两组以上的纬纱与一组经纱交织而成。重组织分为重经组织和重纬组织，重组织可在增加织物的重量、厚度、坚牢度和保暖性的同时，保持外观细洁或者使成本更低，也可使织物正反面显示不同的颜色和组织。

Backed weave fabrics are woven with an extra series of threads, either warp or weft, to add warmth, weight and/or strength, at a single structure which is equally fine on the surface or produced in an economical manner by using an inferior quality of the extra threads. By employing differently colored threads for the face and back a cloth is produced in which the two sides are differently colored, or a cloth may be woven colored on one side and white on the other.

一、经二重组织的构成 /Construction of Warp-Backed Weaves

经二重组织［图 6-1-1（a）］由两组经纱与一组纬纱交织而成，其经纱由形成重叠的两个系统的纱线构成，分别称为表经（阿拉伯数字序号）和里经（罗马数字序号），其某根纱线的截面图分别如图 6-1-1（b）和（c）所示。表经与纬纱交织构成织物正面，称为表组织；里经与同一纬纱交织构成织物反面，称为反组织；表反组织的经组织点用不同的符号表示，以示区别。如果将表经拆去，则显示里组织，如图 6-1-1（d）所示。里组织与反组织呈现底片翻转的

A warp-backed weave as shown in Fig. 6-1-1(a) is composed of two series of warp threads (the face warps and back warps) and one series of weft threads. The face warps (ordered in Arabic numerals) and the back warps (ordered in Roman numerals) overlap in fabric as shown in the transversal section in Fig. 6-1-1(b) or the longitudinal section in Fig. 6-1-1(c). The face weave is composed of the face warps and the wefts while the back weave is composed of the back warps and the wefts. Different marks are used to distinguish the face ends from the back ends. The opposite side of the back weave is the reverse weave. If the face weave is removed, the back weave is visible from front as shown in Fig. 6-1-1(d).

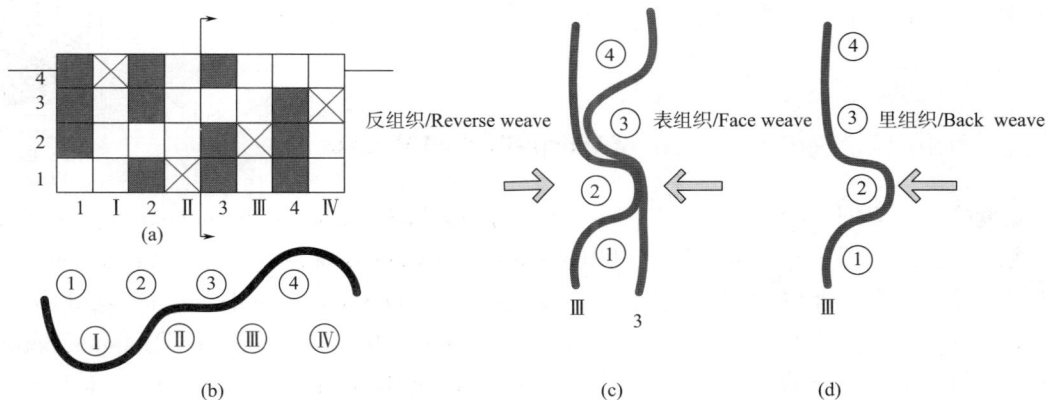

图6-1-1 经二重组织及其截面图
A warp-backed weave and its cross-sections

关系。

因为里组织或者反组织在正面不可见，为了表述表里组织之间的关系，常将表组织与里组织的经纱按照一定的比例间隔排列。图 6-1-2 中的经二重组织就是由表、里组织的经纱（表、里经纱）按照 1:1 比例排列得到的。

A warp-backed weave is arranged in proportion intervals from a face weave and a back weave, therefore, it is actually the combination of face weave and back weave. Fig. 6-1-2 shows a warp-backed weave in which the face warp and the back warp are arranged at the ratio of 1:1.

If the face warps and the back warps in the weave shown in Fig. 6-1-1 are arranged as 1 red and 1 blue, the fabric with high warp sett will appear red in face side and

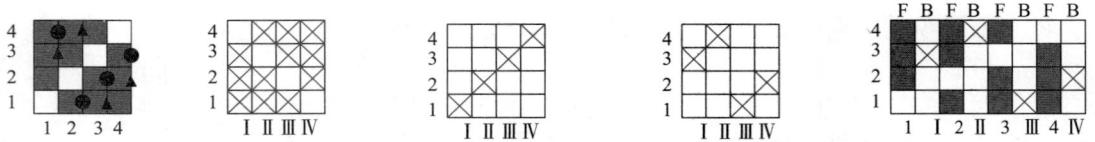

(a) 表组织/Face weave (b) 反组织/Reverse weave (c) 里组织1/Back weave No.1 (d) 里组织2/Back weave No.2 (e) 经二重组织/Warp-backed weave

图6-1-2　经二重组织的组成部分
The composition of a warp-backed weave

如果图 6-1-1 中经二重的经纱排列是 1 红 1 蓝，纬纱为其他颜色，那么该组织织物在经密较大的情况下，正面显示红色，背面显示蓝色，经二重的正反面明显不同，不能按照普通单层组织的配色模纹效果判断其表面效应。因此，在预测一个组织的表面色彩效果时，必须要判断该组织是否为经二重组织等复杂组织，然后才能根据表反组织的配色效果进行预测。

blue in the reverse side. So, the effects of the obverse side and the reverse side are quite different and the color effects can't be expected as the single average weave. Therefore, before predicting the color effects of a fabric, the type of the weave has to be judged first, then, the color effects should be estimated according to the face weave or reverse weave for compound weaves.

Judgement of
Warp-Backed
Weaves

二、经二重组织的判断 /Judgement of Warp-Backed Weaves

按照浮长线原理，同一系统相邻纱线的组织点若浮长差异大，导致高度差异大。在密度较大时，由于相互挤压，会产生重叠。从经纱方向看，如果出现连续的浮长高于（或低于）相邻纱线的所在经浮长高度，则经纱可能产生重叠关系。图 6-1-2 的经二重组织各个组织点所处的经浮长计算值如图 6-1-3 所示。很明显，奇数根位置的经纱都要比其相邻两侧的偶数根位置的经纱的经浮长值大，处于织物的上层，成为表经，与纬纱一起形成表组织；而偶数根位置的经纱处于下层，成为里经，与纬纱一起组成里组织。

A warp-backed weave can be judged by warp floats. If the interweaving points of an end are continuously higher or lower than its neighboring ends due to the difference in the float lengths, the overlapping may happen between the two ends. If all the ends in a weave are hided by their neighboring ends, the weave is probably a warp-backed weave. Fig. 6-1-3 shows the length diagram of the warp floats in a warp-backed weave in Fig. 6-1-2. It is obvious that the values of the float lengths of the odd ends always exceed their neighbors, therefore, the odd ends are naturally led to the upper layer and become the face warps whilst the even ends become the back warps.

3	−3	3	1	3	−3	−1	−3
3	1	3	−3	−1	−3	3	−3
3	−3	−1	−3	3	−3	3	1
−1	−3	3	−3	3	1	3	−3

图6-1-3　经二重组织的经浮长线分布图
The length diagram of the warp floats in a warp–backed weave

三、经二重组织的设计原则 /Principles of Designing Warp–Backed Weaves

（1）表、里组织的选择。表组织要在织物上层，要求经浮长要长，尽量采用经面组织。里组织在织物下层，即经浮长要尽可能短，纬组织点多，多采用纬面组织。里组织与反组织呈现底片翻转的关系，故反组织也是经面组织。即表组织采用经面组织，里组织采用纬面组织，反组织采用经面组织。

（2）里组织的遵循原则。表组织的经浮长应大于里组织的经浮长，且里组织经浮长尽可能配置在表组织经浮长中间，确保表组织可以覆盖重叠在里组织之上，隐藏里组织的经组织点。每一根纬纱要和两种经纱相交织，应使纬纱的屈曲均匀且尽可能小。如果织物有斜向纹路，表、里组织点排列方向相同，确保里组织的短经浮点都能被覆盖重叠。

（3）表、里经纱排列比。表里经纱线密度与密度相同时，表里经排列比可采用1∶1；若仅为增厚增重，里经则可粗、原料较差，排列比为2∶1。

(1) Selection of face weave and back weave. The face weave is at the upper side of the cloth, which requires that the length of the warp float should be longer and higher while the length of the warp float in the back weave is required to be shorter and lower. Therefore, the face weave is mostly warp–faced while the back weave is mostly weft–faced. So, the reverse weave is also warp–faced.

(2) Regulation of the back weave. The short warp floats of the back warp are arranged between two long warp floats of the neighboring face warps to ensure the short warp floats of the back warp threads are concealed. Each filling thread interweaves with the face warps and back warps, therefore, the weft threads should bend evenly and as little as possible to avoid over contraction. If the twill effects appear on the surface, the face and the back are required to have the same directions of diagonal lines to ensure the warp floats of the back weave to be covered by the long warp floats of face weave.

(3) Arrangement of face warp and back warp. If the thickness and the density of the face warps and the back warps are similar, the arrangement ratio of face warps and back warps can be 1∶1. If the fabric is designed only for thickening and weight gaining, coarse and low–grade raw materials can be used for back warps, and the arrangement ratio for face warps and back warps can be set as 2∶1.

(4) Calculation of R_j and R_w. Supposing the arrangement ratio of the face warp and back warp is $m∶n$, the sizes of the

（4）R_j、R_w 的计算。当表经：里经 = $m:n$ 时，表里组织循环根数分别是 R_{jm}、R_{wm}、R_{jn}、R_{wn}，则生成的经二重组织循环根数按照下式计算。

face weave and the back weave are R_{jm}, R_{wm}, R_{jn} and R_{wn} respectively, then, R_j and R_w of the new warp-backed weave are calculated by formulas 6-1-1 and 6-1-2.

$$R_j = \mathrm{lcm} \left[\frac{\mathrm{lcm}(R_{jm}, m)}{m}, \frac{\mathrm{lcm}(R_{jm}, n)}{n} \right] \times (m+n) \qquad (6-1-1)$$

$$R_w = \mathrm{lcm}\,(R_{wm}, R_{wn}) \qquad (6-1-2)$$

四、经二重组织的组织图绘制 /Drawing of Warp-Backed Weaves

以图 6-1-4 中的表、里组织为例，表里经排列比为 1:1 时，绘制经二重组织的组织图步骤如下：

（1）计算 R_j、R_w，圈定组织图范围，并分别为表里经采用不同序号排序。

（2）调整里组织的经纱位置顺序，使之处于表组织对应经纱的长经浮长线之间。首先将里组织第 I 里经移动到第 1、2 表经中间，看里经短浮长是否被表经的长经浮长所覆盖。若是，则第 I 经就是里组织的第 1 根经纱；否则，将第 I 里经移动到第 2、3 表经中间，进行判断；这个过程不断进行，直到所有里经上经浮长被其两侧表经长浮长覆盖为止。这样确定第 I 里经在里组织上的位置。

（3）将里组织的经组织点放置在组织图上的相应位置，得到图 6-1-4 所示的组织图。

若步骤（2）中里组织不动，调整表组织的起始位置，得到如图 6-1-5 所示的组织图。

The face weave and the back weave in Fig. 6-1-4 are taken as example to explain the procedures of drawing the warp-backed weave. The arrangement of the face warp and the back warp is set as 1:1.

Drawing of Warp-Backed Weaves

(1) Calculate R_j and R_w, and determine the whole repeat range on design paper. Number all the end yarns with order of the face and back respectively according to the arrangement ratio of the face warp and back warp.

(2) Adjusting the commencing end of the back weave to ensure that warp floats are concealed by the face ends. Take the short warp float of the end No. I of the back weave, and move them to corresponding positions between end No. 1 and end No. 2 of the face weave. If the short warp floats are hidden or covered by the long warpfloats of the face weave, then the end No. I is the first end in the back weave. Otherwise, the short warp floats move to corresponding positions between end No. 2 and end No. 3 of the face weave and judge whether the short warp floats are hidden or covered by the two yarns of the face weave. And the procedure continues until the commencing end of the back weave is obtained.

(3) Fill in the marks of the back weave according to the sequence of the first end of the back weave.

If the first end of the face weave is to be adjusted, then the warp-backed weave as shown in Fig. 6-1-5 is obtained.

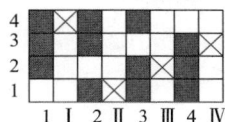

图6-1-4　调整里组织得到的经二重组织
The backed-warp weave by regulating the back warp

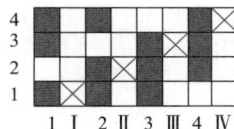

图6-1-5　调整表组织得到的经二重组织
The backed-warp weave by regulating the face warp

五、经二重组织的上机要点 /Looming for Warp-Backed Weaves

（1）穿综。采用分区穿法，提升次数多的表经穿在前综，里经穿后区。后综位置略低于前区的综框，便于经纱的重叠。

（2）织轴。经二重织物一般单轴织造。若表里经纱在原料、强度、缩率等方面显著不同，则采用双轴织造。

（3）穿筘。同一组表里纱穿在同一筘齿中，便于表、里经纱相互重叠。每筘齿穿入数一般为表、里经纱排列比之和的倍数。经纱必须有足够大的密度，否则不能给里经良好的遮盖效果。一般表层经纱紧度不得低于普通类似单层织物的12%，如果里组织松弛，则这个数值是6%。如果表里经排列比是2:1，则表层织物的密度比同等条件下的1:1排列比的小4%～8%。

图 6-1-6 所示为图 6-1-4 中经二重组织的上机图。

(1) Drafting. Divided drafting is employed for warp-backed weaves. Face warps are drawn upon the front heads due to more liftings while the back warps are drawn upon the rear heads which are usually set slightly lower than the front heads for the convenience of overlapping.

(2) Beaming. Usually, the face warps and the back warps are wound from the same beam. If there are significant difference in materials, strength, or rate of contraction between the face warps and the back warps, the back warps will be wound on a separate beam.

(3) Denting. A group of face warps and back warps are threaded in the same dent to help the overlapping of the ends. In order to achieve a well-covered face in a backed cloth, correct settings are very important as without sufficient density of the face threads the binding marks of the back weave cannot be covered no matter how cleverly they are placed. As a general guide it can be stated that in 1 face, 1 back orders of backing the percentage cover of the face yarns could be up to 12% less than in a similar well-constructed single cloth if comparatively short floated weaves are used with firm backing, and about 6% less if longer floated weaves with loose backing are employed. In the order of 2 face, 1 back the reduction in the setting compared with a similar single cloth must be less than for order of 1 face, 1 back and under conditions similar to those stated for the latter could be 8% and 4% less respectively.

The looming plans for a warp-backed weave in Fig. 6-1-4 is shown in Fig. 6-1-6.

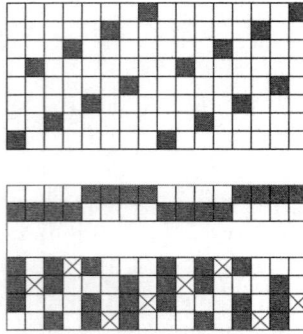

图6-1-6　经二重组织上机图

The looming plans for a warp-backed weave

六、经起花组织 /Extra Warp Figured Weaves

1. 经起花织物的特点 /Features

经起花组织以简单组织为地组织，在织物表面局部采用长浮长线或经二重组织，形成一定的花型。图 6-1-7 所示的经起花组织中，起花部分由经二重组织形成，如 D 部分，表经浮在正面可见，在织物表面形成花型（A 部分），此时表经称为花经；不起花时，花经沉在织物反面成为经浮长线，不可见，如花经上所有的纬组织点（背面则是经浮长线）。简单组织部分为地组织，如图中的 C、E 部分也不起花，使用的经纱称为地经。为了避免花经上出现过长的浮长线，影响织物的服用性能，花经在正面与反面都每隔一定距离与纬纱交织一次，截短在织物正反面的经浮长线，此时的交织点称为接结点，起固结作用。接结点一般不显示，但也可有意作为花型的一部分，如 B 部分的花经上经组织点，因为不能被两侧的平纹地组织点完全重叠遮盖，故隐约可见。采用不同颜色的花经，

In an extra warp figuring fabric, there are one series of ground warp threads, one series of extra figuring warp threads and one series of weft threads. The weave is comprised of the ground weave and the warp-backed weave or long warp floats. In Fig. 6-1-7, the figured portion A is formed by a warp-backed weave where the face warp threads at portion D float on the surface. The warp threads at portion D are called the extra figured threads. On the extra figured threads, the warp will float on the back of the fabric other than the portion A. At the ground area like C and E in the figure, the warps are called ground warp threads. In order to avoid the too long floats which may affect the wear performance of the fabric, the extra warps interweave with the weft threads at interval from the front and back, and cut the long warp floats. The interweaving points are called binding points. The purpose of inserting the binding points is to shorten the length of yarn on the opposite side of the fabric, so as to have a consolidation effect. The binding points are usually concealed by their long float neighbors, however, the points are sometimes purposely designed to be vaguely visible on the surface, such as portion B at the extra figured warp threads. One of the advantages of figuring with extra materials is that bright colors—in sharp contrast with the ground—may be brought to the surface of the cloth in any

可以形成强烈的颜色对比。对于一些轻薄织物，背面的长浮长线可能影响织物效果，可以剪去。如果将花经去除，地部经纬纱仍然正常完整地交织。

　　若花经上的经浮长全部用平纹点，也可形成经起花效应，如图 6-1-8 所示。花经上的背面经浮长线全部在反面，与左右两侧平纹地经形成较大的浮长差异，导致高度差较大，可以被完全重叠覆盖；但正面经组织点与其左右两侧的地经要么形成平纹关系，要么形成并经关系，完全不能重叠。由于其浮

desired proportion. Pleasing color combinations may thus be conveniently obtained. The extra yarn is allowed to float loosely on the back, and is afterwards cut away. It is not applicable to fabrics in which the ground is so light and transparent that the positions of the extra threads on the back can be perceived from the face side. A distinguishing feature of fabrics in which extra materials are employed is that the withdrawal of the extra threads from the fabric leaves a complete ground structure.

Even the plain interweaving points on the extra warp threads can also form the pattern as shown in Fig. 6-1-8. The long floats of the extra warp threads which appear on the reverse side of the fabric are so low that they are completely concealed on the surface. However, the solid square marks of the extra warp thread are either opposed to

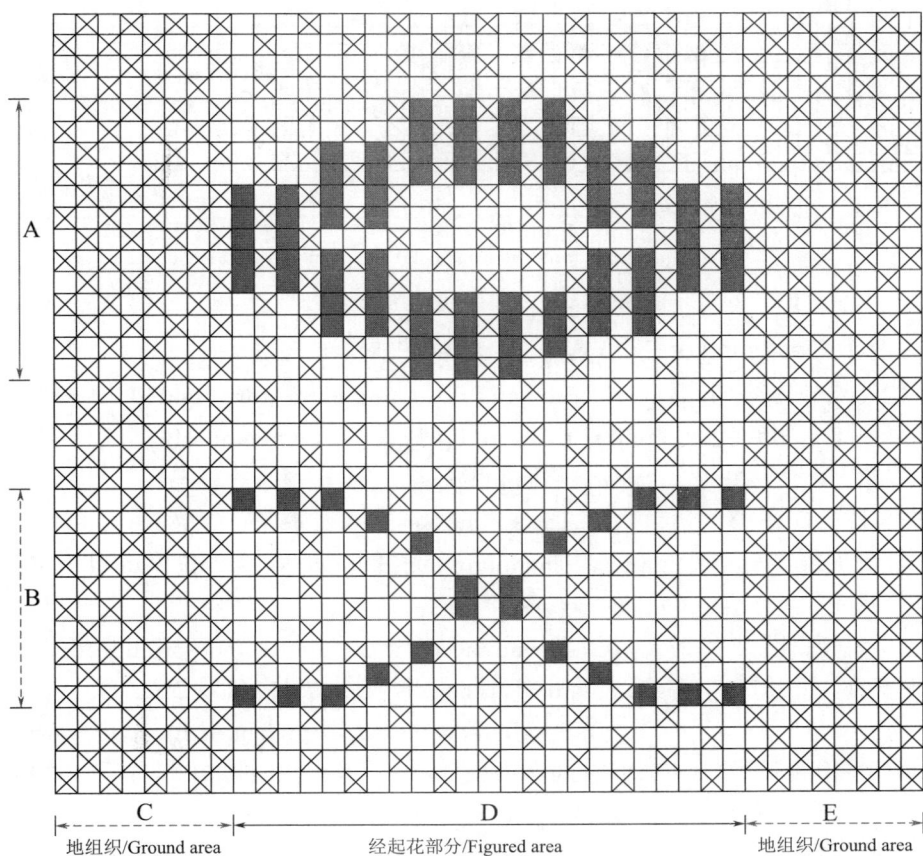

Even the plain interweaving points

地组织/Ground area　　　　経起花部分/Figured area　　　地组织/Ground area

图6-1-7　经起花组织1
An extra warp figured weave No. 1

长短，正面也可显示隐约、含蓄的花纹效应。

or the same as their neighbors which makes them unoble to be overlapped so that the risers of the extra threads can be perceived from the face side to form a vague pattern.

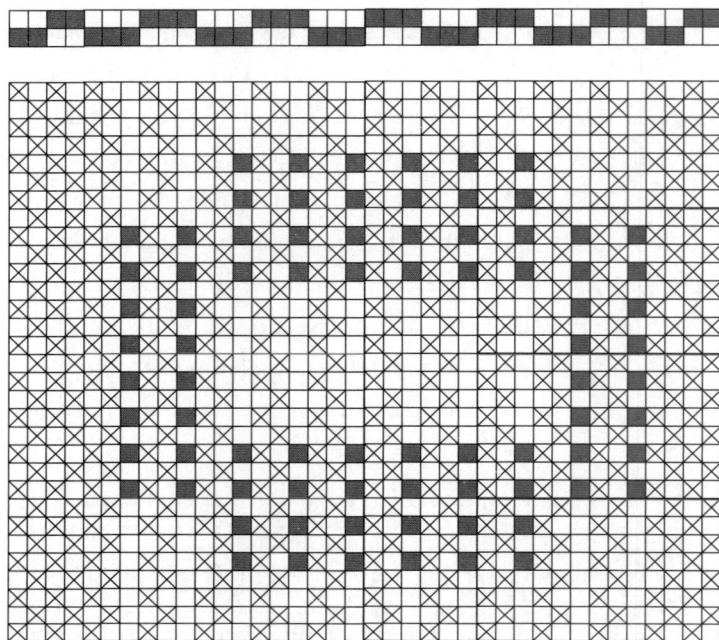

图 6-1-8 经起花组织2
An extra warp figured weave No. 2

2. 经起花织物设计 /Designing of Extra Warp Figured Fabrics

设计经起花织物的设计思路与经二重类似，同时需要结合平纹地小提花织物的设计理念。

（1）地组织的选择。应平整、烘托花型，因此，常用平纹等简单组织。

（2）花经、地经排列比。一般常用的排列比为 1∶1、2∶1 或 2∶2 等。

（3）花型布局与浮长线长度。单轴时，浮长以 3 ~ 5 为好；双轴时，浮长不受限制，但不宜过长。

（4）接结点配置。要求接结点

The idea of designing extra warp figured fabrics is similar to warp-backed weaves, combining the design concept of dobby spots fabric.

(1) The ground weave is preferred to be flat to foil the pattern. Therefore, the plain weave is widely used as the ground weave.

(2) The arrangement of extra figured warp threads and ground threads is usually set as 1∶1, 2∶1 or 2∶2.

(3) Since most of the extra warp figured fabrics are woven with double beams, there are no restriction of the length of the floats. However, too long the floats will damage the strength of the fabric. For the fabric woven with a single beam, the length of floats is preferred to be ranged from 3 to 5.

在织物表面不显露时，接结点应安排在两侧地组织的浮长线中间；若要求接结点显露成花型时，则应安排在地经的纬浮长线之间。

(4) When not exposed on the surface, the binding points shall be arranged between the long floats of their neighboring ground ends. If required to show a pattern, the binding points are arranged between the weft floats of the ground ends.

3. 经起花组织的上机 /Looming for Extra Warp Figured Fabrics

（1）采用分区穿法，地经穿前区，花经穿后区。

（2）花经在地经中并入同一筘。

（3）如有可能，尽量用一个织轴。但因花经与地经交织次数相差较大，一般用两个织轴。

(1) Divided drafting is employed. The ground ends are drawn on the front heals while the extra figured ends are drawn on the rear heals.

(2) The extra figured ends are placed in the same dent with the ground ends overlapped.

(3) A single beam is used as far as possible. However, double beams are mostly used as there is a big difference of the interlacement with fillings between the extra figured ends and the ground ends.

七、经二重组织的其他应用 /Other Applications

经二重组织可以与其他组织联合应用，进一步使织物增厚。图 6-1-9 是某局部重经的厚重凸条组织上机图，凸条与固结组织纬向排列比为 2∶1，在一个组织循环内的 29 根经纱里，有 18 根表经，3 根分隔平纹经纱，4 根芯线和 4 根里经，采用双轴织造。芯纱与里经同一织轴（高张力），其他经纱另一根织轴。除了切割平纹经纱采用 60 英支 /2 棉纱外，纬纱和其余均经纱采用 16 英支 /2 棉纱。

The warp–backed weave can be combined with other weaves to make a heavier fabric. Fig. 6–1–9 demonstrates the looming plans for a Bedford cord weave combining with a warp–backed weave. The long weft floats and the binder weave are arranged at the ratio of 2 ∶ 1. Each cord occupies a total of twenty–nine warp threads, of which eighteen warp threads are "face", three "cutting", four "wadding" and four "back" warp threads. Only two warp beams are necessary to contain the four series of warp threads, namely, one for face and cutting threads (whose rate of contraction during weaving is equal), and one for wadding and back warp threads, which are held at greater tension than face and cutting threads. Two counts of yarn are employed in the production of this example, namely, $60^s/2$ for cutting, and $16^s/2$ for face, wadding and back warp threads (the latter being sized); also $16^s/2$ weft of similar yarn to the warp threads for both face and back picks.

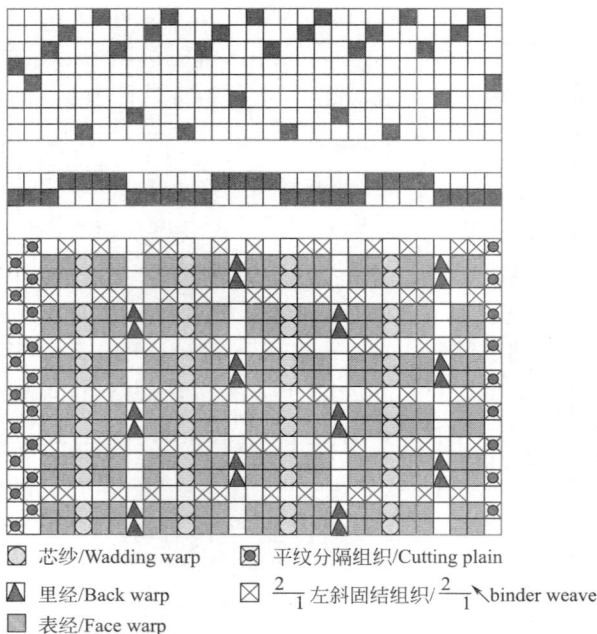

图6-1-9　局部重经的厚重凸条组织上机图

The looming plans for a heavy bedford cord weave combing with a warp-backed weave

第二节　纬二重组织 /Weft-Backed Weaves

一、纬二重组织的构成 /Construction of Weft-Backed Weaves

纬二重组织由两个系统的纬纱（即表纬和里纬）和一个系统的经纱交织而成，织物表面多呈纬面效应。纬二重组织中，表纬与经纱形成表组织，显现在织物正面；里纬与经纱形成里组织，正面不可见；里组织的反面是反组织，在织物背面可见。图 6-2-1 所示的组织中，若表纬显示蓝色，里纬显示红色，经纱是其他颜色，则高纬密织物表面显示蓝色效应，背面显示红色效应。纬二重组织常用于毛毯和工业用滤尘布。

A weft-backed cloth is composed of two series of weft threads (face wefts and back wefts) and one series of warp threads. The face weave is composed of the face wefts and the warps while the back weave is composed of the back wefts and the warps. The opposite side of the back weave is the reverse weave. If the face weave is removed, the back weave is visible from front. If the face weft and the back weft are arranged as 1 blue and 1 red, the weft densely fabric of the weave in Fig. 6-2-1 will appear blue on the face side and red on the reverse side. Weft-backed weaves are usually employed in blankets and filter fabrics.

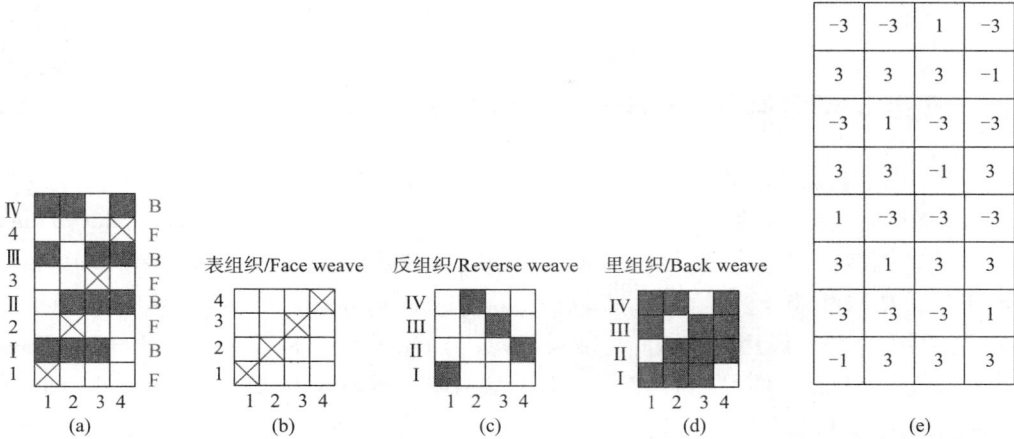

图6-2-1　纬二重组织及其构成
Construction of a weft–backed weave

二、纬二重组织的判断 /Judgement of Weft–Backed Weaves

因里组织在正面不可见，为了表述表里组织之间的关系，将表组织与里组织的纬纱按照一定的比例间隔排列。图 6-2-1 中的纬二重组织［图 6-2-1（a）］是由表组织［图 6-2-1（b）］、里组织［图 6-2-1（d）］的纬纱（表、里纬纱）按照 1:1 比例合成得到的。

按照浮长线原理，从纬纱方向计算各个组织点所在的纬浮长的长度与位置。如果出现连续的浮长高于（或者低于）相邻纱线的所在纬浮长高度，则可能产生纬纱重叠关系。图 6-2-1（e）所示为纬二重组织［图 6-2-1（a）］的纬纱浮长线分布图，可以看出奇数根位置的纬纱都要比其上下相邻两侧的偶数根位置的纬纱的纬浮长值大，处于织物的上层，成为表纬，与经纱一起形成表组织；而偶数根位置的纬纱处于下层，成为里纬，与经纱一

In order to elaborate the construction of the weft-backed weave as shown in Fig. 6-2-1(a), the face weft in the weave as shown in Fig.6-2-1(b) and the back weft in the weave as shown in Fig.6-2-1(d) are arranged at ratio of 1:1.

A weft-backed weave can be judged by the weft floats. If the interweaving points of a weft are continuously higher or lower than its neighboring wefts due to the difference in the float lengths, the overlapping may happen between the two wefts. If all the wefts in a weave are hided by their neighboring wefts, the weave is probably a weft-backed weave. Diagram as shown in Fig. 6-2-1(e) shows the length diagram of the weft floats in the weft-backed weave [Fig. 6-2-1(a)]. It is obvious that the values of the float lengths of the odd wefts are always exceed their neighbors, therefore, the odd wefts are the naturally led to the upper layer and become the face wefts whilst the even wefts become the back wefts.

起组成里组织。

三、纬二重组织的设计原则 /Principles of Designing Weft–Backed Weaves

（1）表组织与里组织的选择。表组织要在织物上层，要求纬浮长要长，尽量采用纬面组织。里组织在织物下层，要求纬浮长要尽可能短，故经组织点多，采用经面组织。里组织与反组织是底片翻转的关系，故反组织也是纬面组织。即表组织采用纬面组织，里组织采用经面组织，反组织采用纬面组织。

（2）里组织遵循的原则。为了使织物正反面具有良好的纬面效应，表纬的纬浮线必须将里纬的纬组织点遮盖住，必须使里纬的短纬浮长配置在相邻表纬的两浮长线之间。若表、里组织点均存在某种排列斜向，则方向要相同，才能满足表纱遮盖里纱的要求。

（3）里纬排列比的选择。取决于表里纬纱的线密度、基础组织的特性以及织机选色装置的条件等。一般常用的排列比为 1∶1、2∶1 或 2∶2 等。如织物正反面组织相同时，如里纬为高特数纱，表里纬排列比可采用 2∶1；若表里纬纱线密度相同，则排列比采用 1∶1 或 2∶2。

（4）R_j、R_w 计算。若表、里经纱排列比为 $m∶n$，表里组织循环根数分别是 R_{jm}、R_{wm}、R_{jn}、R_{wn}，则生成的纬二重组织循环数计算式如下：

(1) Selection of the face weave and back weave. The face weave is at the upper side of the cloth, which requires that the length of the weft float should be longer and higher while the length of the weft float in the back weave is required to be shorter and lower. Therefore, the face weave is mostly weft–faced whereas the back weave is mostly warp–faced. So, the reverse weave is also weft–faced.

(2) Regulation of the back weave. The short weft floats of the back weft are arranged between two long weft floats of the adjacent face wefts to ensure the short weft floats are concealed. If the twill effects appear on the surface, the face weave and the back weave are required to have the same directions of diagonal lines.

(3) Arrangement of the face weft and back weft. The arrangement of the face weft and the back weft are dependent on the thickness of the threads, characteristic of the base weave and the color selector mechanism. The common arrangements are 1∶1, 2∶1 and 2∶2. If the face weave is the same as the reverse weave, and the back weft is coarse and low–grade raw materials, the ratio of face wefts and back wefts can be set as 2∶1. However, if the thickness and the density of the face wefts are similar to that of the back wefts, and the arrangement ratio can be 1∶1 or 2∶2.

(4) Calculation of R_j and R_w. If the ratio of the face warps and back warps is $m : n$, the size of the face weave and the back weave are R_{jm}, R_{wm}, R_{jn} and R_{wn} respectively, then, R_j and R_w of the weft–backed weave are calculated by the formulas 6–2–1 and 6–2–2.

$$R_j = 表组织纬纱循环和里组织纬纱循环的最小公倍数 = lcm（R_{jm}, R_{jn}） \qquad (6–2–1)$$

$$R_w = lcm [\frac{lcm(R_{wm}, m)}{m}, \frac{lcm(R_{wn}, n)}{n}] × （m+n） \qquad (6–2–1)$$

四、纬二重组织的组织图绘制 /Drawing of Weft-Backed Weaves

纬二重组织的组织图绘制步骤如下：

（1）按照 R_j、R_w，圈定组织图范围，并分别为表里纬采用不同序号排序。

（2）调整里组织的纬组织点的位置，使之处于表组织的长纬浮长线之间，确保所有里纬短纬浮长线被遮盖，不在织物正面显示。

（3）将表、里组织的经组织点放置在组织图上的相应位置。首先填入表组织的经组织点，然后根据第一里纬的顺序，依次填入里组织。

图6-2-1中，表、里组织分别是 $\dfrac{1}{3}\nearrow$ 斜纹和 $\dfrac{3}{1}\nearrow$ 斜纹，表里纬排列比为1∶1。因为该顺序下，里组织的纬浮长均被表组织的纬浮长覆盖，故不需要再次调整里纬顺序，直接将表里组织纬纱上的经组织点按照1∶1排列移入新组织，即可得到纬二重组织。

The procedures of drawing a weft-backed weave can be described as following:

(1) Calculate R_j and R_w, and determine the whole repeat range on design paper, then number all the weft yarns with order of the face and back respectively according to the arrangement of the face wefts and back wefts.

(2) Adjusting the commencing weft of the back weave to determine a proper form of back weave. Insert the back weave on back threads only—normal and reversed convention as before—taking care to place a mark of the back weave between two long floats of the face weave thus concealing the binding marks of the back weave by the covering float on the face.

(3) Draw the finial weft-backed weave. Fill the marks in the new weave from the face weave, afterwards, fill the marks from the back weave according to the sequence of the first weft of the back weave.

In designing a weft-backed weave in Fig. 6-2-1, the face weave and the back weave are $\dfrac{1}{3}\nearrow$ twill weave and $\dfrac{3}{1}\nearrow$ twill weave respectively, and the face weft and the back weft are arranged as 1∶1. Since all the weft floats of the back weave are concealed by its neighbors of the face weave, the weft-backed weave is obtained by just inserting the marks of the face weft and the back weft alternately.

五、纬二重组织的上机要点 /Looming for Weft-Backed Weaves

（1）穿综。因经纱方向的交错没有明显的规律，一般采用顺穿法。

（2）穿筘。纬二重织物须有较大的纬密，故经密不宜太大，每筘齿穿入数一般为2～4根。

（3）原料。织造时经纱受外力作用大，故可采用强力较高的原料

(1) Drafting. Straight drafting is usually employed for the fabrics of weft-backed weaves.

(2) Denting. Since there is a comparatively higher weft sett, the warp threads per unit space are less than picks per unit space. 2～4 ends are drawn per dent.

(3) Selection of yarns. Strong materials are required for ends since they subject to the rigorous forces in weaving.

作经纱。如某些毛毯经纱采用棉，纬纱采用毛，经过后整理，毛纱盖住了棉纱。某些棉毯、衬绒织物，经纱采用较细的优质棉纱，而纬纱可用高特数且价廉的棉纱。

（4）投纬。当表里纬纱的纤维材料、线密度、颜色不同时，就须采用多梭箱装置。在纬纱排列比为 2:1 或 1:1 时，织机应该使用双侧多梭箱；而纬纱排列比为 2:2 时，则可用单侧多梭箱装置。新型织机不受表里纬排列比的限制。

For example, some blankets use warp yarns of cotton fiber while filling yarns of wool. After finishing, the wool yarn covers the cotton yarn. For some cotton blankets and lint fabrics, the warp yarn is of finer quality while the filling yarn is of lower count and cheaper cost.

(4) Picking. When the face and back filling are of different fiber materials, thickness and colors, a multi-shuttle box shall be used. If the face filling and back filling are arranged as 2:1 or 1:1, the loom shall use a double shuttle-box; when the alignment ratio is 2:2, a single side multiple shuttle-box device can be used. There is no limit of the arrangement for a shuttleless loom.

六、表里换纬二重组织 /Interchanging Weft–Backed Weaves

如果用甲、乙两种颜色的纬纱并不固定一种专作表纬，另一种不专作里纬，而是按一定的要求在不同区域进行互换，即对于一种纬纱而言，某些区域做表纬，在另外的区域做里纬，这种组织称为表里换纬二重组织。

在图 6-2-2（a）所示的表里换纬二重组织中，纬纱不断在上下层位置变化。前 4 根经纱上，表里纬按照 1:1 排列；中间 4 根经纱上，里表纬按照 1:1 排列；后 4 根经纱上，表里纬按照 2:2 排列。对于此类组织的判断，仍然依据纬浮长线的高低分布来判断，各组织对应的纬浮长长度如图 6-2-2（b）所示。需要注意的是，在表里交换的边界处，可能有极个别组织点所在的浮长线长度没有遵从表层比里层高的规律，但对于整体外观影响很小。

If the wefts of two colors, a and b, are not fixed with one for face weft and the other for back, but is interchanged in different areas according to certain requirements, that is, for one type of filling, some areas for face weft and other areas for back, this arrangement is called interchanging weft–backed weave.

In the interchanging weft–backed weave in Fig. 6-2-2 (a), the wefts change from upper layer to lower layer at different positions. For the first 4 ends, the face weft and the back weft are arranged as 1:1; in the middle 4 ends, the back weft and the face weft are arranged as 1:1; for the last 4 ends, the face weft and the back weft are arranged as 2:2. The weave is also identified by the diagram of the lengths of weft floats shown in Fig. 6-2-2(b). However, at the border of the interchanging, there may be very few interweaving points where the length of the floats does not follow the rule that the face layer is higher than the back layer. However, it has little effects on the appearance in general.

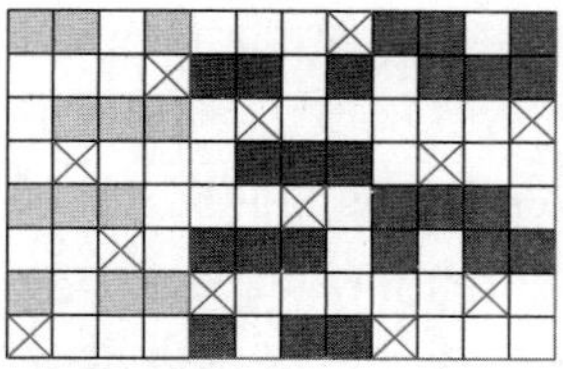

(a) 组织图/The design pattern

-3	-3	1	-1	3	3	3	-3	-3	-3	1	-3
3	3	3	-3	-3	-3	1	-1	3	-3	-3	-3
1	-3	-3	-3	1	-1	5	5	5	5	5	-1
3	-1	3	3	3	-3	-3	-3	1	-1	3	3
-3	-3	-3	3	3	-1	1	-3	-3	-3	1	
2	2	-1	1	-3	-3	-3	1	-1	1	-2	-2
-1	1	-3	-3	-3	5	5	5	5	5	-1	1
-1	3	3	3	-1	1	-3	-3	-3	3	3	3

(b) 浮长线分布图/Diagram of the lengths of weft floats

图6-2-2　表里换纬二重组织
An interchanging weft-backed weave

由于表里纬纱交换，在织物不同部位显示不同颜色，纬二重组织常用来织制彩色条格或者提花毛毯。设计此类组织时，首先将色纬a和b按照1:1的比例排列；然后按照花型，在组织某个颜色区域内，将显示该色的纬纱作为表纬，不显示的色纬作为里纬，按照1:1分布绘制组织图。若某个区域同时显示a和b两种颜色，则色纬按照2:2的排列比绘制组织图，如图6-2-3所示。

The interchanging weft-backed weaves are used to produce the strips, checks or Jacquard blankets. In designing, the colored wefts a and b are arranged as 1:1 firstly. Then, the face weave and the back weave will be inserted a colored section based on 1:1 or 2:2 according to the design pattern. 2 examples are shown in Fig. 6-2-3.

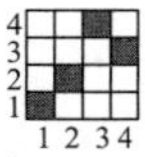

(a) 表组织/ Face weave

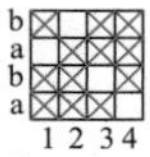

(b) 里组织/Back weave

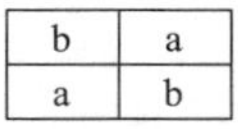

(e) 纹样图案2/Pattern No.2

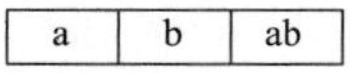

(c) 纹样图案1/ Pattern No.1

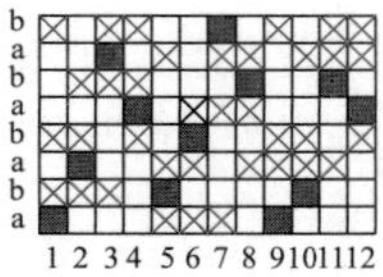

(d)表里换纬二重组织1/
Interchanging weft-backed weave No.1

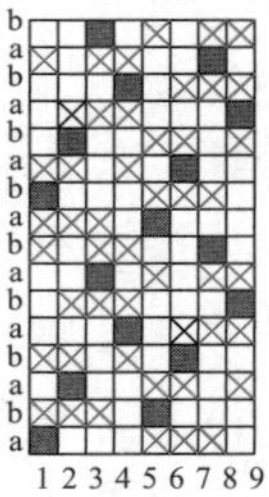

(f)表里换纬二重组织2/
Interchanging weft-backed weave No.2

图6-2-3　表里换纬二重组织图案的组织实现
Design effects and the interchanging backed-weaves

七、纬起花组织 /Extra Weft Figured Weaves

以简单组织为地组织，在织物表面局部采用纬二重组织，使之形成一定的花型，称为纬起花组织。图 6-2-4 所示的纬起花组织中，起花部分由纬二重组织形成，表纬浮在正面可见，在织物表面形成花型，此时表纬称为花纬；不起花时，花纬沉在织物反面成为纬浮长线，不可见，如花纬上所有的经组织点（背面则是纬浮长线）。

An extra weft figured weave is comprised of the ground weave and the weft-backed weave or long weft floats which form the pattern. In an extra weft figuring fabric as shown in Fig. 6-2-4, there are one series of ground weft threads, one series of extra figuring weft threads and one series of warp threads. The figured portion is formed by a weft-backed weave where the visible face weft threads are called the extra figured threads. Where the pattern is undesired, the long weft floats will pass under the warps and be invisible on the face.

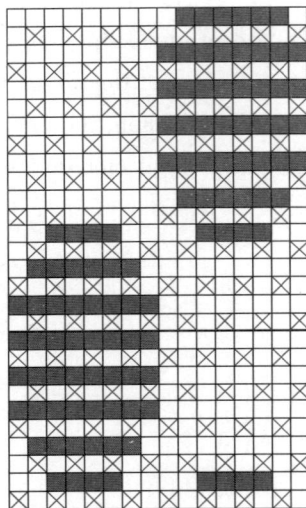

图6-2-4　纬起花组织
An extra figured weft weave

纬起花组织的设计原理与经起花组织类似，要求地布平整，花型突出。根据需要，每隔 4 ~ 5 根地经纱安排一根接结经纱，固结纬浮长线，甚至在接结经旁安排一根地经，抑制正面的过长纬浮长。花纬与地纬的排列比要根据花型需求和织机条件来定。单侧多梭箱织机上，只能采用偶数排列比，如 2∶2 或 4∶2 等。当花纬比例多时，花

Like the extra warp figured fabric, the extra weft figured fabric requires a flat ground and a prominent figure. A binding end is inserted every 4 or 5 ground ends to bind the long weft float as required. Sometimes, another ground end is inserted beside the binding end to prevent an over long float. The arrangement of the extra figured weft thread and the ground weft thread is depended on the pattern and loom. On a single-side multiple box shuttle loom, the ratio can only be set as 2∶2 or 4∶2. If there are more extra figured weft threads, the pattern is prominent but with a low production and stiff hand; if there

型突出，产量低，手感较硬；当花纬比例少时，花型隐约可见。穿综时，地经在前综，起花经在后综，接结经穿在中间综。穿筘时，接结经与相邻花经穿在同一筘齿中。

当连续引入花纬时，织物卷取速度根据地纬计算，卷取机构连续工作。但更常见的是，花纬间隙式引入，一段横条没有花纬，一段有花纬，这时需要间隙式卷纬机构。

Comparison Between Warp–Backed Weaves and Weft–Backed Weaves

are less extra figured weft threads, the pattern is vague. The grounds ends are usually drawn on the front healds, the extra figured ends are drawn on the rear healds while the binding ends are drawn on the middle healds. In denting, the binding ends are usually placed at the same split with their neighboring extra figured ends.

When the extra weft is inserted continuously the take–up speed of the cloth is calculated in terms of ground picks only and the take–up mechanism can run continuously. When, as happens most frequently, the extra weft is introduced intermittently, i.e. when there are bands across the cloth where no extra material is required followed by bands where extra picks are inserted then an intermittent take–up motion is necessary.

八、重经重纬组织的比较 /Comparison Between Warp–Backed Weaves and Weft–Backed Weaves

（1）组织结构。重经组织织物中，经密较大，外观呈双面经效应。织物表面平整、细密。在重纬组织织物中，纬密较大，外观呈双面纬效应。因为纬纱捻度与张力都小，重纬组织织物表面丰满、光亮。

（2）上机织造。与纬二重组织相比，经二重组织具有如下特点：

优点：①因为只有一个纬纱系统，生产效率高，织机速度可以提高；②不需要专门的引纬、多梭箱和卷取机构；③理论上可使用纱线的颜色种数不受限制；④可以得到间断的小花点或者综条带效应。

缺点：①一般需要两个或者多个经轴；②经起花经纱密度大，综丝密度高，同等幅宽的织机织出的组织循环小；③多臂机上织造时，穿综通常比较复杂；④对经

(1) Structure. There are more warps per unit space in a warp–backed fabric, of which both sides are warp–faced and smooth, fine. For a weft–backed fabric, there are more wefts per unit space. Both sides of the weft–backed fabric are softer, fuller, brighter and more loftier owing to the weft containing less twist and being under less tension than the warp.

(2) Looming. Compared with the weft–backed principle, the warp–backed principle has the following features:

Advantages: ① the productivity of a loom is greater because only one series of picks is inserted, and a faster running loom can be used; ② no special picking, box, and uptake motions are required; ③ there is theoretically no limit to the number of colors that can be introduced; ④ in an intermittent arrangement of the extra ends either spotted or stripe patterns can be formed, whereas a similar arrangement in the weft can only be used to form spots (except in special cases) because of the objectionable appearance of horizontal lines.

Disadvantages: ① two or more warp beams may be required instead of one; ② if an ordinary loom and harness are employed a smaller width of repeat is produced by a given size of

纱强力要求高，因此经起花织物不柔软、不丰满、不明亮，花经张力大，花型效果不如纬起花织物；⑤ 经起花织物背面的花经剪除工艺不如纬起花织物的花纬方便。

machine, because the sett of the harness requires to be increased in proportion to the number of extra ends that are introduced in a design; ③ in dobby weaving the drafts are usually more complicated; ④ stronger yarns are required for the figure, and the threads are not so soft, full, and lustrous; extra ends are subjected to greater tension during weaving than extra picks, and, as a rule, there is less contraction in length than in width, and the result is that extra warp effects usually show less prominently than extra weft figures; ⑤ if the extra threads have to be removed from the underside of the cloth, it is more difficult and costly to cut away extra ends than extra picks.

第三节　双层组织 /Double-Layer Weaves

双层组织

如果把两块相互独立、密度较小的相同平纹织物的经纬纱各错一根位置叠放在一起，其交织叠合过程如图 6-3-1（a）所示，由该交织图得到的组织图如图 6-3-1（b）所示。在该组织图中，可以看出一部分经纱（1、2）总是在另一部分纬纱（Ⅰ、Ⅱ）之上，而另一部分经纱（Ⅰ、Ⅱ）总是在某些纬纱（1、2）之下，显然这些经、纬纱线并不交织。

If two separate and open plain fabrics are overlapped and staggered with one thread in both directions, the interlacing diagram and the combined weave are demonstrated in Fig. 6-3-1. As you may see, ends 1, 2 are always passing over the wefts Ⅰ, Ⅱ while ends Ⅰ, Ⅱ are always passing under the wefts 1, 2. Obviously, these ends and wefts actually are not interwoven.

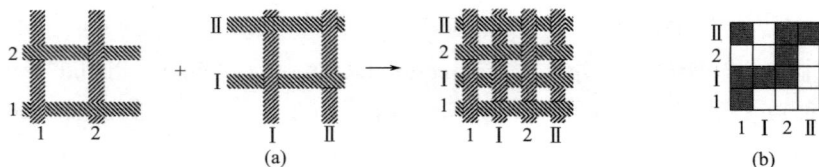

图6-3-1　双层织物交织示意图与组织图
Interlacing diagram and weave of a double-layer fabric

双层组织是由两个系统各自独立的经纱和纬纱分别形成织物的上下两层的组织。双层组织表里两层

A fabric of double-layer weave is woven with two complete independent series of warp and weft threads. The threads of two series override. The threads on the top layer

相互重叠，上层的经纱和纬纱称为表经和表纬，下层的经纱和纬纱称为里经和里纬。上、下两层可以分离，也可以连接在一起。同样的方法，可以形成三层或者四层组织。

are called face warps and face wefts respectively whereas the threads of under layer back warps and back wefts. The two layers are either connected or separated. Similarly, triple-layer weaves or quadruple layer weaves are made.

一、双层组织的结构与形成原理 /Construction and Formation of Double-Layer Weaves

1. 形成双层织物的组织结构沉浮要点 /Principle of Forming Double-Layer Constructions

要形成双层织物，必须满足如下条件：① 投表纬织上层时，里经全部沉在梭口下部不与表纬交织。②投里纬织下层时，表经全部提升不与里纬交织。

以织造图6-3-2所示的双层平纹织物为例。其中，经纱A、C是表经1、2，经纱B、D是里经Ⅰ、Ⅱ。表经穿入1、2片综，里经穿入3、4片综。纬纱a、c是1、2表纬，纬纱b、d是里纬Ⅰ、Ⅱ。

织a纬（表纬1）时，表经1与表经2形成梭口（综片1提起，综片2下沉）与表纬1交织制织上层织物，里经全部下沉。可知，a纬不与经纱B或D交织。

To produce a double-layer fabric, the following conditions are required: ① When being picked, a face filling yarn won't interweave with the back ends which are depressed. ② When being picked, a back filling yarn won't interweave with the face ends which are lifted.

To produce the fabric of the double-layer plain weave in Fig. 6-3-2, the warp threads A and C (or face ends 1 and 2) are drawn on the healds No. 1 and No. 2 respectively while the warp threads B and D (or back ends Ⅰ and Ⅱ) are drawn on the healds No. 3 and No. 4 respectively. The weft threads a and c correspond to face weft threads 1 and 2 respectively while b and d correspond to back weft threads Ⅰ and Ⅱ respectively.

When picking weft a (face weft 1) or c (face weft 2), either the face warp No. 1 or No. 2 is raised while the other warps are lowered. Therefore, wefts a and c do not interweave with warp B or D.

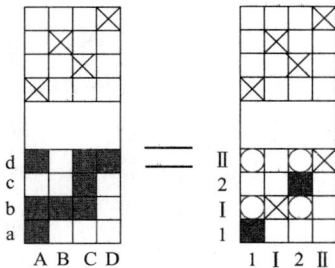

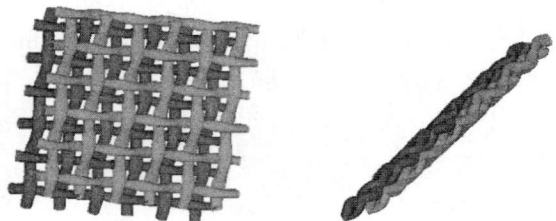

(a) 双层组织上机图/Looming for a double-layer weave　　(b) 双层织物纱线交织立体图/3D interlacing diagram of a double-layer weave

图6-3-2　双层组织
Double-layer weaves

织 b 纬（里纬 I）时，里经 I 与里经 II 形成梭口（综片 3 提起，综片 4 下沉）与里纬 I 交织制织下层织物，表经全部提起。可知，b 纬不与经纱 A 或 C 交织。

织 c 纬（表纬 2）时，表经 1 与表经 2 形成梭口（综片 2 提起，综片 1 下沉）与表纬 2 交织制织上层织物，里经全部下沉。可知，d 纬不与经纱 B 或 D 交织。

织 d 纬（里纬 II）时，里经 I 与里经 II 形成梭口（综片 4 提起，综片 3 下沉）与里纬 II 交织制织下层织物，表经全部提起。可知，d 纬不与经纱 A 或 C 交织。

综框的提综情况见表 6-3-1。

从表中可以看出：①投表纬织上层时，里经必须全部沉在梭口下部，不与表纬交织；②投里纬织下层时，表经必须全部提升，不与里纬交织。

When picking weft b(back wef 1) or d(back wef 2), either the back warp No. I or No. II is lowered while the other warps are raised. Therefore, wefts b and d do not interweave with warp A or C.

The lifting and lowering of the healds are listed in Table 6-3-1.

表 6-3-1　综丝提升顺序 / Lifting and lowering of the healds

投纬/ Weft	第1页综：表经1/ Heald 1: Face end No. 1	第3页综：里经I/ Heald 3: Back end No. I	第2页综：表经2/ Heald 2: Face end No. 2	第4页综：里经II/ Heald 4: Back end No. II
投表纬1/ Face weft No. 1	提升/ Lifted	沉下/ Lowered	沉下/ Lowered	沉下/ Lowered
投里纬I/ Back weft No. I	提升/ Lifted	提升/ Lifted	提升/ Lifted	沉下/ Lowered
投表纬2/ Face weft No. 2	沉下/ Lowered	沉下/ Lowered	提升/ Lifted	沉下/ Lowered
投里纬II/ Back weft No. II	提升/Lifted	沉下/ Lowered	提升/ Lifted	提升/ Lifted

From the above table, the following arguments hold: ① while the face wefts are inserted in weaving the upper layer, the back warps are all depressed in the lower part of the shed without interweaving with the face wefts; ② while the back wefts are inserted in weaving the lower layer, all the face warps are lifted and without interweaving with the back wefts.

2. 双层组织的判断 /Judgement of the Double-Layer Weaves

根据浮长线原理，双层组织在经、纬两个方向均存在相邻纱线的重叠关系。以图 6-3-2 为例，经向的连续经浮长线长度图如图 6-3-3（a）所示。显然，奇数根的经纱均处于上层。同理，该组织在纬纱方向的连续纬浮长线长度图如图 6-3-3（b）所示。显然，奇数根的纬纱均处于上层。

3. 组织图的设计步骤 /Designing of Double-Layer Weaves

（1）表、里层组织的确定。双层织物的上、下两层是各自独立的，两层组织的关系不如二重组织

According to the principles of float, the overlapping of the threads must occur in both warp and weft directions. Take the weave in Fig. 6-3-2 as example. The diagrams of the lengths of the warp floats and the weft floats are shown in Fig. 6-3-3(a) and (b) respectively. Obviously, the warps and the wefts at the odd order interweave at upper layer.

(1) Determine the Face and Back Weaves. The two layers of the fabric are independent each other, therefore, the selection of the face weave and the back weave is

3	-3	3	1
-1	-3	3	-3
3	1	3	-3
3	-3	-1	-3

-3	1	-3	-3
3	3	-1	3
-3	-3	-3	1
-1	3	3	3

(a) 经向/Warp direction　　　　(b) 纬向/Weft direction

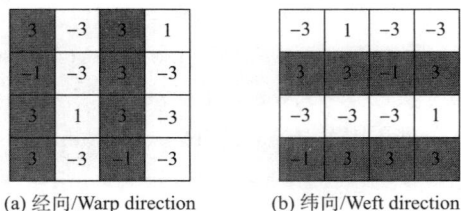

图6-3-3　浮长分布图

The float length diagram of the weave in Fig. 6–3–2

那样严格。表、里两层组织可以是经面组织也可以是纬面组织，但要求表、里组织交错次数应接近。常用的表、里组织有平纹、斜纹、重平、方平、四枚破斜纹等组织。

（2）确定表经与里经的排列比。如果表、里经线密度相同，各层紧度相似，表经与里经的排列比取1:1或2:2。如果表经细、里经粗（此时紧度仍然相同），或表层紧密、里层稀疏（此时线密度相同），表经与里经的排列比可采用2:1。

（3）确定表纬与里纬的投纬比。表纬与里纬的投纬比与纬纱的线密度及织物紧度有关，其规律与表里经排列比类似。此外，还与织机的多梭箱装置有关。若使用有梭织机、单侧多梭箱，投纬比必须都是偶数；若投纬比中有奇数，必须采用双侧多梭箱。若使用无梭织机，则表里纬投纬比无此限制。

（4）确定组织循环。双层组织的表里经（纬）比 $m:n$ 时，则循环根数计算式参见经二重组织的 R_j 和纬二重组织的 R_w。

4. 双层组织图的绘制 /Drawing of a Double–Layer Weaves

（1）计算组织循环根数，用不同符号在方格纸上标出表里经和表

optional. However, the weaves are expected to have similar intersections in a repeat. The ordinary weaves for choosing are plain, twills, ribs, hopsacks, satinette, etc.

(2) Determine the Arrangement of Face and Back ends. Usually, the arrangement of face and back ends is 1:1 or 2:2. If the fine threads/dense structure and the coarse threads/open structure are used alternately, 2:1 is preferred.

(3) Determine the arrangement of face and back picks. The selection of the ratio is mainly depended on the thickness of wefts and the covering factor of the fabric in weft direction, which is similar to the warp threads. In addition, the ratio is confined to the weft thread selecting mechanism. For a shuttle loom with multiple shuttle boxing mechanism at single side, the numbers in ratio must be all even numbers; if there are any odd numbers, two side multiple shuttle boxing mechanism is used. For a shuttleless loom, there is no limitation.

(4) Calculating R_j and R_w. Supposing the arrangement ratio of the face warp (weft) and back warp (weft) is $m:n$, then, R_j and R_w of the new double–layer weave are calculated like R_j of warp–backed weaves and R_w of weft–backed weaves.

Drawing of a Double–Layer Weaves

(1) Calculate R_j and R_w, and number the order of the face and back threads respectively in different symbols on

里纬的排列序号。

（2）用不同的符号在表经和表纬相交处的方格内填表组织（实心方块■），在里经和里纬相交处的方格内填绘里组织（交叉线 ×）。

（3）在投里纬织下层时，表经必须全部提升，不与里纬交织。在绘组织图时要注意表经与里纬相交处的方格内，必须全部填入特殊的经组织点符号（空心圈○，表示分层点或提综符号）。

5. 双层组织的上机要点 /Looming for Double-Layer Fabrics

双层组织的上机穿经、穿筘、织轴等织造要求与经二重组织类似。采用分区穿法，一般表经穿在前区，里经穿在后区。同一组的表里经穿入同一筘齿中，以便表里经上下重叠。图 6-3-2 所示为双层组织的上机图。

6. 双层组织织物的分类与连接方法 /Connections of Double-Layer Fabrics

双层组织必须将其两层按照某种形式连接起来，才具有价值。上下层可以仅仅是稍加连接，很容易判断出是两个独立的织物，也可以是紧密接结为类似复杂的单层结构。双层组织连接方式有：织物边部连接（图 6-3-4）、沿花纹边缘连接（图 6-3-5）、自身接结（图 6-3-6）和外加纱线连接（图 6-3-7）。根据连接方式的不同，双层组织可分为管状织物组织、双幅织物组织、袋织物组织、表里换层组织和接结双层组织等。

design paper.

(2) Fill the marks (solid square ■) for intersecting points of warp over weft at the squares where the face warps and the face wefts cross, and fill another kind of marks (cross lines ×) for intersecting points of warp over weft at the squares where the back warps and the back wefts cross.

(3) Fill the third kind of marks (loop ○) for lifters at the squares where the face warps and the back wefts cross since all the face warp threads are raised and won't interweave with the back weft thread picked.

The looming for the double-layer fabric is similar to that for the warp-backed weave in drafting, denting and beaming. Divided drafting is used and the face ends are drawing on the front healds while the back ends are drawn on the back healds. The face ends and the back ends are placed in the same split of reed to help overlapping. The looming plans for a double-layer weave is shown in Fig. 6-3-2.

Two layers of the double-layer weave have to be connect to embody its value. The two layers may be only loosely connected together in which case each may be readily identified as a different entity or they may be so intricately stitched or tied together that they appear to form a complex single structure. The two layers can be connected by edge-stitching (Fig. 6-3-4), interchanging stitching (Fig. 6-3-5), self-stitching (Fig. 6-3-6) and center-stitching (Fig. 6-3-7). Several fabrics are produced by double-layer weaves including tubular fabrics, double-width fabrics, bag-shape fabrics, interchanging double-layer fabrics, stitched double-layer fabrics. etc.

(a) 两边连接：管状组织/Double-edge stitching: Tubular weave

(b) 单边连接：双幅组织/Single-edge stitching: Double-width weave

(c) 两边连接：三幅组织/Double-edge stitching: Treble-width weave

图6-3-4　边部连接
Edge-stitching

图6-3-5　表里换层
Interchanging stitching

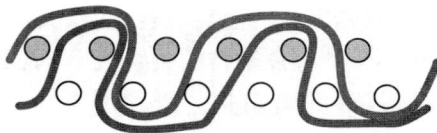

图6-3-6　自身接结
Self-stitching

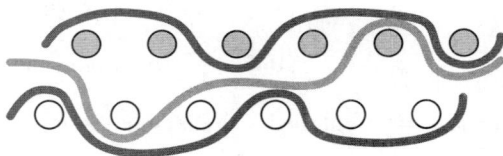

图6-3-7　外纱接结
Center-stitching

二、管状织物 /Tubular Weaves

1. 管状织物定义 /Description

连接双层组织的两边缘构成管状织物，其幅宽一般在 10cm 以上，或周长一般在 20cm 以上（图 6-3-8、图 6-3-9）。

By connecting the two edges of a double-layer weave with picks of weft, a tubular fabric is formed. The fabric is of cylindrical form having a width of 10cm or more or a circumference of 20cm or more(Fig. 6-3-8, Fig. 6-3-9).

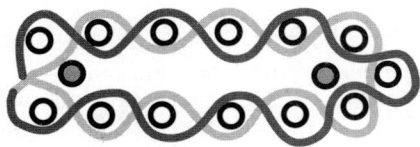

图6-3-8　管状织物结构图
Structural diagram of a tubular weave

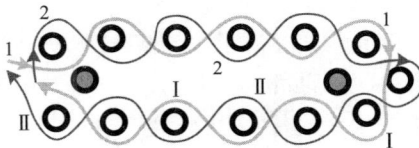

图6-3-9　管状织物交织顺序图解
Order of picking of a tubular weave

其特点如下：

（1）管状织物纬纱既作表纬

The features of the weave are described as following:

(1) The series of weft threads works as both face wefts

239

又作里纬，往复循环于表里两层之间。

（2）管状织物的表里两层只在两侧边缘连接，中间分离。

（3）表里两层的经纱呈平行排列，纬纱呈螺旋形状态。

2. 管状织物组织设计要点 /Designing of Tubular Weaves

（1）表、里组织的选择。表组织和反组织应相同（注意：表组织与反组织不是"底片翻转的关系"，表组织与其背面组织是"底片翻转"的关系）。如果将管子沿某根经纱剪开，然后展平，可知表组织与里组织必须是相同的组织。

如将 $\dfrac{1}{3}$↖ 织物折叠后，形成上下不等大小的双层织物，类似管状织物被剪开。表组织为左斜，里组织与表组织是典型的底片翻转关系，故里组织为 $\dfrac{3}{1}$↗；而里组织与反组织也是底片翻转关系。故两次翻转后，表反组织相同。实际双层组织中，并不展示反组织。造成人眼认为表反组织不同的原因是，看表反组织的视线方向是相反的：表组织是正面从上而下看，而反组织是从背面看，即从下而上看织物。

故管状织物的表、里组织必须采用同一组织的正反面。为保证折幅处组织连续，应保证纬向飞数 S_w 为常数，常用的组织有平纹、斜纹和缎纹原组织和纬重平组织。若对折幅处连续性要求不严格，也可以采用方平、破斜纹组织。

（2）表、里纬纱排列比：必须为 1:1。

and back wefts.

(2) The fabric is actually a hollow fabric connecting at the edges.

(3) The ends in two layers are parallel one another while the wefts are in spiral shape.

(1) Selection of the Face Weave and Back Weave. The face weave and the back weave must adopt both sides of the same weave (Note: The face weave is not reversed to the reverse weave, however, the face weave is reversed to the weave at the other side of the fabric). If a tubular fabric is cut along the warp direction and expanded to flatten status, you may see that the upper layer and the front of the bottom layer (actually the back weave) are just reversing.

If a $\dfrac{1}{3}$↖ twill fabric is folded , a two-layer fabric is formed with different size in two layers, which could simulate the cut tubular fabric. The face weave is $\dfrac{1}{3}$↖ while the back weave, reversing to the face, is a $\dfrac{3}{1}$↗ twill. Since the revere side of the back weave is reversing to the back weave again, therefore, the reverse weave is just the same as the face weave.

In order to ensure the continuity of folding edges, the weave with a constant step number S_w should be used, such as the basic weaves (plain weave, twill weave, satin weave) and weft rib weave. If the continuous structure at the fold is not strictly required, hopsack and broken twill weaves can be used as the face or back weave.

(2) The arrangement of the face weft and back weft must keep the ratio of 1:1.

(3) Determination the Size of Weave Repeat. The calculation method of the weave repeat for tubular weaves is the same as that for double weaves.

(4) Determine the Total Number of Ends for Tubular Fabrics. In order to ensure the continuity at the connecting

（3）组织循环根数的确定方法与普通的双层组织一样。

（4）总经根数 M_j 的计算与修正。为了确保边部连续，织物的总经根数不能随意增减。计算过程分为以下两步。

①估算总经根数。根据管状织物的用途和要求，确定管状织物的直径 D（cm），计算管子幅宽 W（cm）。若管状织物单层经密是 P_j（根 /10cm），则：

②双层组织的总经根数修正。首先根据下式计算管状织物中的组织循环个数 Z。

双层组织的总经根数 M_j 必须严格按照下式来修正：

第一纬从左向右投里纬，或第一纬从右向左投表纬，取"+"；第一纬从右向左投里纬，或第一纬从左向右投表纬，取"–"。里组织第一纬的起点应根据管状截面图确定。

3. 管状织物组织图的绘制 /Drawing of Tubular Weaves

管状织物组织图的绘制基本方法与双层组织一样，不同之处在于需要根据总经根数和投纬方向绘制管状织物截面图，调整里组织第一纬的交织顺序，确保在连接处连续。方法如下：①截面图中，上层表经的数量与下层里经的数量任意，只

edges, the total number of ends cannot be increased or decreased at will. There are two steps to obtain an accurate number.

①Estimate the total number of the ends. Determine the diameter D (cm) of the tubular fabric according to the purpose and requirements, then calculate the width W (cm) of the fabric by the Formula (6–3–1). If the warp density of a single layer of tubular fabric is P_j (ends /10cm), then the number of the ends is estimated by Formular (6–3–2).

$$W=\pi \times D/2 \qquad (6\text{–}3\text{–}1)$$

$$M_j=2W\times \frac{P_j}{10}=\pi \times D \times P_j/10 \qquad (6\text{–}3\text{–}2)$$

② Correction of total number of the ends. The number Z of the weaves in a tubular fabric is calculated by the following formular.

$$Z=\frac{M_j}{R_j} \qquad (6\text{–}3\text{–}3)$$

The accurate number of the ends in the tubular fabric is corrected by the following formula.

$$M_j=R_jZ\pm S_w \qquad (6\text{–}3\text{–}4)$$

Operator "+" is used when insert the back filling from left to right or the face filling from right to left; "–" is used when insert the back filling from right to left or the face filling from left to right. The commencing weaving point of the first back weft is determined by the sectional diagram based on the first face weft.

Drawing of Tubular Weaves

The principle of drawing a double–layer weave is adopted in drawing the tubular weave. However, there is a slight difference for the latter. The sectional diagram of the tubular fabric is drawn firstly according to the total number of ends, then the interlacing order of the first weft in the back weave is adjusted to ensure continuity at the edges. The method to adjust the commencing point by sectional diagram is described as

要确保两种之和等于总经根数即可。为了绘制方便，采用尽量少的组织循环；②若第一表纬从左向右投纬，则在确定第一里纬时，根据最左一个循环的纬纱的规律，确定第一里纬的沉浮次序；③若第一表纬从右向左投纬，则在确定第一里纬时，根据最右一个循环的纬纱的规律，确定第一里纬的沉浮次序。

　　例：已知管状织物的表、里组织均为5枚2飞经面缎纹。若基础组织个数为5，从左向右投第一纬。该织物组织图的绘制过程如下：

　　（1）首先绘制表组织，里组织按照表组织底面翻转得到，分别如图6-3-10（a）（b）所示。

　　（2）计算总经根数。从表纬投第一纬，从左向右，纬向飞数为3，故总经根数 R_j=55-3=22。

　　（3）画管状织物截面图。因总经根数是22，故上、下层各11根经纱。若第一根纬纱从表纬1开始，自左向右投纬，交织规律是一下四上；故第二纬（即里纬第一纬）从右向左，交织规律为一上四下。根据里纬Ⅰ与最左方的5根下层经纱的交织截面顺序［图6-3-10（c）］，找到里组织中与该交织顺序相同的那个纬纱，作为第一里纬，调整里组织顺序，如图6-3-10（d）所示。需要说明的是，若上层10根纱线，或者上层12根纱线，均不影响第一里纬的顺序。

　　（4）按照绘制双层组织的方法，绘制管状织物的组织图，如图6-3-11所示。

　　管状织物表组织为 $\dfrac{2}{2}\nearrow$ 斜

following: ① ensure that the total number of the face warps and the back warps in the sectional diagram is the constant that is just calculated according to Formula (6–3–4); in the sectional diagram, the number of the threads on the upper layer and the bottom layer are set optionally as soon as the sum is the number of the total ends; ② if the first face filling is inserted from the left to right, the interlacing order of the first back filling shall be determined according to the left–most back fillings; ③ if the first face weft is picked from the right to left, the interlacing order of the first back filling shall be determined according to the right–most back fillings.

　　Example: In a tubular weave, both the face weave and the back weave are $\dfrac{5}{2}$ satin, and the first face filling thread is picked from the left. To show 5 weave repeat units in the looming plans, the procedures are illustrated as following:

　　(1) Draw the face weave, and the back weave is obtained by reversing the face weave as shown in Fig. 6–3–10(a)(b).

　　(2) Calculate the number of total ends in the sectional diagram. Since the first face filling is inserted from the left to right, thus, R_j=55-3=22.

　　(3) Draw the sectional diagram of the tubular weave. Place 11 ends in the upper layer and another 11 ends in the lower layer. Pick the first face weft thread from the left by the interlacing of 1 down and 4 up. The second weft (or the first back weft Ⅰ) is inserted from the right to left by the interlacing order of 1 up and 4 down. The 2nd filling (the 1st back filling) in the sectional diagram as shown in Fig. 6–3–10(c) determines the order of the interlacement of the first back weft (from the left, the interlacing of the left–most 5 ends). Therefore, the back weave is adjusted as shown in Fig. 6–3–10(d). It should be noted that it won't change the ultimate result if the number of upper ends is set as 10 or 12 as long as the number of the total ends in the sectional diagram is 22.

　　(4) The weave plan can be drawn by the method for an average double–layer weave(Fig. 6–3–11).

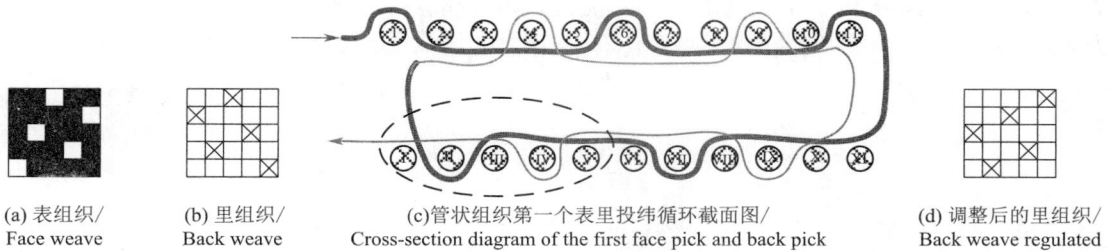

(a) 表组织/
Face weave

(b) 里组织/
Back weave

(c)管状组织第一个表里投纬循环截面图/
Cross-section diagram of the first face pick and back pick

(d) 调整后的里组织/
Back weave regulated

图6-3-10 管状织物基本组织构成
Constitution of the tubular weave

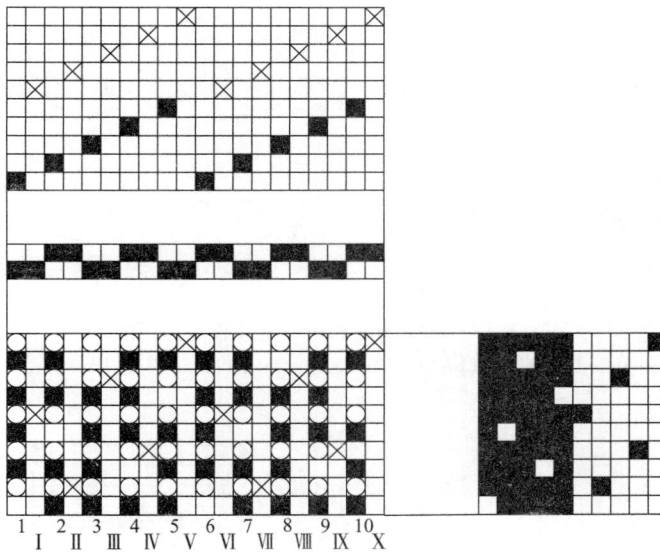

1 2 3 4 5 6 7 8 9 10
I II III IV V VI VII VIII IX X

图6-3-11 管状织物组织图
The looming plans for a tubular weave

纹，表、里组织如图 6-3-12（a）（b）所示。若基础组织个数为 4，从右向左投第一纬，故总经根数为 4×4+1=17。当上层经纱根数为 8 时，管状织物截面图如图 6-3-13（a）所示。根据该图调整，右侧第一里纬的交织规律均如图 6-3-12（c）所示。若上层经纱根数为 9 时，则第一表里纬投纬截面如图 6-3-13（b）。可见，上、下层经纱总和一定时，上层根数变化不影响最终结果。该组织的投纬完全循环的截

If the face weave of the tubular weave is a $\frac{2}{2}\nearrow$ twill, and the face weave and the back weave are shown in Fig. 6-3-12(a) and (b) respectively. The first face filling is inserted from the right to left. If 4 weave repeats are shown in a sectional diagram, the total number of ends is calculated as $R=4 \times 4+1=17$. When the number of the ends in upper layer is 8, the sectional diagram of the first back weft is drawn as shown in Fig. 6-3-13(a), which is the base to regulate the back weave as shown in Fig. 6-3-12(c). If the number of ends in upper layer is 9, then the sectional diagram of the first back weft is drawn as shown in Fig. 6-3-13(b). It can be concluded that the

面图如图 6-3-14 所示，上机图如
图 6-3-15 所示。

changing of the number of ends in upper layer won't change the outcome. The complete sectional diagram for picking is shown in Fig. 6-3-14 and the looming plans are shown in Fig. 6-3-15.

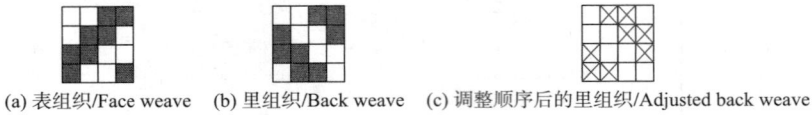

(a) 表组织/Face weave　(b) 里组织/Back weave　(c) 调整顺序后的里组织/Adjusted back weave

图6-3-12　管状组织的表里组织
Face weave and back weave

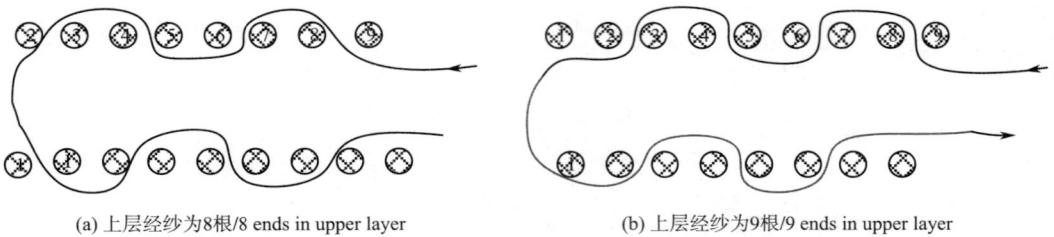

(a) 上层经纱为8根/8 ends in upper layer　　(b) 上层经纱为9根/9 ends in upper layer

图6-3-13　第一表纬和第一里纬投纬示意图
Sectional diagram of picking for the first face weft and first back weft

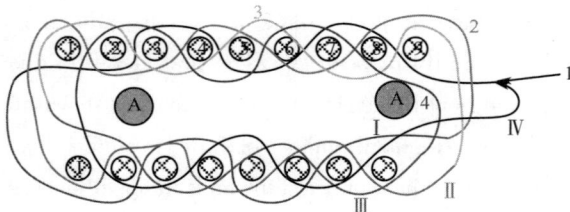

图6-3-14　一个完整的投纬循环截面图
A complete sectional diagram for picking

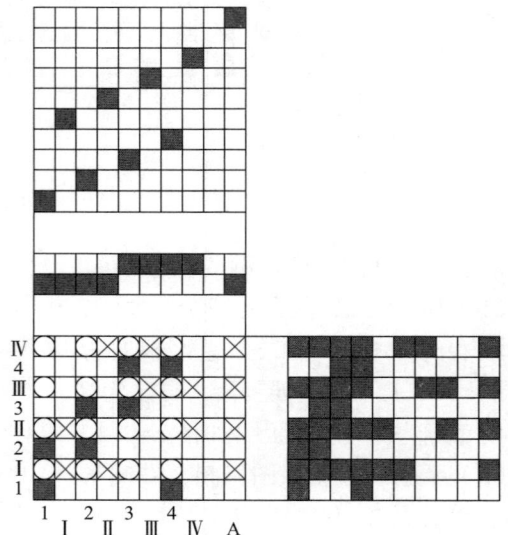

图6-3-15　管状织物上机图
The looming plans for a tubular weave

244

4. 管状织物上机要点 /Looming for Tubular Fabrics

（1）穿综。管状织物一般采用分区穿法（将表经穿入前综），偶尔采用顺穿法。

（2）穿筘。每组表里经穿入同一筘齿中。在织造时，由于纬纱的收缩，会引起织物两侧经密偏大。为了防止这种现象，并保证折幅处织物平整，对不同厚度织物，穿筘时采用不同方法。

①轻薄型管状织物。逐渐减少边部筘齿穿入数，中间为4，边部为3、2，如图6-3-16所示。

(1) Drafting. Divided drafting is usually used for the tubular fabrics, the face ends are drawn on the front healds, and the back ends on the rear healds. Straight draft is occasionally employed for the fabrics.

(2) Denting. The face ends and the back ends in same group shall be drawn in the same dent. The crimp of the wefts may cause a high warp-sett at the selvedge of the fabrics, which should be avoided. To keep the fabric flat in the folded area, some measures are taken in denting.

① For light tubular fabrics. The ends per dent at the fold edge of the tubular are reduced gradually, e.g., 4 ends/middle dent, 3 or 2 ends/dent at the fold edge. The schematic denting diagram is shown in Fig. 6-3-16.

2	2	2	3	3	3	4	4	4	4	4			4	4	4	4	4	4	3	3	3	2	2

边部/Side dent　　　　中部/Middle dent　　　　边部/Side dent

图6-3-16　穿筘示意图
Denting diagram

②中厚型管状织物。在两侧边缘处的内侧，各穿入一根张力较大的特线A，如图6-3-14所示，其作用是防止经纱向内侧收缩，保证经密均匀。特线的结构性质类似于芯线，夹在表里层中间，不与织物交织。特线单独穿入一片综（后综），单独穿入一个筘齿中。投入表纬时，特线下沉；投入里纬时，特线提升。织物下机时，可将特线抽出。

③织物经密很大、纱线粗。此时，经纱张力大，可使用内撑幅器来替代特线。

（3）织机。管状织物用一组纬纱循环往复交织于表里两层，纬纱不可剪断，故只能使用有梭织机或者织带机织制，不需要多梭箱装置，但必须注意投纬方向。

② For medium tubular fabrics. At the inner edge of the folded edge, a special tighten coarse thread A is introduced as shown in Fig. 6-3-14. The coarse thread itself is drawn on a single heald (rear heald) and occupies a single dent. The coarse yarn is lowered when a face filling is inserted but lifted if a back filling is inserted, i.e., the coarse thread A doesn't interweave the fillings. The coarse thread will be removed when the fabric is off loom.

③ For high warp-sett and coarse tubular fabric. The ends are high tensioned when the tubular fabric is of high warp sett and produced of coarse threads. An inner temple is used to expand the tubular fabric on loom.

(3) Loom. The weft yarns that serve as the face and back yarn interweave with the face warps and back warps without cutting, therefore, the fabric can only be woven by shuttle looms or tape looms. The single-box shuttle loom is sufficient for the fabrics, but a correct direction of picking should be guaranteed.

（4）经轴。管状织物中，表里经采用相同原料、同一种纬纱、表里组织相同，纱线屈曲相同，故表里经卷在同一织轴上。

(4) Beam. In tubular fabrics, the materials and the weaves for the face and the back are the same, therefore, the rates of contraction are similar to the face warps and the back warps. Both the face ends and the back ends are wound on the same beam.

三、双幅织物 /Double-Width Weaves

1. 双幅织物定义 /Description

双幅织物的设计原理类似于管状织物，但仅在一侧连接，如图6-3-17所示，该结构使得在窄幅有梭织机上织造宽幅织物成为可能。为了使双幅织物在连接一侧的经纱不至于过分收缩，导致经纱密度过大，可以添加一根较粗的纱线作为特线A。为了方便织造，布边侧加一根缝线B，将上、下布边连接，织物下机后将缝线拆除。若轮流连接两侧边缘，可织成三幅或三幅以上的宽幅织物，原理如图6-3-4（c）所示。

The double-width structure as shown in Fig. 6-3-17 enables the designer to produce woven fabrics of wider width on a narrow shuttle loom. The principle of designing the double-width weaves is similar to the tubular fabrics. The difference is in that only one side is connected and the other side is separated. A tighten coarse thread A is usually applied to prevent the fabric contracting at the folding edge. At the other side, a stitching thread B which connects the two layers is used to keep a flat and even fabric on loom. Thread B will be removed when the fabric is off loom. Similarly, the treble-width fabric can also be woven by alternately connecting the side of the fabric as shown in Fig. 6-3-4(c).

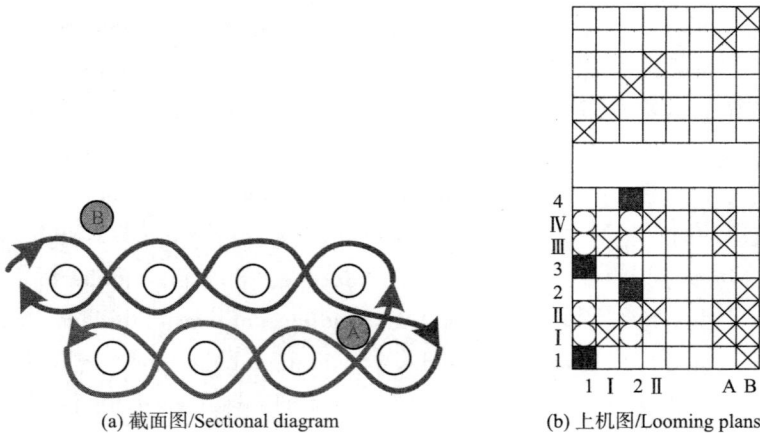

(a) 截面图/Sectional diagram

(b) 上机图/Looming plans

图6-3-17 双幅织物组织
The double-width weave

246

2. 双幅织物组织设计要点 /Designing of Double-Width Fabrics

设计双幅织物组织时，需考虑以下因素：

（1）基础组织的选择。与管状织物组织的设计方法相同，采用同一种简单组织的正、反面作为表组织和里组织。

（2）表、里经排列比。可以为1:1或者2:2。

（3）表、里纬投纬比。根据双幅织物组织的形成过程，表、里纬的投纬比必须是2:2；投梭顺序为：表、里、里、表。

（4）计算 R_j 和 R_w。组织循环大小的计算方法与双层组织相同。

The following factors are required to consider in designing a double-width fabric.

(1) Choice of base weaves. The face weave is just reversal to the back weave. Usually, a simple weave is a good choice for them. Weft rib weaves are also used.

(2) The arrangement of face end and back end is usually chosen as 1:1 or 2:2.

(3) The arrangement of face pick and back pick must be 2:2. The picking order is usually set as 1 face weft, 2 back wefts and 1 face weft.

(4) Calculation of R_j and R_w. The size of the new double-width weave is calculated like R_j of the warp-backed weave and R_w of the weft-backed weave.

3. 双幅织物组织图的绘制 /Drawing of Double-Width Weaves

类似管状织物，为保证折幅处纹路连续，虽然里组织第一纬的起点也应根据双幅织物组织的截面图确定，但截面图上经纱根数是组织循环的倍数，故里组织与表组织严格底片翻转，无需调整位置。

若双幅织物采用的表、里组织分别如图6-3-18所示，第一表纬的投梭方向从左到右，使用特线和缝线。表组织采用 $\frac{1}{2}\nearrow$ 斜纹组织，故里组织为 $\frac{2}{1}\nwarrow$ 斜纹组织。第一表里纬和整个纬纱循环的截面图分别如图6-3-19和图6-3-20所示，经纱使用了3个循环，其上机图如图6-3-21所示。

To ensure the continuity of the fabric at the fold edge, drawing of a double-width weave is similar to that of a tubular weave in that the order of interlacing of the first back weft is determined by the sectional diagram of picking the wefts. However, the number of the total ends will not affect the looming plans. It is unnecessary to regulate the back weave which is strictly reversal to the face weave.

Fig. 6-3-18 shows the face weave of $\frac{1}{2}\nearrow$ twill and the back weave of a double-width weave. Supposing that a tighten coarse thread and a stitching thread are applied in weaving and the first face weft is picked from the left to right, the sectional diagram of the fabric for the first back weft and all the wefts of 3 weave repeats are shown in Fig. 6-3-19 and Fig. 6-3-20 respectively, and the looming plans are shown in Fig. 6-3-21.

4. 双幅织物上机要点 /Looming for Double-Width Weaves

（1）穿综方法。采用顺穿法或分区穿法。如用分区穿法，表经穿前区，里经穿后区。地部（不含

(1) Drafting. Divided drafting or straight drafting is applied for the double-width fabric. If divided drafting is used, the face ends are drawn on the front heads, and the

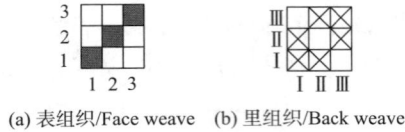

(a) 表组织/Face weave   (b) 里组织/Back weave

图6-3-18   双幅织物的表里组织
The face weave and back weave of a double-width weave

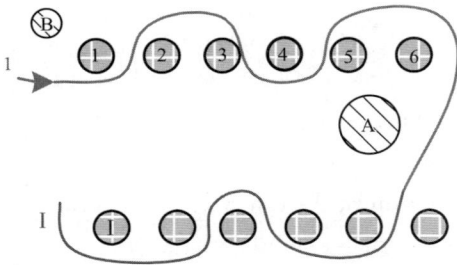

图6-3-19   第一表里纬与经纱交织的截面图
Sectional diagram of the weft 1 and I

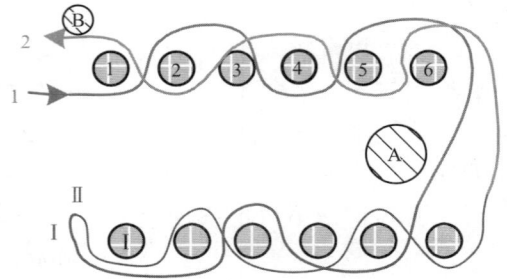

图6-3-20   整个纬纱循环交织的截面图
Sectional diagram of all the wefts in a weave repeat

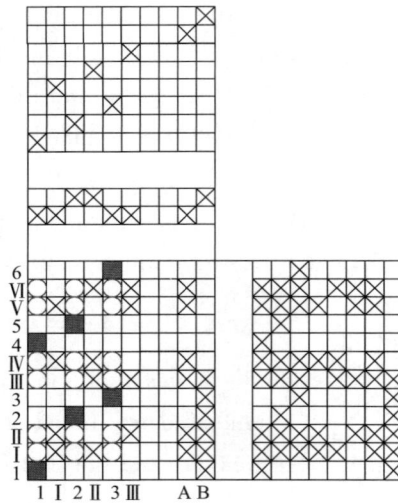

图6-3-21   双幅织物上机图
The looming plans of the double-width weave

边）所需综数为基础组织的组织循环经纱数的 2 倍。特线与缝线均需单独用一片综。

（2）穿筘。同一组表里经穿入同一筘齿中。在折幅侧应逐渐减少每筘齿穿入数或加一根特线。特线

back ends rear healds. The number of the healds for the body (selvedge excluded) is 2 times of the warp repeat of the base weave. The tighten coarse thread and the stitching thread should be drawn upon the separate heald.

(2) Denting. The face ends and the back ends which are overlapped in a group shall be placed in the same dent.

单独穿一筘齿，织物下机后将特线抽出。在布边侧每筘齿穿入数应多（增加布边牢度）。缝线也要单独穿一筘齿。

（3）织轴与梭箱。因表、里经纱的织缩率一致，故可用单织轴织机制织双幅织物。表、里纬纱的排列比虽为 2∶2，但因表、里纬纱原料相同，所以可用单梭箱织机制织。因双幅织物有一侧要将表、里两层连接在一起，故必须使用有梭织机才可制织。

（4）其他。在管状织物折边内侧，经常放置一根粗纱线——特线（图 6-3-21 中的 A 纱）。在织造表纬时，特线全部沉下；在织里纬时，特线全部提升。因此，特线并不与织物交织，下机后抽走。在织物的另一侧，缝线（图 6-3-21 中 B 纱）在织物上将两层织物缝在一起，使其平整。缝线的纬纱循环应该是双幅组织纬纱循环根数的约数或倍数，半个缝线纬纱循环提升，另外半个下沉。缝线在下机后拆除。

The ends per dent at the fold edge of the fabric are reduced gradually. More ends of the selvedge are drawn in a dent to increase its strength. The tighten coarse thread and the stitching thread should be drawn into the separate dent split.

(3) Beam and Shuttle-Box. Since the rates of the contraction of the face ends and the back ends are similar, the double-piece fabrics can be woven on a single-beam loom. The arrangement of the face and back filling is 2∶2, therefore, the single-box loom can be applied to produce the fabrics. To keep the continuity of the yarn in one selvedge, only the shuttle loom applies.

(4) Others. At the inner selvedge of the fold edge, a tighten special coarse thread A (Fig. 6-3-21) is introduced. The coarse yarn is lowered as inserting a face filling but lifted as inserting a back filling, i.e., the coarse yarn doesn't interweave with the fillings. The coarse thread is draw off after off loom. At another side of the selvedge, a stitching thread B (Fig. 6-3-21) is used to connect the two layers of fabric. The stitching thread B itself is drawn in a single shaft (rear) and occupies a single dent. The size of the stitching repeat is the factor of the multiple of the double-width weave. The stitching thread will be lifted in half of a whole repeat in the weft direction and lowered in the other half repeat. The stitching thread is also removed after the fabric is off loom.

Double Interchanging Weaves

四、表里换层织物 /Double Interchanging Weaves

1. 表里换层织物定义 /Description

如果双层织物的上、下两层纱线粗细、颜色、组织不同，在某些区域交换上、下层位置，形成表里换层织物。在层交换位置处，形成花纹图案轮廓，同时将双层织物连接成一个整体，两层中间形成空心袋状。有些织物中，经纬方向全部换层；而有些织物中，仅仅在单向换层。

A double layer cloth in which two layers are different in color of thread, number of threads in a unit space will show some pattern on the place where the two layers interchange. In a double interchanging weave, the two layers are so woven as to interchange with each other and the layers are joined only at pattern changes. The spaces between the two layers of cloth are called pockets. In some cases, the fabrics are completely interchanged whereas in others only the warp or weft threads interchange.

表里换层的目的可能是为了在精细织物上改善热绝缘性能，也可能是为了美观需要，在织物正反面因为纱线颜色改变而获得不同的精美图案。

The purpose of the construction may be entirely utilitarian, such as the improvement of the thermal insulation value of a fabric in which a fine, smart face appearance is necessary; or, it may be aesthetic in intention for which purpose the existence of two series of threads in each direction improves the capacity for producing intricate effects dependent upon either color, or structural changes.

2. 表里换层织物组织图的绘制 /Drawing of Double Interchanging Weaves

表里换层双层织物的设计步骤如下：

（1）设计纹样图。

（2）选择基础组织。常用简单组织为表、里组织，如平纹、$\dfrac{2}{2}$斜纹、$\dfrac{2}{2}$方平等组织。

（3）确定经纬纱排列比。在表里换层组织中，经纬纱需按纹样要求换层，在某一位置为表经、表纬，在另一位置就为里经、里纬。为了避免混淆，在表里换层组织中称为色经 A、色经 B 或色纬 A、色纬 B。常用的色经排列比有 1:1、2:1、2:2 等；色纬排列比有 1:1、2:1、2:2、2:4 等。

（4）确定组织经、纬循环（一个花纹循环）。表里换层组织的经、纬循环是表里层基础组织循环的整倍数。

（5）根据纹样图分割组织图区域。在不同区域，按照经、纬纱色排顺序填充表组织和里组织。

（6）在不同区域填充分层点符号。在各个区域的表经和里纬交错点处，填入符号"○"。

若设计的纹样图如图 6-3-22 所示，共占用 16 根经、纬纱。选用的表、里组织分别如图 6-3-23 所示，色经色纬的排列比均为 1:1。

The procedures to design interchanging double weaves can be described as following:

(1) Design the effect pattern.

(2) Choose the face and back weaves, usually simple weaves such as the plain weave, $\dfrac{2}{2}$ twill weave, $\dfrac{2}{2}$ basket weave.

(3) Determine the arrangement of threads. Threads change their layers according to the desired pattern. A thread is a face thread at one position, but the back thread at another position. Therefore, color A or color B is a preferred name other than face or back thread. Usually, the arrangement for ends is 1:1, 2:1 or 2:2, for filling is 1:1, 2:1, 2:2 or 2:4.

(4) Calculate the pattern repeat. The repeat of an interchanging double weave is the integral multiple of the base weaves.

(5) Divide the weave into several areas according to the pattern diagram. First, the areas are arranged in order according to the thread, and only the marks of the basic weave are filled in. Then, according to the design, the face warps and face wefts are determined.

(6) Fill in the symbols "○" of all the lifters (virtual interleaving points) at places where the face warps intersect with the back wefts.

Supposing the designed effect pattern is shown in Fig. 6-3-22, there are 16 warps and 16 wefts. The two base weaves are shown in Fig. 6-3-23. The arrangement for the colored warp and colored weft is listed as following: 1 white and 1 red for warps, designated as A and B; and 1 white and

因为表里换层，故将 16×16 的组织图等分成 4 份，色经排列均为 1 白 1 红，用 A、B 表示；色纬排列也是 1 白 1 红，用数字 1、2 表示，如图 6-3-24 所示。将表、里组织分别按照 1:1 的表里经、表里纬的顺序填充在整个组织图中。

由于在左下角要显示红色区域，故 B 经应该是表经，2 纬应为表纬，故 A 经应为里经，1 纬应为里纬。在表经 B 和里纬 1 对应的组织方格中填入虚拟组织点符号"○"。在左上角，因为要显示白色区域，故 A 经是表经，2 纬是里纬，因此要在 A 经和 2 纬对应的组织方格中填入符号"○"。右上角和左下角一样，右下角与左上角一样，按照相同的方式填入符号"○"。绘图过程如图 6-3-24 所示。

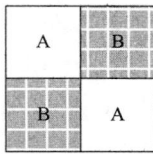

1 red for wefts, designated as 1 and 2 in Fig. 6-3-24. The whole new weave of 16×16 will be divided into 4 parts. The two base weaves are arranged alternately in the weave in different parts of the new weave.

According to the effect pattern, to display a red effect in the bottom-left corner, end B is required to be the face warp and the weft No. 2 is a face weft, therefore, the warp A must be a back end and the weft No. 1 a back weft. The squares where the face end B cross the back weft 1 must be the lifters and filled with the mark "○". At the upper-left corner, a white effect is displayed if the end A is face and the weft No. 2 is the back, thus, the squares where the warp A and the weft No. 2 cross should be filled with mark "○" to denote they are lifters. The upper-right corner and the lower-right corner are processed by the same way. The whole procedures are shown in Fig. 6-3-24.

图6-3-22 图案的纹样
The effect pattern

图6-3-23 表里换双层织物的两种基础组织
Two base weaves of the double interchanging weave

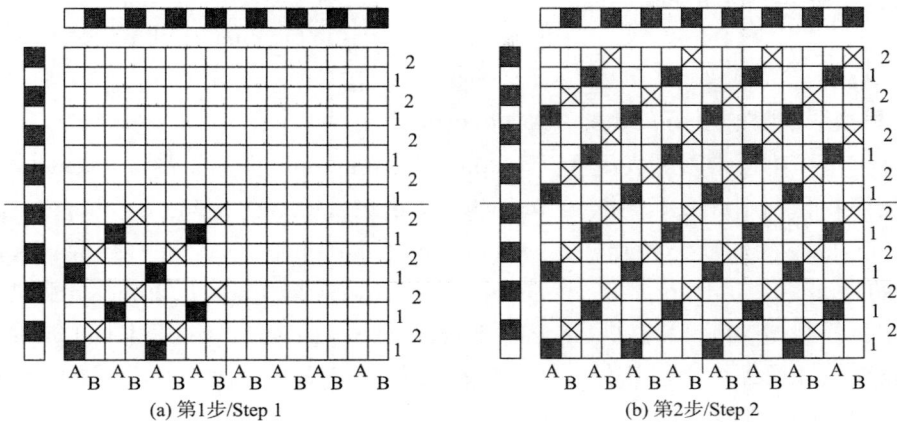

(a) 第1步/Step 1

(b) 第2步/Step 2

图6-3-24

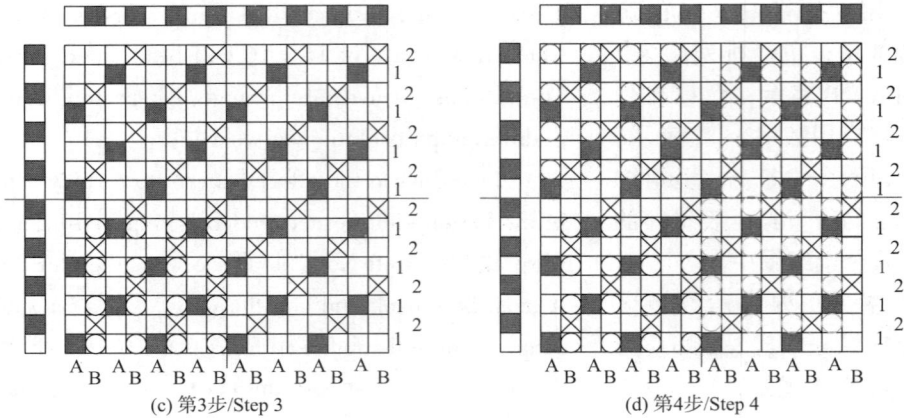

(c) 第3步/Step 3　　　　　　　　　(d) 第4步/Step 4

图6-3-24　接结经组织绘制步骤

Procedures of drawing a center-warp stitching weave

3. 表里换层组织的判断 /Judgement of Double Interchanging Weaves

表里换层组织的判断与双层组织判断的原理相同。在每个花纹内部纱线重叠分层，但在花纹或者换层处，没有明显的重叠关系。

The principle of judging a double interchanging weave is similar to a double-layer weave. Inside each component of the pattern, the threads are overlapped. However, the threads might not overlap at the contour of the pattern.

4. 表里换层织物上机要点 /Looming for Interchanging Double Weaves

表里换层织物采用分区间断穿法。图 6-3-24 所示组织图的穿综顺序为：（5、1、6、2）×2，（3、7、4、8）×2。穿筘时，同一组表里经穿入同一筘齿中。因为采用多色纬，有梭织机需要多梭箱装置。纹板数等于一个花纹中的纬纱循环数。

The looming for interchanging double fabrics is similar to double weaves. Combination of divided drafting and grouped drafting is used and the grouped warp threads are drawn in the same dent split. The ends for the weave in Fig. 6-3-24 are drawn at the order of (5, 1, 6, 2) × 2, (3, 7, 4, 8) × 2. If the fabric is woven on a shuttle loom, the multiple shuttle-box mechanism is required. The number of pattern lags equals to the number of the color repeat in weft direction.

5. 表里换层织物的应用 /Other Applications

精纺牙签条织物通常采用表里换层的双层织物，利用纱线不同捻向的排列，形成不同的反光特点，形成精细的条带效应。图 6-3-25 所示组织中，经纱 1、2 和纬纱 I、II 采用 S 捻股线，而经纱 I、II 和纬纱 1、2 采用 Z 捻股线，最终织物的反光效果如图 6-3-26 所示。

Interchanging double weaves are sometime used in fancy worsted fabrics. By placing the threads in different twist direction, the reflection to the light results in a stripe effect. In Fig. 6-3-25, ends 1, 2 and picks I, II are of folded thread of S twist while ends I, II and picks 1, 2 are of folded thread of Z twist. The effect of the fabric is shown in Fig. 6-3-26.

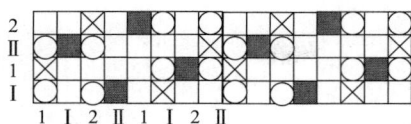

经纱/Ends 1, 2　　纬纱/ Picks Ⅰ, Ⅱ Z×S
经纱/Ends Ⅰ, Ⅱ　　纬纱/ Picks 1, 2 S×Z

图6-3-25　牙签条组织与纱线配置
Configuration of weaves and threads

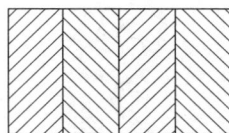

图6-3-26　织物的反光效果
Reflection of the fabric

五、接结双层织物 /Stitched Double Weaves

依靠各种连接方法，使分离的表里两层构成一个整体的组织，称为表里接结双层组织。接结后，两层织物牢固地接在一起，不容易被分开。对接结方法的一般要求是不破坏织物结构，不影响表、里层外观效果。但是由于织物强力和牢度的原因，对于织物结构有一定影响也是允许的。里层的作用主要是增加织物的厚度，对原料纱线要求不高；表层的纱线质量要好一些，以增强织物外观效果。

In a multi-layer woven fabric, each layer is attached to some other layers by means of stitching. Although there are still two distinct fabrics formed one above the other, they may be so closely united that the separation of the two layers is impossible. Usually, stitches should not interfere with the fabric structure, especially on the face and back layers. However, for some special reasons, such as fabric strength and firmness, stitches may be allowed to interfere with the fabric structure. There is no special requirement for the yarn materials of back layer if it is used only for increasing the weight or thickness of the fabric. However, the fine yarns are applied to the face layer to enhance the appearance of the fabric.

（一）接结方法分类 /Types of Stitching

根据连接上下两层的纱线来源，接结方法可分为以下五种，即①里经接结法或"下接上"接结法；②表经接结法或"上接下"接结法；③联合接结法；④接结经接结法；⑤接结纬接结法。这五种接结方法如图 6-3-27 所示。其中，前三种也称自身接结法，分类如图 6-3-28所示。

According to the source of the stitching threads, the types of stitching are classified as ① up-going warp stitching or down-going weft stitching, ② down-going warp stitching or up-going weft stitching to the warp, ③ warp-stitching up and down or double-stitching, ④ center-warp stitching, and ⑤ center-weft stitching. These types of stitching are described in Fig. 6-3-27 respectively. There are other ways of classifying the stitching which are described in Fig. 6-3-28.

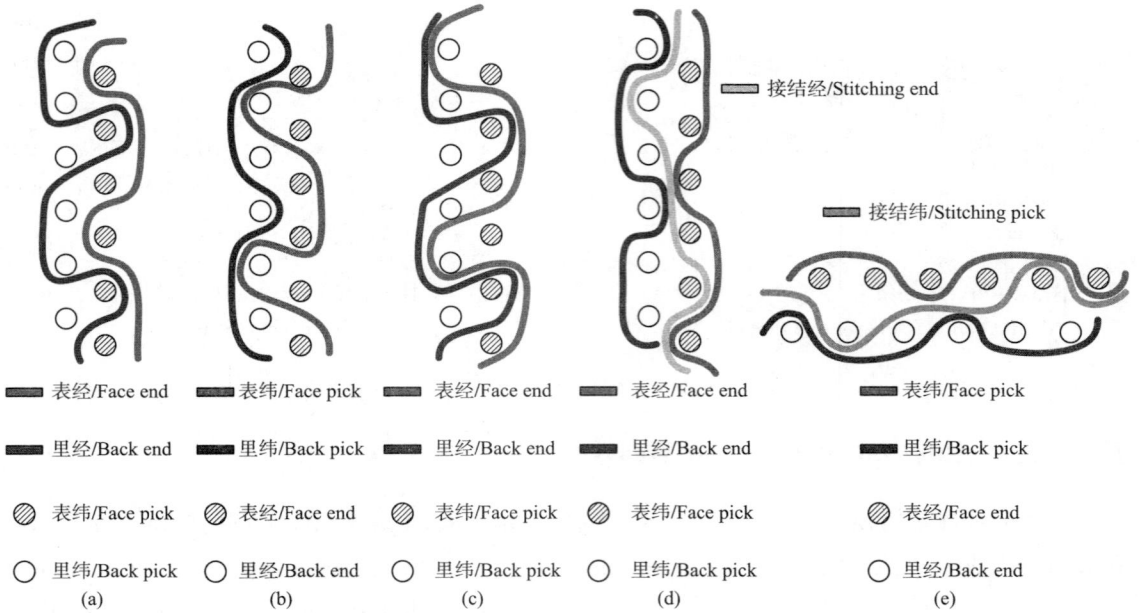

▬ 表经/Face end	▬ 表纬/Face pick	▬ 表经/Face end	▬ 表经/Face end	▬ 表纬/Face pick
▬ 里经/Back end	▬ 里纬/Back pick	▬ 里经/Back end	▬ 里经/Back end	▬ 里纬/Back pick
⊘ 表纬/Face pick	⊘ 表经/Face end	⊘ 表纬/Face pick	⊘ 表纬/Face pick	⊘ 表经/Face end
○ 里纬/Back pick	○ 里经/Back end	○ 里纬/Back pick	○ 里纬/Back pick	○ 里经/Back end
(a)	(b)	(c)	(d)	(e)

图6-3-27　接结方法示意图
Various ways of stitching

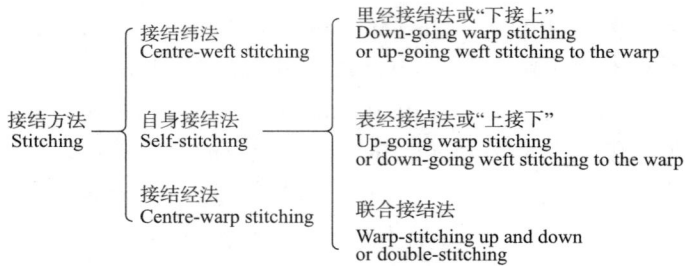

图6-3-28　接结方法分类
Ways of classifying the stitching

1. 里经接结法 /Up-Going Warp Stitching

里经提起与表纬交织，形成接结。如果在投表纬时，部分里经提起，某层上纱线就会与另外一层织物的纱线交织。图 6-3-29 中（a）是表组织，（b）是里组织，接结组织如图 6-3-29（c）所示，其安排确保里经与表纬接结点被表组织的长经浮线所覆盖。组织图绘制过程如图 6-3-30 所示，首先确定双层

In this stitching, the back ends are lifted to interweave with the face fillings. If a proportion of the back warp threads are raised when a face pick is inserted, the threads of one layer interweave with the threads of the other layer. Fig. 6-3-29 shows the 3 base weaves of the up-going warp stitched weave. The stitching weave as shown in Fig. 6-3-29(a) should be arranged in the manner that its warp floats be covered and concealed by the long floats of the face weave. When drawing the weave, the size of the stitched weave is

组织循环大小，后面的步骤共有四步：①在表经表纬处填入表组织；②在里经里纬处填入里组织；③在里经表纬处按照接结组织填入符号"▲"；④在表经里纬交叉处填入分层点符号"○"。需要指出的是，第一步与第二步可以互换次序，第三步与第四步也可以互换次序。

calculated firstly. Then, 4 steps are used to draw the up-going warp stitched weave. ① Step 1: fill in the marks of the face weave ; ② Step 2: fill in the marks of the back weave; ③ Step 3: fill in the marks "▲" for stitching points ; and ④Step 4: fill in the marks "○" for lifters. It should be noted that the order of the Step 1 and Step 2 can be exchanged, so does Step 3 and Step 4 (Fig. 6-3-30).

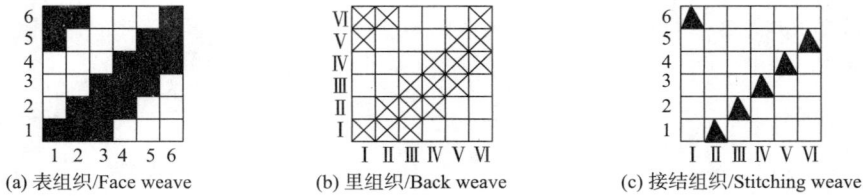

(a) 表组织/Face weave　　(b) 里组织/Back weave　　(c) 接结组织/Stitching weave

图6-3-29　接结组织的基础组织
The elements of the up-going warp stitching weave

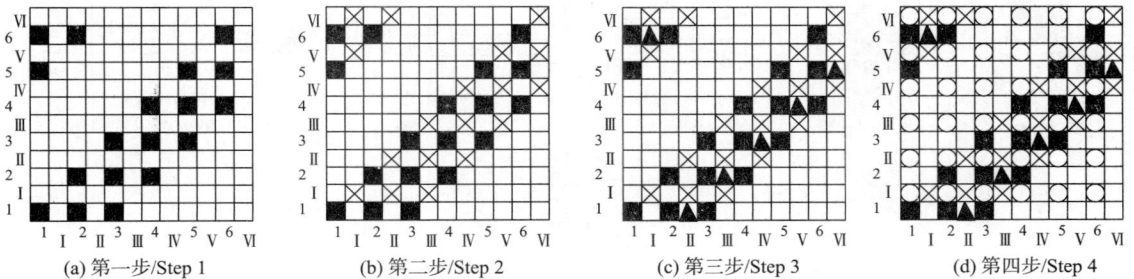

(a) 第一步/Step 1　　(b) 第二步/Step 2　　(c) 第三步/Step 3　　(d) 第四步/Step 4

图6-3-30　绘制里经接结双层组织步骤
The procedures of drawing an up-going warp stitching weave

2. 表经接结法 /Down-Going Warp Stitching

当里纬织入时，部分表经下沉，某层纱线与另一层纱线交织的接结方式称"上接下接结法"。表经下沉与里纬交织，形成接结。因此，部分分层点消失。图6-3-31（a）是表组织，（b）是里组织，接结组织如图6-3-31（c）或（d）所示，其安排确保里纬与表经接结点被表组织的长纬浮线所覆盖。需要注意的是：该点应为纬组织点，为

If a proportion of the face warp threads is left down when a back pick is inserted, the threads of one layer interweave with the threads of the other layer, which is the principle of "down-going warp stitching". The face ends are dropped and down-going, interweaving with the back fillings. Therefore, some lifters are canceled. The base weaves are shown in Fig. 6-3-31. In drawing the weave, the weft stitching points are covered and concealed by the long weft floats of the face weft. Since the point is a weft over warp, a special symbol "□" is used. In drawing the weave,

了区分，用"□"表示。在绘制组织图时，首先确定双层组织尺寸，分别在表经表纬处填入表组织，在里经里纬处填入里组织，在接结的表经里纬处按照接结组织填入符号"□"，最后在其他表经里纬交叉处填入分层点符号"○"。最终的组织图如图 6-3-31（e）所示。特别需要注意的是，在纹板图中，符号"□"对应位置的纹板不应植入纹钉或被提起。

the size should be determined firstly. Then, the face weave is filled where the face end interlaces with the face pick, and the back weave is filled where the back end interlaces with the back pick. At the square where the face end crosses the back pick, fill the symbol "□" if it is the stitching weaving point or symbol "○" otherwise. The final down-going warp stitched weave is shown in Fig. 6-3-31(e). It should be noted that the squares in the lifting plan corresponding to the symbol "□" in weave keep empty.

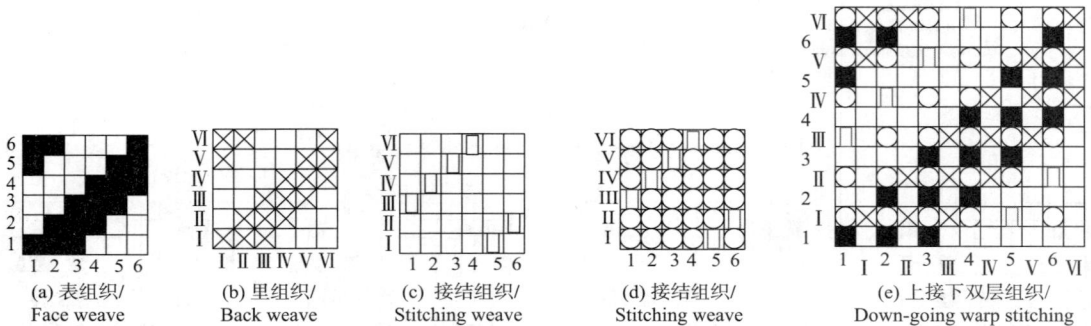

图6-3-31　上接下双层组织
Down-going warp stitched double-layer weave

3. 联合接结法 /Double Stitching

采用这种方法接结时，里经提起与表纬交织；同时，表经下沉与里纬交织，共同形成接结。图 6-3-32（a）是表组织，（b）是里组织。这时，同时设计"下接上"的里经表纬接结组织［图 6-3-32（c）］和"上接下"的表经里纬接结组织［图 6-3-32（d）］。最终设计的组织图如图 6-3-32（e）所示。

In double stitching, parts of the back ends are lifted to interweave with the face fillings while some face fillings drop and interweave with the back fillings. Fig. 6-3-32 describes the face weave, back weave, up-going warp stitching weave, down-going warp stitching weave and the final double stitched weave.

Center-Warp Stitching

4. 接结经接结法 /Center-Warp Stitching

采用附加的接结经与表里纬纱交织，把分离的两层织物连接起来。当表、里纱在原料、细度和颜

A third series of threads is introduced in the warp direction whose entire function is to stitch the two separate layers of cloth together. The stitching is used for cloths in

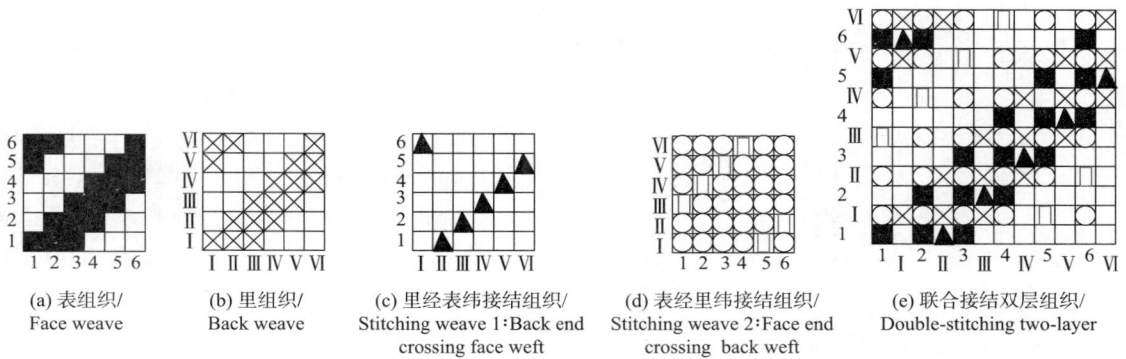

| (a) 表组织/ Face weave | (b) 里组织/ Back weave | (c) 里经表纬接结组织/ Stitching weave 1：Back end crossing face weft | (d) 表经里纬接结组织/ Stitching weave 2：Face end crossing back weft | (e) 联合接结双层组织/ Double-stitching two-layer |

图6-3-32 联合接结组织
A double stitching two-layer weave

色等方面差异很大时，一般另引入一根较细的接结纱线，在两层织物间来回穿梭，增加两层织物之间的结合力。由于两层织物并不交织，接结经纱大多数情况下处于两层之间，仅在交织点处接结，故织物接结牢固程度不如自身接结双层织物，手感较柔软。

在设计接结经双层织物时，需遵循以下原则：①在不需要接结处，接结经纱在表纬之下、里纬之上；②与表纬交织时，必须被两侧表经长浮线遮盖；③在与里纬交织时，必须要被里经在背面的经长浮线所覆盖。

若表、里组织分别如图 6-3-33（a）（b）所示的 $\frac{2}{2}\nearrow$ 斜纹，表经、里经、接结经排列比为 2：2：1。根据以上原则，则设计的接结经与表纬、里纬的接结组织如图 6-3-33（c）（d）所示。

which there is a great difference either in the thickness or the colors of the face and back threads. As a rule, the center threads are finer than either the face or backing threads. The center threads lie between the face and the back cloth and for the purpose of stitching oscillate at regular intervals between the face and the back thus achieving the required inter-layer cohesion. In this system the threads of one fabric do not interweave with those of the other fabric; the center threads oscillate between one and the other, and lie between them when not employed for tying. The two fabrics are less firmly united than with the self-stitching, and the cloth has a softer and fuller handle.

In designing the interlacing of the center-warp with face wefts and back wefts, the following principle needs to be obeyed: ① where no ties occur the center warp lies between the face and the back fabric and, therefore, must be lowered on the face picks and raised on the back picks; ② in tying to the face cloth the center ends are raised over the face picks where these are absent from the face, i.e., where they are covered by two adjacent floats of the face warp; ③ in tying to the back cloths the center ends are lowered on the back picks where these are absent from the underside, i.e., where they are covered on the underside of the back cloth by two adjacent floats of the back warp.

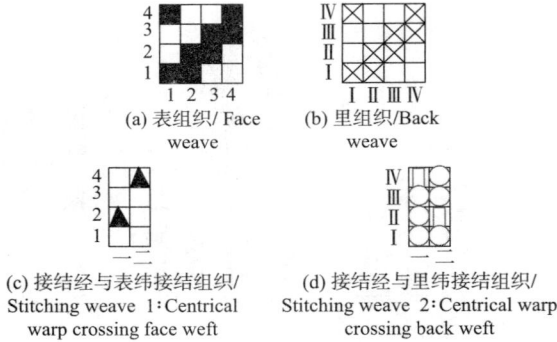

(a) 表组织/ Face weave

(b) 里组织/Back weave

(c) 接结经与表纬接结组织/ Stitching weave 1：Centrical warp crossing face weft

(d) 接结经与里纬接结组织/ Stitching weave 2：Centrical warp crossing back weft

图6-3-33 基础组织
Base weaves

Fig. 6-3-33(a)(b) shows the face weave ($\frac{2}{2}\nearrow$ twill) and the back weave ($\frac{2}{2}\nearrow$ twill) of a center-warp stitched weave. The face warp, back warp and stitched warp are arranged in proportion of 2:2:1. Based on the above principles, the stitching weaves of the center warp with the face weft and back weft are designed as shown in Fig. 6-3-33 (c)(d) respectively.

接结经组织的 R_j 等于双层组织的 R_{jd} 加上 1 个经纱循环内的接结经纱数，R_w 等于双层组织的 R_w。因此，R_j=8+2=10，R_w=8。因为一个经纱循环内，有 2 根接结经纱，故在 2 根表经 2 根里经后排列 1 根接结经纱。图 6-3-34 分别展示了填入表组织、里组织、接结组织、表经里纬分层点的过程，图 6-3-35 是该组织的上机图。

织造时，接结经双层组织采用分区穿法，接结经纱穿入后面综框，必须单独卷绕在另外一根织轴上。

The size of the weave repeat in warp direction R_j is calculated as the sum of the warp size of the double-layer weave repeat and the number of the center stitching warps, and the size of the weave repeat in weft direction R_w keeps the same as the double-layer weave. Therefore, R_j=8+2=10, R_w=8. The ends are arranged in 2 face ends, 2 back ends, 1 stitching end. Fig. 6-3-34 demonstrates how the weave is drawn step by step and Fig. 6-3-35 shows the loom plans for the weave.

Obviously, the divided drafting is suitable for the center-warp stitching and the stitching ends are drawn on the rear shafts. An extra beam is required for the stitching warp being released from.

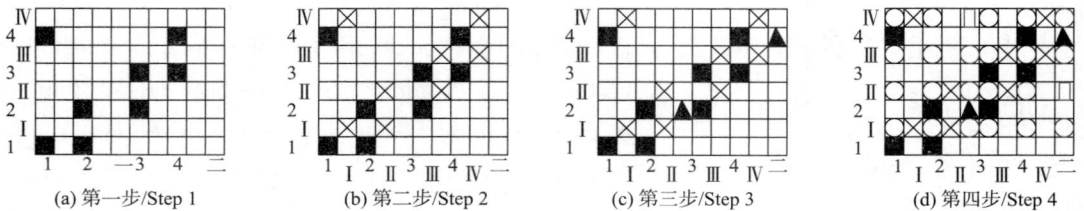

(a) 第一步/Step 1

(b) 第二步/Step 2

(c) 第三步/Step 3

(d) 第四步/Step 4

图6-3-34 接结经组织绘制步骤
Procedures of drawing a center-warp stitching weave

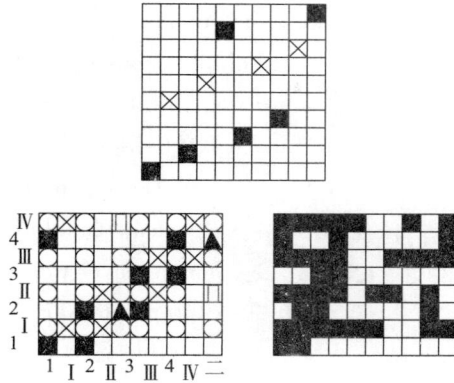

图6-3-35 接结经组织上机图
The looming plans for a center–warp stitching weave

5. 接结纬接结法 /Center–Weft Stitching

接结纬接结法是采用附加的较细接结纬在两层织物间来回穿梭，与表里经纱交织，把两层织物连接起来。接结纬接结法因为要多卷取，织造效率低，一般不常用。若采用接结经另装经轴困难，考虑使用接结纬接结法。

在设计接结组织时，需要遵循如下原则：①在不需要接结处，接结纬纱在表经之下、里经之上；②与表经交织时，必须被两侧表纬长浮线遮盖；③在与里经交织时，必须要被里纬在背面的纬长浮线所覆盖。

若表、里组织分别如图 6-3-36（a）（b）所示，表纬、里纬、接结纬排列比为 1:1:1。根据以上原则，则设计的接结纬与表经、里经的接结组织如图 6-3-36（c）（d）所示。

接结纬组织的 R_w 等于双层组织的 R_{wd} 加上 1 个纬纱循环内的接结纬纱数，R_j 等于双层组织的 R_j。因此，$R_w=8+4=12$，$R_j=8$。因为一个纬纱循环内，有 4 根接结纬纱，故纬纱排列方式是 1 表 1 里 1 接结。图 6-3-37

A third series of threads is introduced in the weft direction whose entire function is to stitch the two otherwise separate layers of cloth together. The center threads lie between the face and the back cloth and for the purpose of stitching oscillate at regular intervals between the face and the back. The stitching is not very often used as it reduces the rate of cloth production due to the take–up of the center weft picks that do not contribute to the length of cloth being produced. Occasionally the center weft is applied if the mounting of an extra beam required by the center warp threads presents a particular difficulty in respect of the control or access to the warp yarns.

When designing a center–weft stitching the following procedures need to be observed: ① where no ties occur the center weft lies between the face and the back cloth. To achieve this on center weft picks the face ends are raised and the back ends are lowered; ② to achieve a face fabric stitch a face end must be dropped on a center pick at a point at which it is absent from the surface, i.e. when it is covered by two adjacent floats of the face weft; ③ to achieve a back fabric stitch a back end is raised on a center pick at a point at which it is absent from the underside of the cloth, being covered by two adjacent floats of the back weft.

Fig. 6-3-36(a)(b) shows the face weave and the back weave of a center–weft stitched weave. The face weft, back

分别展示了填入表组织、里组织、表经里纬分层点、接结纬与表经交织点、接结纬与里经交织点的过程，图 6-3-38 是该组织的上机图。

weft and stitched weft are arranged in proportion of 1:1:1. Based on the above principles, the stitching weaves of the center weft with face warp and back warp are designed as shown in Fig. 6-3-36(c)(d) respectively.

The size of the weave repeat in weft direction R_w is calculated as the sum of the weft size of the double-layer weave repeat and the number of the center stitching wefts, and the warp size of the weave repeat R_j keeps the same as the double-layer weave. Therefore, $R_w=8+4=12$, $R_j=8$. There are 4 stitching weft threads in a weave repeat, and the weft threads are arranged as 1 face filling, 1 back filling and 1 stitching filling. Fig. 6-3-37 demonstrates how the weave is drawn step by step and Fig. 6-3-38 shows the loom plans for the weave.

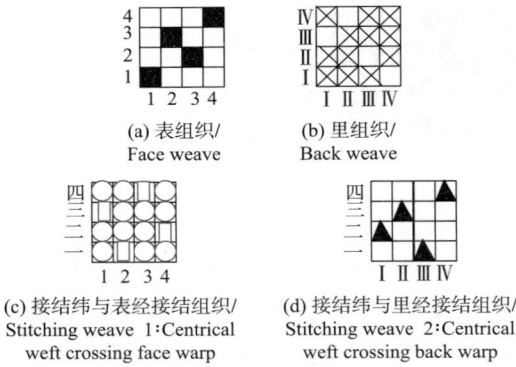

图6-3-36 基础组织
Base weaves

(a) 表组织/Face weave
(b) 里组织/Back weave
(c) 接结纬与表经接结组织/Stitching weave 1:Centrical weft crossing face warp
(d) 接结纬与里经接结组织/Stitching weave 2:Centrical weft crossing back warp

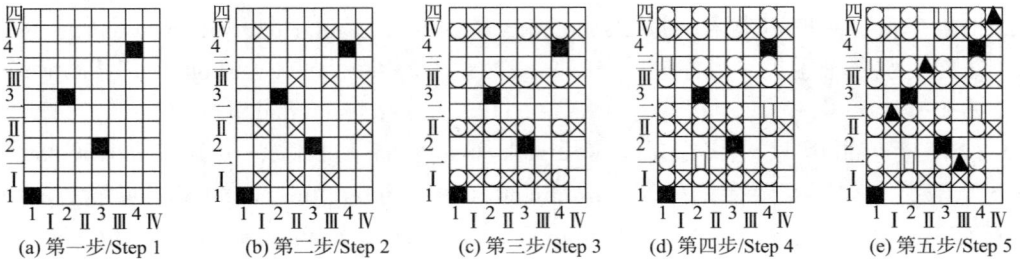

(a) 第一步/Step 1 (b) 第二步/Step 2 (c) 第三步/Step 3 (d) 第四步/Step 4 (e) 第五步/Step 5

图6-3-37 接结纬组织绘制步骤
Procedures of drawing a center-weft stitching weave

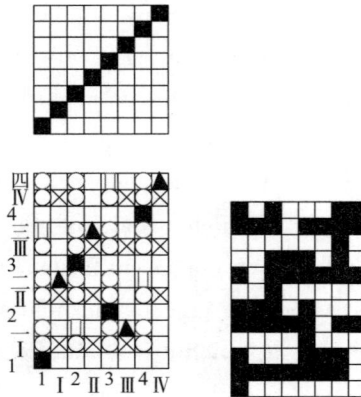

图6-3-38 接结纬组织上机图
The looming plans for a center-weft stitching weave

（二）接结双层组织组织图的绘制 /Drawing of Stitching Double Weaves

（1）选择表里基础组织。先确定表层组织，然后根据织物要求确定里层组织。

（2）确定表里经、纬的排列比。排列比要根据织物的用途、表里层组织、线密度、密度等情况来确定。常用表里经的排列比有 1:1、2:1、3:1 等，常用表里纬的排列比有 1:1、2:1、3:1、2:2、4:2 等。

（3）确定接结组织。用来接结的纱线与上下两层组织的交织点，称为接结点。接结的纱线可能是表层纱，也可能是里层纱，甚至是外来纱线。接结点在上下两层的配置分布规律称为接结组织。确定接结组织的原则：①在一个组织循环内，接结点的分布要均匀。②接结点不要在织物表面显露，因此接结点应安排在两侧长浮线之间（接结点是经组织点，应位于左右表经长浮线之间；接结点是纬组织点，应位于上下表纬长浮线之间）。③若表组织为斜纹组织，接结点的分布方向应与表组织的斜纹方向一致。表组织为经面组织，选用"里经接结法"有利于接结点的遮盖。表组织为纬面组织，选用"表经接结法"有利于接结点的遮盖。表组织为同面组织，选用"里经接结法"为好，因为一般经纱比纬纱细而牢，接结点易遮盖且接结牢固。④接结组织的经纬纱循环数应与表里组织经纬纱循环数相等，或是其约数或倍数。

（4）确定组织循环。经、纬循环数的确定可参照经、纬二重组织

(1) Choose the Base Weaves. The face weave is determined according to the usage of the fabric first, then the back weave is set accordingly.

(2) Determine the arrangement of the face thread and the back thread. The arrangement of threads should be designed by considering the usages, base weaves, linear density of the yarns, fabric settings. Usually, the ratio is set as 1:1, 2:1 or 3:1 for face ends and back ends or 1:1, 2:1, 3:1, 2:2 or 4:2 for face fillings and back fillings.

(3) Determine the Stitching Weave. The stitching point is the square where a stitching thread cross the face or back layer. The stitching thread may be a face thread, or a back thread, or an extra thread. The stitching weave is the distribution of the stitching points of the two layers. The following rules should be followed for determining the stitching weave: ① The stitching points are evenly distributed in a weave repeat. ② The stitching points are concealed in the appearance, therefore, they must be covered by the long floats bilaterally from the face weave. ③ If a twill weave is chosen for the face, then the direction of the stitching weave should be in line with. If the face weave is warp-faced, it is better to use up-going warp stitching while for a weft-dominated face weave, it is better to use down-going warp stitching. If the face weave is double-faced, an up-going warp stitching is preferred since a fine and strong end is easily concealed. ④ The size of the stitching weave repeat is equal to the multiple or factor of the face weave or back weave.

(4) Calculate the size of the stitching double weave repeat. If a self-stitching is used, the calculation of the repeat is similar to a normal double weave. However, if a center-warp stitching or a center-weft stitching is used, the number of the stitching warps or wefts should be counted.

的计算方法。在使用接结经（纬）
接结时，应另加上接结经的根数。

（三）接结双层组织的上机要点 /Looming for the Stitched Double Weaves

（1）经轴。表里经纱缩率一致，单经轴可以织造；否则，采用多经轴织造。接结导致经纱张力比为接结经纱大时，接结的经纱的送经量要增大。采用自身接结时，因为采用相同的纱线和组织，即使单经轴也可以顺利织造。接结经需单独使用一个经轴。

（2）梭箱。表里纬纱不同原料、不同色需多梭箱；接结纬需多梭箱。

（3）穿综。与重经组织相同，提升多在前区。自身接结的双层组织所用综框个数不仅与表、里组织大小有关，还与接结组织相关。

（4）穿筘。按照排列比，里经需与表经同筘，且尽可能放入中间。例如，表∶里 =2∶1 时，按照 1 表、1 里、1 表方式穿筘。

（5）纹板图。与一般纹板图类似，但是表经里纬的接结点符号为"□"，实际是纬组织点，纹板图对应位置应为空白，不植入纹钉。

(1) Beaming. Single beam is sufficient if the tensions of face ends and back ends are similar; otherwise, multiple beams are used. The stitches put tension on the warp threads, hence, other things being equal, the series used for tying requires to be longer than the unstitched series, and two warp beams are therefore necessary. By employing double stitching, however, and using similar yarns and weaves of equal firmness for the two fabrics, a perfect double cloth can be woven with only one warp beam. Center-warp stitching ends need an extra beam.

(2) Shuttle-box. For the face and back fillings of different materials and colors, a multiple shuttle-box mechanism is required. The device is also necessary for manufacturing of center-weft stitching fabrics.

(3) Drafting. Divided drafting is applied to stitched double weaves. The number of healds require in a self-stitched double cloth depends not only on the respective sizes of repeats in the face and the back cloths but also on the order of stitching.

(4) Denting. The ends are placed in the dent according to yarn group. The back ends are drawn in the same dent with the face ends of a group and it is better to place the back ends in the middle of the face ends, eg, face∶back=2∶1, and the ends are drawn in dents according to the order of 1 face,1 back and 1 face.

(5) Weaving plan. The symbol "□" that denotes the stitching points of the face ends and the back wefts in the weave design is actually the weaving point of "weft over warp", which cannot be drawn as mark at the corresponding square in the weaving plan.

（四）接结双层组织的判断 /Judgement of the Stitched Double Weaves

接结双层组织的判断方法与普通双层组织相似，按照浮长线的长度判断，在经纱方向和纬纱方向都形成重叠。

图6-3-39所示为待判断的组织图及其经纱浮长线分布图和纬纱浮长线分布图。可以发现，在经纱方向，第1、3、5经总是比其相邻的第2、4、6经高，应该是处于织物的上层；在纬纱方向，第1、3、5纬也不低于与其相邻的第2、4、6纬，也应居于织物的上层。故正常双层组织中，表经（奇数根）里纬（偶数根）交织点应该是经组织点，而里经（偶数根）表纬（奇数根）交织点应该是纬组织点。图6-3-40（a）中，发现表经与里纬交织点都是经组织点，故都是正常分层点；但图6-3-40（b）中，有三处地方表纬、里经交织，分别是第1纬（表纬）第6经（里经）、第3纬（表）第2经（里）、第5纬第4经为经组织点，故这三个点必然是接结点，属于下接上法接结双层组织。

The judgement of the stitched double weave is similar to that for normal double-layer weave. According to the lengths of the floats, the warps and the wefts will get overlapped both lengthwise and crosswise.

The diagrams of the lengths of the warp floats and the weft floats are shown in Fig. 6-3-39(b)(c) as regard to the weave shown in Fig. 6-3-39(a). It is observed that the warp floats of the ends No. 1, No. 3 and No. 5 are always exceed that of their neighbors, i.e., ends No. 2, No. 4 and No. 6, which means that the former warps are at the upper layer. Similarly, weft threads No. 1, No. 3 and No. 5 are also located at the upper layer. Therefore, for a normal double-layer weave, the crosses where the odd warps (face warps) and the even wefts (back wefts) intersect are all warp-up points while the crosses where the even warps (back warps) and the odd wefts (face wefts) are all warp-down points. By observing the weave as shown in Fig. 6-3-40(a), it is normal for all the intersecting points of the face warps interlacing the back wefts are of warp-up. However, three squares where the face wefts intersect the back warps, i.e. weft No. 1 interlacing warp No. 6, weft No. 3 interlacing warp No. 2, and weft No. 5 interlacing warp No. 4 are found filled with marks. It is obvious that the weave is formed by up-going warp stitching. The stitching weave of the face wefts crossing the back warps is drawn in Fig. 6-3-40(b).

(a) 组织图/Weave pattern

5	-4	5	2	5	-4
-1	-4	5	2	5	-4
5	2	5	-4	5	-4
5	2	5	-4	-1	-4
5	-4	5	-4	5	2
5	-4	-1	-4	5	2

(b) 经向浮长分布图/Lengths of warp floats

-1	1	-3	-3	-3	1
3	3	-3	-3	-3	3
-3	-3	-3	1	-1	1
-3	-3	-3	3	3	3
-3	1	-1	1	-3	-3
-3	3	3	3	-3	-3

(c) 纬向浮长分布图/Lengths of weft floats

图6-3-39 待判断组织

A weave for judgement

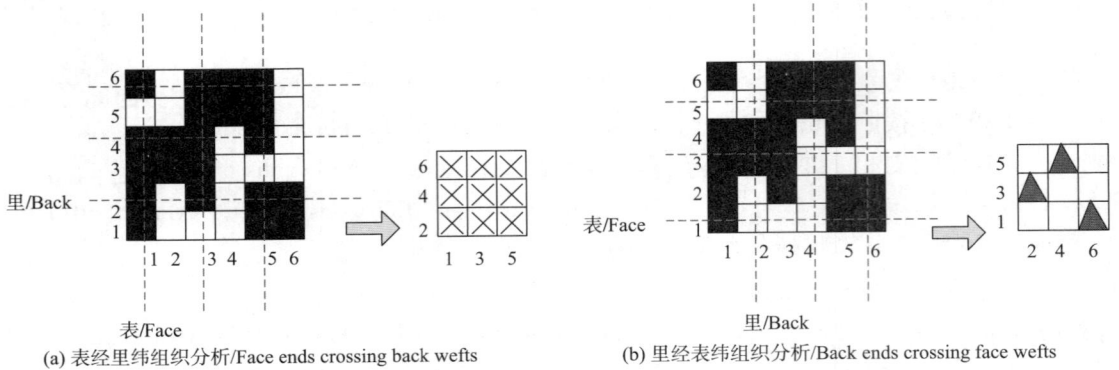

(a) 表经里纬组织分析/Face ends crossing back wefts   (b) 里经表纬组织分析/Back ends crossing face wefts

图6-3-40　接结双层组织判断

Judgement of a stitching duble weave

习题 /Questions

1. 判断习题图 6-1 中各组织点处纱线所在的层数。

2. 试作 $\dfrac{4}{1}\nearrow$ 斜纹为表组织，$\dfrac{4}{1}$ 斜纹为反组织，经纱排列比为 1:1 的经二重组织的上机图和经向截面图。

3. 已知纬二重组织的表组织和反组织均为 $\dfrac{1}{3}$ 破斜纹，表里纬纱排列比为 1:1，试作织物上机图及纬向截面图。

1. Judge the layers of each interweaving point in the weaves.

2. Draw the looming plans and the longitudinal section diagram of the warp-backed weave under the following conditions: a $\dfrac{4}{1}\nearrow$ twill weave as the face weave, a $\dfrac{4}{1}$ twill as the reverse weave; the order of warping is 1 face and 1 back.

3. Draw the looming plans and the transversal section diagram of the weft-backed weave under the following conditions: both the face weave and reverse weave are the $\dfrac{1}{3}$ satinette; the order of wefting is 1 face and 1 back.

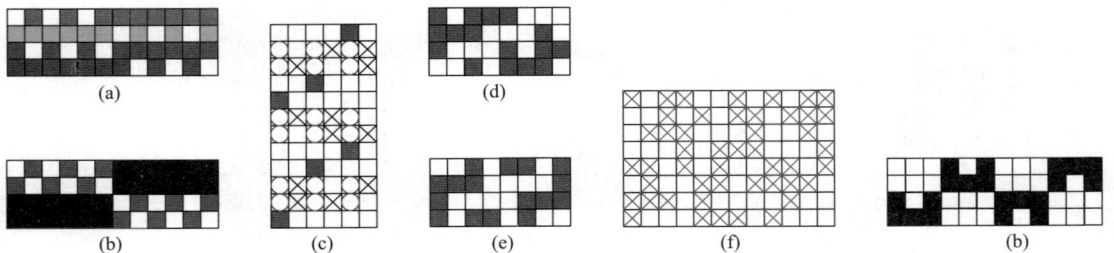

(a)　(b)　(c)　(d)　(e)　(f)　(b)

习题图6-1

4. 某纬二重织物，表组织为 $\dfrac{2}{4}\nearrow$，里组织自选，表、里纬纱排列比为 1:1，试作出组织图。

5. 某纬二重织物表里组织均为 $\dfrac{1}{3}$ 破斜纹，纬纱采用两种颜色制造，排列比为：1A:1B。该织物 $R_w=16$，$R_j=4$。在织物下半部分显 B 色，上半部分显 A 色。试绘该织物的上机图。

6. 某纬二重织物表组织均为 $\dfrac{2}{2}\nearrow$ 斜纹，里组织均为 $\dfrac{1}{3}$ 斜纹，表里纬排列比为 2:2。试绘该织物的上机图及纬向横截面。

7. 试作 $\dfrac{3}{1}\nearrow$ 为基础组织的管状织物上机图及纬向截面图。

8. 某管状织物基础组织为 $\dfrac{2}{2}$ 斜纹组织，要求经纱总数大于 2 倍基础组织经纱循环数，织造时第 1 纬自左向右投入表层，试作该织物的上机图及纬向截面图。

9. 以 4 枚不规则缎纹为基础组织，试作双幅织物的上机图及纬向截面图。

10. 试作以平纹为表、里层的基础组织，经纬纱排列比均为 2A:2B 的表里交换双层组织。花纹如图 $\begin{array}{|c|c|}\hline A & B \\\hline B & A \\\hline\end{array}$，每小方格分别由 4 根表、里经纱和 4 根表、里纬纱组成。

11. 以 4 枚不规则缎纹为基础组织，试作表里换层组织及其经纬

4. Draw the design of the weft-backed weave under the following conditions: the face weave is set as $\dfrac{2}{4}\nearrow$; the order of wefting is 1 face and 1 back; the back weave is chosen by the designer.

5. In a weft-backed weave, both the face weave and the back weave are the $\dfrac{1}{3}$ satinette，and the order of wefting is 1A and 1B. The weave repeats on 16 picks and 4 ends. The bottom half part of the weave shows the effect of color B and the upper part shows the effect of color A. Try to design the Looming plans.

6. Draw the looming plans and the transversal section diagram of the weft-backed weave under the following conditions: a $\dfrac{2}{2}\nearrow$ twill weave as face weave，and a $\dfrac{1}{3}$ twill as reverse weave; the order of wefting is 2 face and 2 back.

7. Draw the looming plans and the transversal section diagram of the tubular weave based on a $\dfrac{3}{1}\nearrow$ twill.

8. A $\dfrac{2}{2}$ twill based tubular weave is started by picking the face thread from the left to right. Draw the looming plans and the transversal sectional diagram requiring the number of the ends is larger than 2 times of the size of the base repeat.

9. Try to draw the looming plans and the transversal sectional diagram of the double-width weave based on a satinette.

10. Draw the design of an interchanging double-layer weave under the following conditions: both the face and the back are plain weaves; the order of warping is 2A:2B; the pattern is shown as $\begin{array}{|c|c|}\hline A & B \\\hline B & A \\\hline\end{array}$, in which each square represents 4 face threads and 4 back threads.

11. Draw the design and the longitudinal section diagram and the transversal section diagram of an

向截面图，其纹样如图 $\boxed{\begin{array}{cc} A & B \\ B & A \end{array}}$ ，每区分别由表、里各 8 根经纱和纬纱组成，表、里经与表、里纬排列比均为 1:1。

12. 某表里换层双层组织的基础组织均为平纹，表、里经及表、里纬排列比均为 1 白 1 红，其纹样如习题图 6-2 所示，试作该组织图。

interchanging double-layer weave under the following conditions: both the face and the back are satinettes, the order of warping is 1A1B, the order of wefting is 1A1B, and the pattern is shown as $\boxed{\begin{array}{cc} A & B \\ B & A \end{array}}$, in which each square represents 8 face threads and 8 back threads.

12. Draw the double interchanging weave under the following conditions: both the face and the back are plains; the order of warping is 1white, 1red. The design is shown as below.

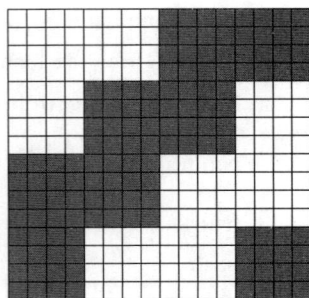

习题图6-2

13. 已知表组织为 $\dfrac{4}{2}\nearrow$ ，里组织为 $\dfrac{2}{1}\nearrow$ ，表、里经纬纱排列比均为 2:1，试作"下接上"双层接结组织的上机图和经纬向截面图。

14. 以平纹为表组织， $\dfrac{2}{2}\nearrow$ 为里组织，表、里经纬纱排列比均为 1:1，选用八枚缎纹作为"下接上"的接结组织，绘制其上机图和经纬向截面图。

15. 已知表组织为 $\dfrac{2}{1}\nearrow$ ，里组织为 $\dfrac{1}{2}$ 斜纹，接结组织也为 $\dfrac{1}{2}$ 斜纹，表、里经纬纱排列比均为 1:1，试作"下接上"双层接结组织的上

13. Draw the looming plans for the up-going warp stitching double weave, the longitudinal section diagram and the transversal section diagram under the following conditions: face weave $\dfrac{4}{2}\nearrow$, back weave $\dfrac{2}{1}\nearrow$; the order of warping is 1face, 1back; the order of wefting is face, 1back.

14. Draw the looming plans for the up-going warp stitching double weave , the longitudinal section diagram and the transversal section diagram under the following conditions: face weave—plain weave, back weave $\dfrac{2}{2}\nearrow$, binder weave—8 -shaft sateen; the order of warping is 1face, 1back; the order of wefting is face, 1back.

15. Draw the looming plans for the up-going warp stitching double weave , the longitudinal section diagram and the transversal section diagram under the following conditions: face weave $\dfrac{2}{2}\nearrow$ twill , back weave $\dfrac{1}{2}$ twill , binder weave

机图和经纬向截面图。

16. 某双层织物，表里组织均为 $\frac{2}{2}$↗，采用"上接下"法，表、里经纬纱排列比均为1:1。接结组织自选，绘制该织物组织图。

17. 某双层织物，表、里组织分别如习题图6-3所示，采用"上接下"法，表、里经纬纱排列比均为1:1。接结组织自选，绘制该织物组织图。

$\frac{1}{2}$ twill; the order of warping is 1face, 1back; the order of wefting is face, 1back.

16. Draw the down–going warp stitching double weave under the following conditions: face weave $\frac{2}{2}$↗ twill, back weave $\frac{2}{2}$↗ twill, and binder weave is selected by the designer; the order of warping is 1face, 1back, the order of wefting is face, 1back.

17. Draw the down–going warp stitching double weave under the following conditions: face weave and back weave are shown as below, and binder weave is selected by the designer; the order of warping is 1face, 1back; the order of wefting is face, 1back.

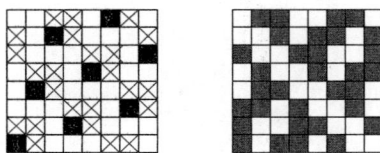

习题图6-3

18. 某双层织物，表面组织及里组织均为 $\frac{2}{2}$↗，采用"联合接结法"，表、里经纬纱排列比均为1:1，试作该织物上机图及经纬向截面图。

19. 判断习题图6-4中的组织及各个组织点所在纱线的层数。如果是接结双层组织，说明其表、里组织及接结方法。

20. 下面的缎背华达呢组织属于缎纹变化组织，分类时属于假经二重组织，经密是纬密的1.5倍。组织正反面分别如习题图6-5所示，分析其外观特点及成形原因。

18. Draw the double stitching weave, the longitudinal section diagram and the transversal section diagram under the following conditions: face weave $\frac{2}{2}$ ↗ , back weave $\frac{2}{2}$ ↗ ;the order of warping is 1face, 1back; the order of wefting is face, 1back.

19. Judge the type of the weave and mark the layer of each interweaving point located. If the weave is of a stitched double layer, describe the face weave, the back weave, and the method of stitching.

20. The satin–back gaberdine is belonged to the derivative of a satin weave, and it can also be classified as pseudo warp–backed weave. Usually, 1.5 times of ends are placed in the same space as the picks. The face and the back are show as below. Try to analyze the structure and the appearance of the weave.

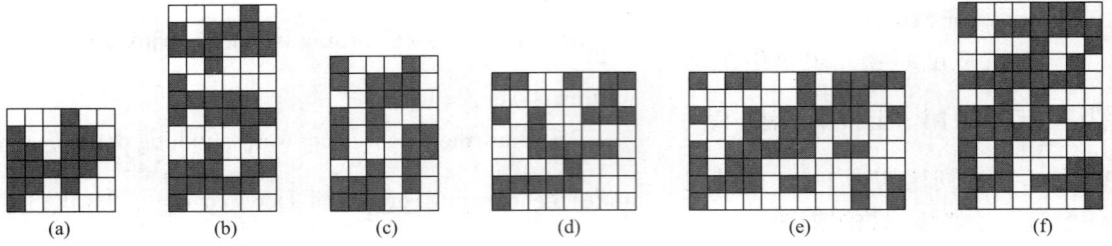

习题图6-4

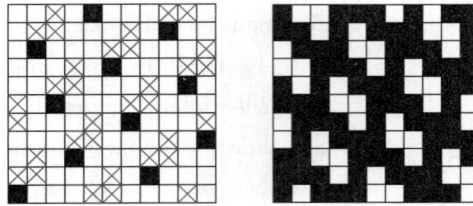

习题图6-5

第七章　起毛起圈织物 /Fleecy Fabrics

第一节　起毛起圈组织的类型 /Classification of Fleecy Fabrics

在织物表面引入装饰性的高度效果的立体态纤维或者纱线，可形成起毛（起绒）起圈织物，特别满足柔软、保暖、吸水性方面的用途。起毛起圈可以通过超细纤维的原纤化产生，或者使用花式纱线如雪尼尔形成，或者对毛织物进行缩呢整理得到，或者利用针织、簇绒、静电植绒和机织生产方式得到。起毛起圈织物柔软、厚重、结构紧密、耐用。机织物生产中，起毛主要有两种方式：开毛（包括割绒和起圈）、拉毛（拉绒）。根据起毛系统，织物又可分为经起毛织物与纬起毛织物。纬起毛织物一般是棉织物，纬密大，在纬纱上割绒或者拉绒形成，包括帝国绒、广东棉（斜纹绒）、仿鼹鼠皮、海狸绒（斜纹绒）、平绒和灯芯绒。本章主要介绍用割绒法生产的起毛起圈织物结构。

The fleecy fabrics introduce a decorative third dimension, creating an effect of depth. The construction is especially used when softness, warmth, and absorbency are desired. The fabrics can be manufactured by fibrillation of super fine fibers, introducing chenille yarn, felting of wool fibers in finishing, knitting, tufting, electrostatic flocking and weaving. They are comparatively soft, heavy and compact textures of great durability. For woven fabrics, there are two subtypes: pile (cut pile and loop pile) fabrics and napped fabrics. According to the yarn series to be napped or cut, the fleecy fabrics can also be classified as warp pile fabric and fustian. Fustian is the general term to refer to heavy fabrics with a high proportion of filling yarns usually cotton and often napped or cut. It comprises of imperial, canton, moleskin, beaverteen, velveteen, and corduroy. In this chapter, pile fabrics are main introduced.

一、割绒起圈织物 /Pile Fabrics

割绒起圈织物表面具有密集的直立纤维。通过特定结构，首先将纱线垂直于布面，然后切断起毛纱或者形成毛圈得到。割绒织物的地

The pile fabric is a special type of fabric having a dense upright fibrous surface obtained by a special form of construction which permits pile yarns or pile loops to be projected and then completely cut through or kept intact,

组织纱线没有被切断，提供织物强力，形成绒毛附着的基础。

此类织物包括地毯、雪尼尔织物、灯芯绒、丝绒、长毛绒、挂毯、毛巾、羊毛绒、纬平绒（棉绒）。该织物有多种分类形式。

1. 割绒与起圈织物 /Cut Pile Fabrics and Uncut Pile Fabrics

通过切断绒经、绒纬浮长线产生的起绒织物称为割绒织物，如切割绒经纱的丝绒、长毛绒、威尔顿地毯等，切割绒纬纱的灯芯绒、纬平绒等。毛圈可以剪开，也可以不剪开。有些丝绒品种仍然保留丝圈。毛巾则是典型的未切断毛圈结构的织物。

生产丝绒时，当绒经浮在起绒杆（作用类似纬纱）上后，从织物中抽出起绒杆。若采用普通起绒杆，则形成未切断的线圈；若起绒杆末端有刀片，则割断纱线，形成绒面，原理如图 7-1-1 所示。毛巾的毛圈形成原理如图 7-1-2 所示，由单独卷绕在织轴上的松弛毛经纱利用特殊打纬机构与特定的组织共同形成的。

whilst ground yarns, which are not cut form the basis of fabric strength and pile support.

Pile fabrics include carpet, chenille, corduroy, velvet, plush, rugs, terry, velour, velveteen. There are various ways to classify the fabrics.

The cut pile fabrics are produced by cutting the warp floats or weft floats. These fabrics include velvet, plush, Wilton carpet by cutting warps and corduroy, velveteen by cutting wefts. The loops may be cut or closed sheared, or left uncut if a loop surface is desired. In some varieties of velvet, the loops are kept. Terry is a typical structure with an uncut pile (loop pile).

In weaving a velvet, pile loops are produced by weaving an extra series of yarns over wires (act as a pick) that are then drawn out of the fabric. Plain wires leave uncut loops; wires with a razor-like blade produce a cut-pile surface as shown in Fig. 7-1-1. Terry loops are formed by beating the extra slack ends which are unwound on a separate beam at a special reed motion combining a special weave, as shown in Fig. 7-1-2.

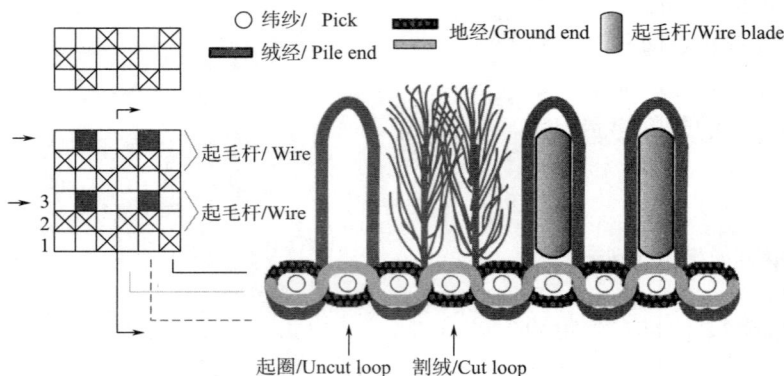

图7-1-1 丝绒形成原理

The principle of forming a velvet pile by a wire

毛经（绒经）纱线轮流与双层织物交织也可以得到起毛织物。沿着幅宽方向用刀子将织物剪开，形成两块割绒织物，原理如图7-1-3所示。灯芯绒表面绒纬浮长线被割断，形成了直立绒毛，如图7-1-4所示。

Pile fabrics can also be made by producing a double-cloth structure woven face to face, with an extra set of yarn interlacing with each cloth alternately. The two fabrics are cut apart by a traversing knife, producing two fabrics with a cut-pile face. The principle of making pile fabrics by face to face is illustrated in Fig. 7-1-3. In corduroys fabric, long filling floats on the surface are slit, causing the pile to stand erect as shown in Fig. 7-1-4.

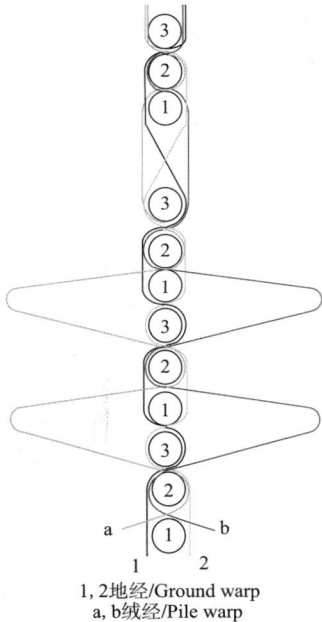

图7-1-2　毛巾毛圈形成原理
The principle of forming a terry loop

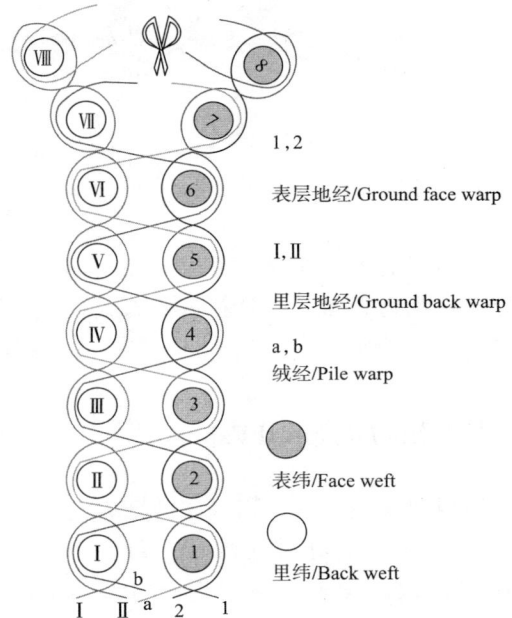

图7-1-3　双层织物起毛形成原理
The principle of forming a pile by a double-fabric

2. 纬起毛与经起毛织物 /Weft Pile Fabrics and Warp Pile Fabrics

当经纱作为毛纱时，则是经起毛织物，如丝绒、长毛绒和毛巾。长毛绒织物的绒毛高度通常超过3mm。若由纬纱作为毛纱时，则是纬起毛织物，如灯芯绒和纬平绒。纬平绒织物，绒毛短而均匀地分布在织物上，织物表面非常平整；而灯芯绒的毛绒形成与经纱平行的纵向条带。纬平绒和灯芯绒的差异在

Warp pile fabrics, i.e., velvet, plush and terry are produced by an extra warp yarn. In a plush fabric, the pile is more than 3 mm high. Weft pile fabrics, such as corduroy and velveteen, are produced by an extra weft yarn. In plain velveteens, the short and even tufts of pile are distributed uniformly over the fabric, thereby forming a perfectly level surface; but in corduroys, the tufts of pile are caused to develop a ribbed or corded formation, with the cords produced lengthwise or parallel with warp threads.

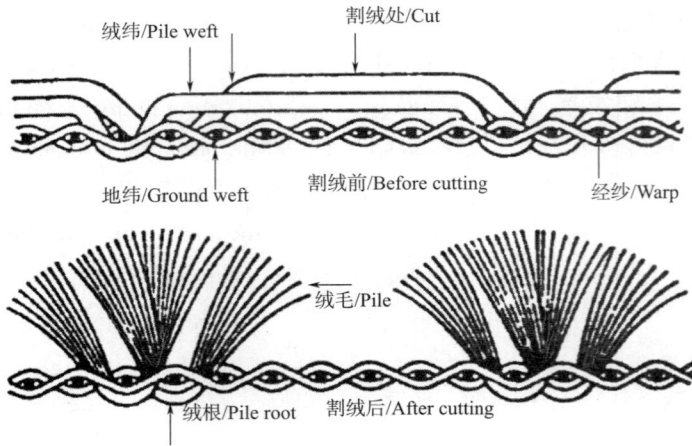

绒纬/Pile weft　　割绒处/Cut

地纬/Ground weft　　割绒前/Before cutting　　经纱/Warp

绒毛/Pile

绒根/Pile root　　割绒后/After cutting

图7-1-4　灯芯绒形成原理

The principle of forming a corduroy

于，纬平绒的绒纬浮长尽可能均匀地分散在织物表面，而灯芯绒的绒纬仅在几组经纱上。

Velveteen differs from corduroy in the arrangement of the floats of the pile filling, which are scattered as regularly as possible instead of being confined to groups of warp yarns.

二、拉绒织物 /Napped Fabrics

拉绒织物通过将织物表面纤维拉成长短不齐的纤维而形成，纤维在织物双面突起，通过对低捻纱线进行拉绒、刷绒和起绒工序产生。绒毛一般呈现条纹状或其他图案，通过剪毛工序，绒毛高度均匀。法兰绒、毛毯和大衣呢等不同面料的绒毛高度不同。通过起绒机高速回转的滚筒上起毛钢丝或刺果的撕扯，将织物结构中的纤维拉断起毛。除了这种刺果起绒方法外，其他类似刷绒、搓呢工艺也可以将织物表面的纤维拉成垂直于织物平面。烤花大衣呢通过绒纬拉毛产生，绒毛在织物表面产生斜条或波纹图案。

Napped fabric is formed by shredding surface yarns of a fabric. An array of fiber ends protruding from the surface of one of both sides of the napped fabric and gives the fabric a fuzzy appearance. Produced by napping, raising and brushing the fibers of the loosely twisted yarn, the nap may be in stripes or other patterns or may be chopped or sheared to the obtain the uniform length. The length of the nap varies in the different fabrics that are given this type of finish, e.g., flannel, blanketing, overcoating. In these fabrics, part of the fibre from the basic structure of a textile material is raised to the surface by means of revolving cylinders covered with metal points or teazle burrs. Apart from teazling, other means like brushing, or rubbing also may be employed for producing such a layer of protruding fibers on the surface of fabrics. Elysian overcoat is a thick woolen overcoating with a heavy nap formed from extra filling yarns floated on the surface. The nap makes a pattern of diagonal or wavy lines.

Napped Fabrics

第二节　灯芯绒 /Corduroy Weaves

灯芯绒也称条绒，是一种有纵向平行起绒条纹的纬起毛织物，如图 7-2-1 所示。灯芯绒织物条子圆润、丰满、柔软、厚实、结实、弹性好、光泽柔和。由于地部被绒毛遮盖，与外界摩擦时，耐磨性能好。灯芯绒常按宽度来分类。在同一块面料中，条纹一般等宽，偶尔条子粗细变化。通常条子宽度在 1.25 ~ 4mm，也可以更细或者更粗些。

Corduroys are also called cords. Corduroy weaves as shown in Fig. 7-2-1 are filling-pile fabrics with ridges of pile (cords) running lengthwise parallel to the selvage. The cords are round, plump, soft, thick, solid, resilient and soft sheen. The fabrics are relatively resistant to wearing because the grounds areas are covered by the pile surface when they are rubbing with the outsides. The cords are usually of the uniform width in the same fabric, but sometimes they are variegated. In general, the width of each cord is in the range of 1.25~4 mm, and may be less than

图7-2-1　灯芯绒织物
Fabrics of corduroy

一、灯芯绒组织的构成 /Construction of Corduroy Weaves

灯芯绒由一个系统的经纱和两个系统的纬纱（绒纬和地纬）组成，或者说由地组织与绒纬组织构成。地组织通常采用平纹、三页或四页斜纹或其他简单组织。绒纬有规律地间隔排列在 2 ~ 6 根连续经纱上（由条绒宽度和特点决定）。在绒根

1.25mm or wider than 4 mm.

Corduroys are composed of one series of warp threads and two series of weft threads, the ground and the pile. Or, they consist of a foundation texture, usually based upon a plain or three-end or four-end twill or other simple weaves, containing tufts of pile disposed at regular intervals on from two to six contiguous warp threads (according to the width

273

之间的浮长正中间将绒纬割断，在地布上绒毛竖直伸出，形成纵向条绒。

图 7-2-2 展示了灯芯绒结构的形成原理，图 7-2-2（b）（c）分别是开毛前、后的横向截面图。图 7-2-3 是该结构对应的组织分解图。从图中可以看出，灯芯绒组织的经纱由地经与压绒经部分（简称绒经）组成，但均属于一个纱线系统。绒经是在割绒后仍然与起绒纬纱交织在一起的经纱。灯芯绒有 2 个纬纱系统——绒纬和地纬纱，其排列比为 2:1，并以平纹为地组织。将图 7-2-3（a）所示灯芯绒组织分解为地组织和绒纬组织。地组织是地经、地纬交织的组织，而绒纬组织是所有经纱与绒纬交织的组织，由浮长线与绒根组成。绒根组织则是绒经和绒纬交织的组织。

and character of cord required). The cuts are made right up the center of the space between the pile binding points, with the result that the tufts of fibers project from the foundation in the form of cords or ribs running lengthwise of the fabric.

Fig. 7-2-2 shows the schematic of the construction of a corduroy fabric, and the cross-sections of the structures before and after cutting are shown in Fig. 7-2-2(b)(c). Fig. 7-2-3 is the decomposition diagram of the corresponding weave. It can be seen from the figures that the ends in a corduroy comprises of ground ends and pile ends, but they act as one series. Pile ends are those that still interlaced with the pile filling yarns after fustian cutting. However, there are two filling series—pile fillings and ground fillings, which are arranged in the proportion of 2:1. Plain weaves are used as ground. The whole weave as shown in Fig. 7-2-3(a) is decomposed into the ground weave, the interlacing of ground ends and ground fillings, and the pile weave where the pile fillings and all the ends intersect. And again, the pile weave comprises of the binding weave and floats of pile filling. The binding weave is the pattern that the pile ends and pile fillings interlace.

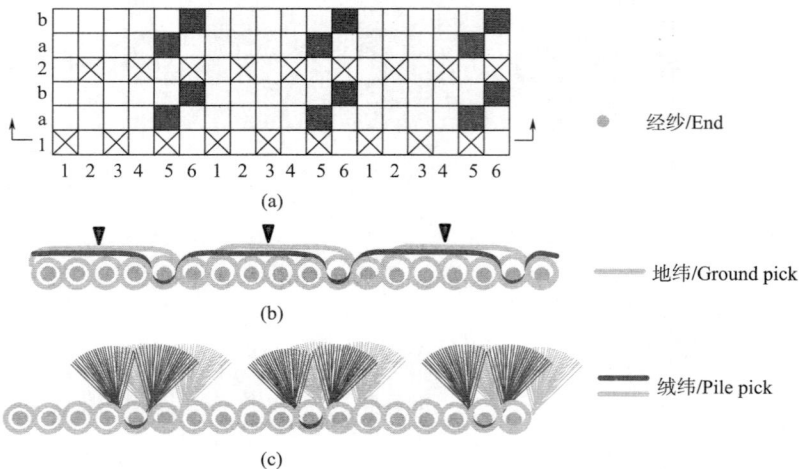

图7-2-2　灯芯绒结构示意图
Schematic of the construction of a corduroy fabric

为了避免绒毛脱落，绒毛应该被经纱或纬纱夹紧。如果经纱密度

To void the fraying out of the tufts of the pile, the pile should be secured by the ground warp threads or

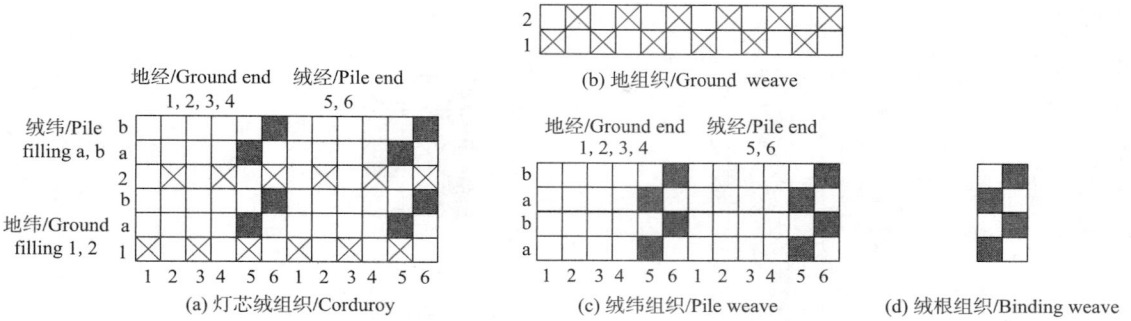

图7-2-3　灯芯绒组织分解图
The constitution of a corduroy weave

大，那么纬密就小，绒毛密度不足。实际上，灯芯绒织物的纬向紧度为140% ~ 180%，经向紧度为50% ~ 60%。如要增大纬纱密度，则要选择恰当的织物组织。灯芯绒地经部分实际就是凸条组织，结构上属于纬二重组织，这种设计使得纬纱之间可以重叠形成两层打紧。但是绒经部分，特别是固结组织，仅仅是普通的单层组织，正常情况下，无法达到如此大的紧度。因此，采用加大绒经张力的措施，使得绒经（如第5经纱）纵向结构由图 7-2-4（a）所示的情况转变为类似图 7-2-4（b）所示的结构，绒经屈曲波减小，纬纱屈曲波增大，但是纬纱占的空间从 L_{w1} 减小到 L_{w2}，使得打纬更加紧密些。或者，可采用斜纹等可密性更大的组织。与此同时，灯芯绒织物的纬向织缩大，约20%甚至更高，给织造带来困难。

ground weft threads. If the fabric is high in warp sett, there will be less filling inserting and result in a low density of pile population per area. Generally, the range of the weft percentage cover is between 140%~180% while the range of the warp percentage cover is only between 50%~60%. To increase the weft density, suitable weave structure is required. The part of the ground ends in a corduroy is actually a Bedford cord weave which belongs to the weft-backed weave theoretically. The weaves must be so selected that successive picks can be beaten-up one on top of another. The pile fillings can overlap the back fillings and a high weft density is achieved. However, the part of pile ends, especially for the binding weave, is usually single structure, and can't achieve such a high weft density in normal looming conditions. Measures have to be taken to tight the pile ends in loom and expel the pile ends tend to be straight as shown in Fig. 7-2-4. Take end No. 5 as example, the longitudinal section of the pile end changes from the structure as shown in Fig. 7-2-4(a) to the structure as shown in Fig. 7-2-4(b) where the filling threads bend more but occupy a space decreased from L_{w1} to L_{w2}, and more fillings are allowed to insert in the structure of the fabric by beating-up. Other binding weaves like twill weave allows more fillings to be inserted. Meanwhile, the cross-over contraction is as high much as 20%, which makes weaving difficult.

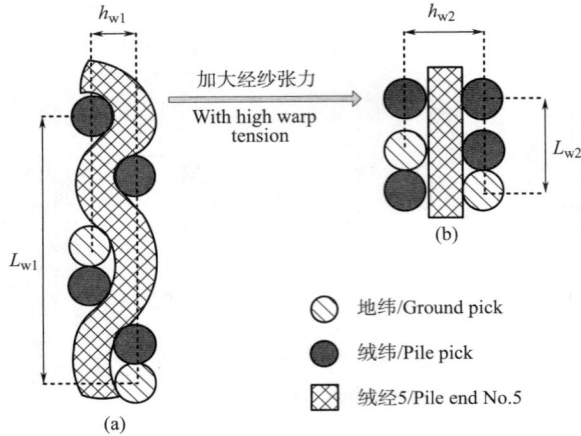

图7-2-4　绒经5结构变化示意图
Schematic diagram of the structure changing at pile end No. 5

二、割绒工序 /Fustian Cutting

在割绒前，需要对织物表面进行硬挺处理，确保割绒在切割轨道上精确控制，切割顺利。织物背面用黏合剂，确保绒毛在地组织上不脱落。割绒后，经过横向刷绒和退捻处理，绒毛耸立在织物表面，形成纵向条绒。最后，织物经烧毛、染色处理。

在割绒机上，薄型圆刀片以等间隔（取决于条绒宽度）安装在轴上高速旋转。灯芯绒在整个织物幅宽方向上，由与条子数量相等的薄型圆刀片同时切割。圆盘刀片穿入纬浮长形成的"跑道"，在浮长中间切割纬纱，绒毛在压绒经处凸起，形成圆润的条子。

图 7-2-5 是割绒原理图。金属导针插入纬浮长"跑道"到割绒点处，其作用是：①引导纬浮长线形成"跑道"，在割绒时张紧浮长，便于割绒；②保证可移动的刀片位

Before fustian cutting, the cloth is prepared for the operation by stiffening the surface float in order to define the cutting races more precisely and to ensure crisper cutting. The back of the cloth is also treated by an application of an adhesive to ensure that the tufts during cutting are not plucked out from the ground structure. The fabrics after cutting undergo a crosswise brushing operation and are then brushed and untwisted in pile finishing, and then the piles are made to stand up on the surface to form a longitudinal pile. Finally, the fabrics are singed and dyed.

All cords across the entire width of cloth are cut simultaneously by means of a corresponding number of thin sharp-edged steel discs, placed at regular intervals (coinciding with the width of cord) upon a mandrel, which extends across the machine and revolves with considerable velocity in the direction indicated. As a fustian knife is thrust along each successive "race" the floats of weft are severed at or near the center, thereby producing tufts of pile which rise on each side of binding warp threads and form the characteristic rounded cords of pile.

The principle of cutting is shown in Fig. 7-2-5. The

Fustian Cutting

于"跑道"中心。

guide wires are inserted in the "races" at the cutting points where cloth approaches the knives. Guide wires serve the functions of ① guiding floats of weft forming "race" to the knives, and tautening them as they are cut; and ② keeping the knives (which are not fixed rigidly, but are placed somewhat freely upon the mandrel) in the center of each "race".

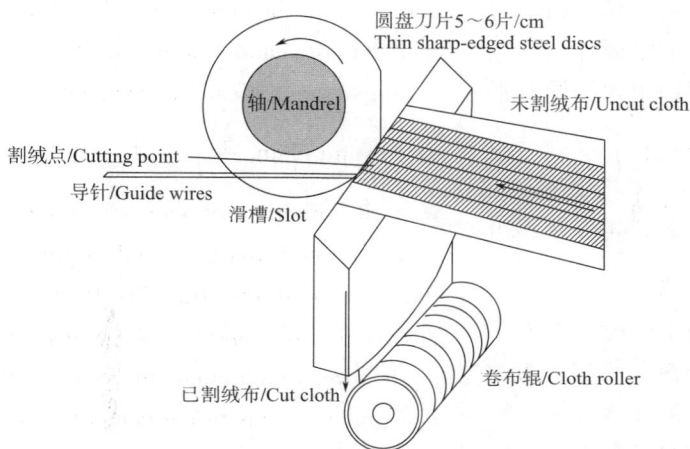

图7-2-5　割绒原理图
The principle of cutting

三、灯芯绒织物设计要点 /Key Points of Designing Corduroy Weaves

设计灯芯绒织物时，需要考虑以下几点：①地组织和绒纬组织的选择；②绒纬与地纬的比例。这些与织物经纬密度因素共同影响条绒的宽度、高度、密度和牢度。

In constructing the corduroy, the chief points to note are ① the weaves that are used for the ground and pile respectively; and ② the ratio of pile picks to ground picks. These factors, together with the ends and picks per unit length of the cloth, influence the length, height, density, and fastness of the pile.

1. 织物密度 /Fabric Sett

灯芯绒织物的密度与绒毛高度和织物牢度有关，纬纱密度比经纱密度大得多，绒毛密度较大。在其他条件相等的条件下，若经纱密度较高，则绒头短而结实，织物紧实但手感不佳。

The sett of corduroy fabric is related to the pile height and fabric fastness. The weft density is much higher than the warp density, thereby the pile is dense. Other conditions equaling, if with a higher warp density, the pile is short and firm, and the fabric is tight but hard to feel.

2. 地纬纱与绒纬纱的排列比 /The Arrangement Ratio of Ground Weft and Pile Weft

绒纬纱与地纬纱的排列比为 2:1、3:1、4:1 或 5:1，通常为 2:1 和 3:1。

The pile wefts are arranged in the proportion respectively of two, three, four, five to each ground pick. Usually, the ratio is 2:1 or 3:1.

3. 地组织的选择 /Choice of the Ground Weaves

在织物中，地组织承受外力并固定绒毛。地组织的选择与织物的手感、纬密、绒毛的牢固性、割绒方便性有关。常用地组织结构有平纹组织、$\frac{2}{1}$斜纹组织、$\frac{2}{2}$斜纹组织、$\frac{2}{2}$纬重平组织、$\frac{2}{2}$经重平组织和平纹变化组织等。

It is the ground weave that withstands external forces and anchors the piles. The choosing of ground weaves is related to the handle, weft density, firmness of piles, fustian cutting of the fabrics. The common structures for the ground weaves are plain, $\frac{2}{1}$ twill, $\frac{2}{2}$ twill, $\frac{2}{2}$ weft rib, $\frac{2}{2}$ warp rib, variations of plain weave, etc.

采用平纹做地组织时，布面平整，手感粗糙、硬挺、坚牢。布的正面不容易脱毛，绒纬纱容易被导针挑起，有利于割绒。但交织次数多，对纬密的增加有限制。在背面，绒根突出，绒毛容易被外力磨掉。

If a plain weave is used as ground, the fabric touches flat, hard, stiff and firm. The front of the fabric won't be depilated easily, and the pile weft is easily be uplift by the guide wire, which is conducive to cutting. However, since more intersections in the construction, the increase of weft density is limited. On the reverse side, the knuckle of the tuft is prominent and it is easy to rub off the pile when exposed to external forces.

斜纹地组织［图 7-2-6（a）］比平纹松，因此可密性大，要得到同样的绒毛牢度，必须加大地纬密度。在绒纬与地纬排列比相同的情况下，能织入更多纬纱，绒毛密集。背面绒毛被地纬浮长保护，改善了脱毛现象。而且，斜纹地灯芯绒更柔软，但不如平纹地组织坚固、平整，故增加了割绒的困难。

A twill foundation weave as shown in Fig. 7-2-6(a) is looser than a plain weave, and therefore, not only permits, but in order to maintain the same firmness of pile, requires a large number of ground picks to be inserted. Hence, with the same ratio of piles to ground picks, more pile picks can be put in and a denser piles formed. The piles are protected by the float of ground weft on the reverse side thereby improving depilation. Also, a cloth with a twill ground is softer and more flexible. However, the fabric is not as firm and flat as plain ground, which may result in difficulty in fustian cutting.

如果在采用平纹地组织时，改变绒经与地纬的交织规律，如图 7-2-6（b）所示，绒根在背面的上下两侧是纬浮长线，凸出在背面，对绒根形成保护，减少绒毛的脱落；同时，绒纬浮长下地部保持平整，不影响割绒的顺利进行。此时地组织称为双经保护平纹变化组织。

When a plain weave is used as ground, if the interlacing at the pile ends and the ground filling is changed as shown in Fig. 7-2-6(b), the binding points of the pile filling at the reverse side are hidden by their neighboring weft floats. Thereby, the knuckle of the tufts is protected in chafing or rubbing and depilation is relieved. Meanwhile, the variated plain ground keeps flat, which is beneficial to fustian cutting.

(a) 斜纹组织/Twill weave　　　(b) 双经保护平纹变化组织/Variated plain

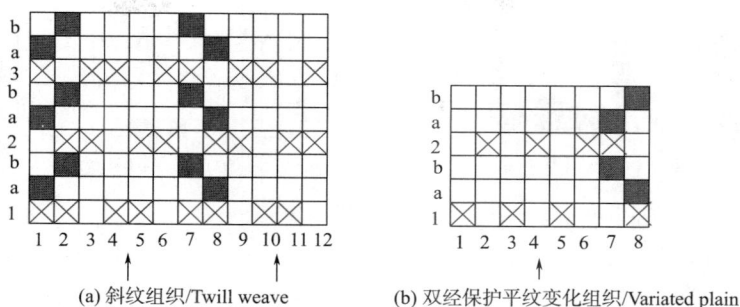

图7-2-6　灯芯绒固结组织
The ground weaves of corduroys

Pile Weaves

4. 绒纬组织 /Pile Weaves

绒纬组织是经纱绒纬的交织规律，很大程度上决定了绒毛牢固程度和绒毛长度。

（1）绒根的固结方式。绒根决定了绒毛的固结方式。如果绒纬仅仅绕过一根经纱，则该固结方式称为松毛固结，或者 V 型固结，如图 7-2-7（a）所示。图 7-2-3 中组织采用 V 型固结。因为交织次数少，绒纬密度大，绒毛丰满，但牢固程度较差，容易脱毛。

若不需要绒毛密度大，或者需要较长的绒毛时，可以通过增加绒纬交织次数的方式提高绒毛固结程度，此固结方式称为紧毛固结。每根绒纬通过与连续 3 根经纱交织，提高了牢度，此时根据其形状称 W 型固结，如图 7-2-7（b）所示。由于交织紧密，使得纬密增加困难，这种固结方式导致布面绒毛不够丰满，但绒毛耐磨程度大大提高。图 7-2-8 是采用该固结方式的阔条灯芯绒组织。

（2）绒纬浮长。绒毛的长度、宽度和密度一定条件下取决于经纱密度和浮长。但是，绒纬浮长也起

The pile weave is the intersection of ends and pile wefts. It affects the fastness of the piles and the height of the tufts to a great extent.

(1) Mode of binding the piles. The mode of binding the piles is determined by the binding weave. The pile which is looped under and held by only one warp thread is called "loosen" pile, or is V–form pile as shown in Fig. 7–2–7(a). In the weave as shown in Fig. 7–2–3, the piles are V– form pile. Due to less intersections, the high density in weft piles and richness of tufts can be achieved but with less fastness, and there is a tendency to have tufts fraying out.

If it is desired to introduce fewer picks per cm, or to make a long pile, the necessary firmness can be secured by interweaving the pile picks more frequently and thus making what is termed a fast pile or a lashed pile. Each tuft of pile is secured by interlacing with three consecutive warp threads, then the pile called W–bound on account of its shape as shown in Fig. 7–2–7(b). The firmer interweaving renders it more difficult to insert a larger amount of weft, and it is generally recognized that in a fast pile the richness of the cloth will suffer, but there is the advantage that the greater firmness gives the cloth better wearing qualities. Fig. 7–2–8 shows a corduroy weave for a broad rib with lashes pile.

(2) Length of the pile weft floats. The length, width and density of pile are determined under certain conditions of warp density and the length of the float. The longer the float

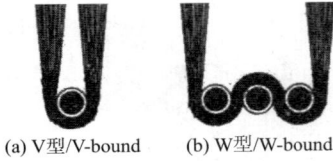

(a) V型/V-bound   (b) W型/W-bound

图7-2-7　灯芯绒绒根固结方式
Two types of binding the piles

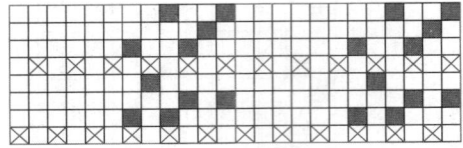

图7-2-8　阔条灯芯绒组织
A corduroy with a broad rib

到很大作用。绒纬浮长越长，绒毛越高，绒条越宽，地组织越裸露。绒毛的高度（mm）=50 × 绒纬浮长 /10cm 内经纱根数。

（3）绒根分布。绒根分布直接影响织物外观，关系到绒条的圆润与露底，对细条和阔条灯芯绒非常重要。若每束绒毛长短差异小，绒条平坦；若绒根占中间的绒毛高，两边的绒毛短，则绒条圆润。由于绒纬在割绒时与两侧的绒根不等距，导致绒毛长短不齐，使得绒条圆润。图 7-2-9 很好地展示了整个原理。图 7-2-10 就是绒根不规则分布的典型例子。由于浮长长短不一，割绒后，绒毛长度不同，形成的绒条就非常圆润。

of pile weft, the higher the pile, the wider the cords, and the bolder the grounds. The height of the piles (mm) =50 × the number of ends covered by the pile wefts / the number of ends per 10 cm.

(3) Distribution of the pile binding points. The positions of the pile binding points affect the aesthetic appearance of the corduroy related to the round of the cords or bare construction, especially for the fine cords or broad cords. The same lengths of the neighboring tufts of pile will form a flat cord. However, the difference in the lengths causes the cords to have a rounded formation, as the long side of the tufts forms the center, and the short side the outer parts of the cords. The rounded or convex formation of cords in corduroys is entirely due to floating weft being cut at unequal distances on each side of binding points thereby causing each complete tuft to be formed with a long and short tuft, which is easily understood on examining Fig. 7-2-9. The irregular method of binding, exemplified in Fig. 7-2-10, is for the purpose of producing a variety of different lengths of floats, which, after cutting, will produce various lengths of pile, and thereby develop cords having a much rounder formation.

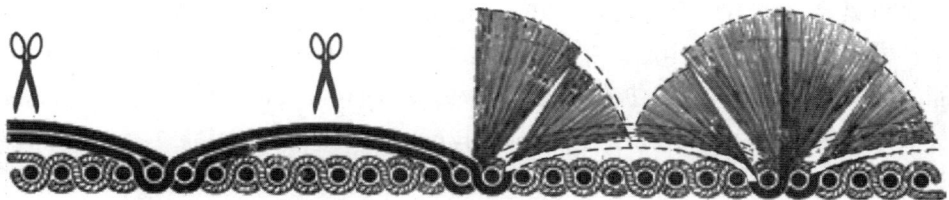

图7-2-9　形成圆润绒条机理
Mechanics of forming a round cord

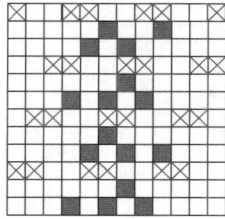

图7-2-10 圆润条子的灯芯绒组织
Round cords

由于在绒纬浮长中心割绒，在图7-2-11所示的两个灯芯绒组织中，尽管绒根分布不同，但是得到的绒条圆润效果基本相同。图7-2-11（a）中，每个条子的绒根一样，浮长相等，故割绒点肯定总与浮长中心有点偏移（而导致绒毛长度不同）。而图7-2-11（b）中，绒根对称，但相邻两个条子长短浮长交替出现，尽管可以准确地在中心割绒，但是割绒后仍然是一长一短。无论如何，割绒后，一侧绒毛总比另一侧长，条绒圆润。

在设计阔条灯芯绒时，单纯增加绒纬浮长，并不能使绒毛耸立，反而会导致露底。此时，要求绒根分布均匀，绒根散开分布（压绒经根数多），所占空间宽；绒根以中间多、两边少分布为宜，如图7-2-8所示。细条灯芯绒仅半数绒纬起绒，另一半绒纬用作相邻条子的绒根，如图7-2-12所示。因此，可以用绒根的分布结合浮长长度判断条子的粗细。

For the two weaves as shown in Fig. 7-2-11, the result of the roundness of the cords is practically the same whichever method of binding is adopted, because the floats are cut in the middle of the space between the pile binding points. The pile binding may be the same in each cord, as shown in Fig. 7-2-11(a), and in this case all the pile floats are equal. All floats will be cut a little out of the center. The plain binding weave of the pile picks may be reversed in alternate cords, as shown in Fig. 7-2-11(b), in which case the design extends over the width of two cords, and each pile pick forms alternately a long and a short float. They will all be cut exactly in the center; yet, in both instances, each complete tuft will be formed with a long and short tuft with precisely similar results. Consequently, in either case, one side of each tuft is longer than the other side, and round cords are obtained.

In designing a corduroy with broad cords, it is not an appropriate scheme to simply increase the float length of the pile fillings because the long tufts of the pile can't be projected on the surface, instead, a bare foundation appears. In this case, it is better to evenly scatter the binding points and have more pile ends so that the pile ends occupy more space. More than half of the binding points are arranged in the middle with the rest points at each side, as shown in Fig. 7-2-8. For the fine cords, only half the pile wefts are cut while the other half work as the binding points of the neighboring cord as shown in Fig. 7-2-12. Therefore, the width of the cords can be judged by the distribution of the pile roots and the length of the pile wefts.

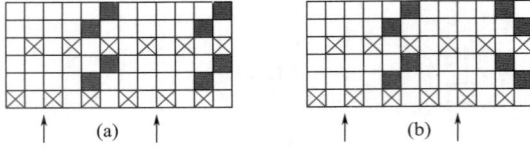

图7-2-11　条绒效果相同的2个组织
Two corduroys with the same effect

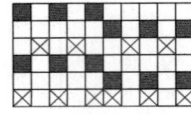

图7-2-12　细条灯芯绒组织
The corduroy weave with fine cords

（4）绒根与地组织的配合。如果绒根经组织点与其相邻的地纬的经组织点在同一梭口，则绒根会受到保护，如图 7-2-6（b）和图 7-2-13（b）所示，同时，有助于打紧纬纱，提高纬纱密度。因此，同样是 V+W 型固结，图 7-2-13（b）所示的组织优于图 7-2-13（a）所示的组织。表 7-2-1 所示为各种灯芯绒组织设计时的主要参数选择范围。

(4) Combing the ground and the binding weave. If at the same shed with the riser of its neighboring ground weft, the binding point will be protected and easily be beaten tight as shown in Fig. 7-2-6(b) and Fig. 7-2-13(b). More fillings can be inserted. Therefore, the weave as shown in Fig. 7-2-13(b) is superior to the weave as shown in Fig. 7-2-13(a) in spite of the same V+W binding. Table 7-2-1 shows the selection of main parameters for different corduroys.

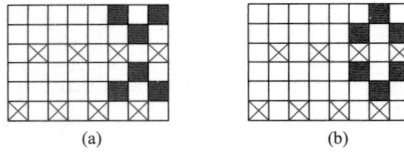

图7-2-13　绒根与地组织配合
Ground and binding weave

表 7-2-1　各种灯芯绒组织设计时的主要选择范围 /The selection of main parameters for different corduroys

类型/Type	超细/Ultra fine cord	细条/Fine cord	中条/Medium cord				粗条/Wide cord			阔条/Broad cord	
条绒宽度/每英寸条子数量/cords /per inch	>18	13~18	8 ~ 12				5 ~ 7			<5	
地组织/Ground weave	平纹/Plain	平纹/Plain	平纹/Plain	纬重平双经/Weft rib+double ends	纬重平单经/Weft rib+single end	斜纹/Twill	平纹/Plain	纬重平双经/Weft rib+double ends	纬重平单经/Weft rib+single end	变化平纹/Plain derivaties	变化纬重平/Weft rib
固结方式/Mode of binding	W	W	V, W	V	W	V	W, W+V	W, W+V	W, W+V	W, W+V	W, W+V
绒纬浮长/Length of the pile weft	3	3 ~ 4	5 ~ 6				7 ~ 8			>8	
毛地比/Ratio of the pile weft and ground weft	3 : 1	2 : 1	2 : 1, 3 : 1				3 : 1, 4 : 1			4 : 1	

四、灯芯绒组织设计 /Designing of Corduroy Weaves

灯芯绒组织的设计步骤如下：

（1）计算组织循环根数 R_j 和 R_w。R_j 由绒纬纱长度和绒根的分布长度确定，R_w 由地纬与绒纬的排列比及地组织确定。V 型固结至少需要 2 根压绒经；W 型固结至少需要 3 根压绒经。

R_j = 绒纬浮长 + 压绒经根数

R_w = 地组织的纬向循环根数 × 绒纬地纬的排列比之和

（2）画出一个组织循环，标记绒经、地经、绒纬和地纬的顺序。

（3）地纬与经纱交织处，填入地组织；绒纬与绒经交织处填入绒纬组织；绒纬与地经的交织处，是绒纬浮长，最后会被割绒，不填任何符号。

例：设计一个灯芯绒组织，要求浮长是 4，绒毛采用 V 型固结，地组织采用平纹，毛地比是 2:1。

①计算 R_w = lcm [lcm (2,2)/2, lcm (2,1)/1] × (2+1)=6。

②浮长为 4，V 型固结，取绒经根数 2，故 R_j =4+2=6。

③确定 6×6 的组织图范围，标记纬纱的顺序。在地纬与所有经纱交织的地方按照平纹绘制组织图。

④V 型固结，在每一绒纬上，与所有的压绒经上只填绘一个经组织点。

整个过程如图 7-2-14（a）所示。为了使组织图更易理解，③和④中的经组织点最好用不同的标记。如果浮长为 6，则 R_j=6+2=8，得到图 7-2-14（b）所示组织。如采用双经保护变化平纹，则地组织

The followings illustrate the procedures of designing corduroy weaves.

(1) Calculate of R_j and R_w. The repeat number of the warps R_j is determined by the length of the pile wefts and the length occupied by the pile roots while the weft repeat number of the ground weave and ratio of the ground wefts and pile wefts determine the R_w of the corduroy weave. There are at least 2 pile ends for V–form pile while 3 pile ends for W–form pile.

R_j =The length of float of the pile wefts + the number of the pile ends

R_w = The repeat size of ground weave in weft direction × sum of arrangement ratios of pile fillings and ground fillings

(2) Determine the full repeat range of the corduroy weave, and mark the order of pile ends, ground ends, pile fillings and ground fillings.

(3) Fill the ground weave where the ground fillings and ends cross; fill the pile weave where the pile wefts and pile ends cross; keep the squares where the pile wefts and ground ends cross blank since they are all weft floats to be cut later.

Example: Design a corduroy weave meeting the following requirements: the float length of the pile weft is 4, V–bound pile, plain ground, and the arrangement of pile weft to ground weft is 2:1.

① R_w=lcm[lcm (2,2)/2, lcm (2,1)/1] × (2+1)=6.

② The length of weft float = 4, V–binding, therefore, the number of pile ends is 2, R_j=4+2=6

③ Draw 6×6 squares for the repeat range of the corduroy weave, and mark the order of pile ends, ground ends, pile fillings and ground fillings. Draw the plain weave where the ground fillings cross the ends.

④ Due to V–binding, draw only one mark on each pile filling in the binding weave.

The whole procedure is illustrated in Fig. 7-2-14(a). For a better understanding of the weave, it is recommend

略有变化，得到图 7-2-14（c）所示组织。

that the risers in step ③ and ④ should be drawn in different marks. If the float length of pile weft is 6, then R_j =6+2=8, and the weave as shown in Fig. 7-2-14(b) is obtained. If varied plain with double-end protection is required for the ground, then the weave as shown in Fig. 7-2-14(c) is designed.

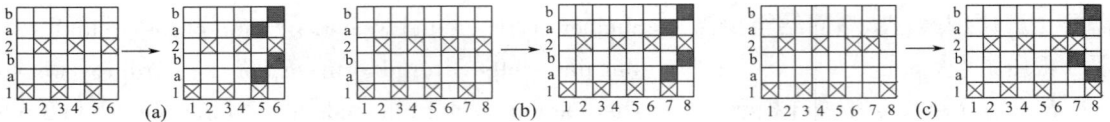

图7-2-14 灯芯绒组织图画法
The designing of corduroys

五、灯芯绒上机要点 /Looming for Corduroy Weaves

由于纬纱密度大，引起较大的打纬阻力、经纱张力和摩擦力。因此，经纱大多采用股线。如果使用单纱经纱，则必须充分进行上浆。纬纱使用单纱。

纬纱的纤维长度应该均匀。纬纱要求毛羽少、棉结杂质少，且均匀一致。如果杂质或棉结过多，会严重阻碍导针的通过，损伤表面，降低切割效率，还会导致破洞。在不影响织造生产的前提下，适当降低纬纱捻度，这样在割绒和整理过程中，纱线容易退捻并保持绒毛圆润。

图 7-2-15 分别显示了特细条灯芯绒、细条灯芯绒、中条灯芯绒、粗条灯芯绒、阔条灯芯绒织物的上机图，箭头指向割绒方向。通常，绒经纱穿入后面的综框，并保持较高的张力，加大纬纱密度和绒纬弯曲程度，以便提高摩擦力，减少掉毛的现象。由于纬向缩率大，

A high filling density is required, and a great weft beating-up action resistance during weaving should be presented, which leads to a great tension and friction on the warp yarns. Therefore, plied yarns are mostly used for warp yarns. If a single warp is used, sizing must be done adequately. Single yarns are used for weft yarns.

The fiber for the weft yarn should be of uniform length. The weft yarn requires less hairiness, kneps or impurities, and should be uniform. The impurities or kneps will seriously hinder the passage of the guide needle, damage the surface, and reduce the efficiency of cutting and result in the tearing holes. The twist of the weft yarn should be reduced properly without affecting the weaving production, so that the pile will be easy to untwist and loosen during cutting and finishing, making the pile round.

The looming plans for corduroy weaves of ultra-fine cord, fine cord, medium cord, wide cord and broad cord are shown in Fig. 7-2-15 respectively. The arrows in the diagram point the direction of cutting. Generally, the pile ends are drawn on the back healds and maintain high tension, so that the pile wefts are allowed to be inserted and are bent more to increase the anchorage, therefore, reduce

布边需要特别设计。

the fraying out of the tufts. Due to high weaving contraction cross-wise, the selvage has to be specially designed.

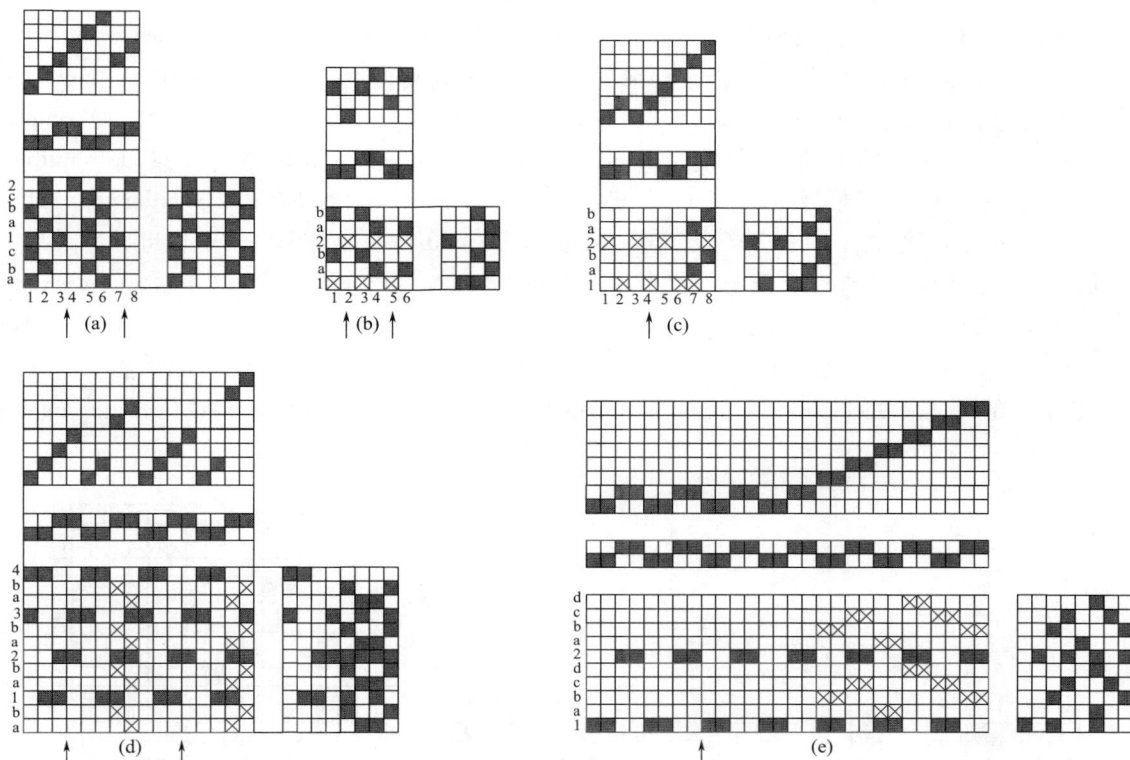

图7-2-15　不同粗细条绒织物的上机图

The looming plans for corduroys

Figure Corduroy Weaves

六、花式灯芯绒 /Figured Corduroy Weaves

在普通灯芯绒的基础上，通过改变绒纬的交织规律、绒根的分布，可以得到花式灯芯绒。需要指出的是，在细条绒上的起花装饰效果要多于宽条与阔条灯芯绒。

Figured corduroys can be obtained by changing the interlacing of pile fillings and binding weave on the basis of common corduroys. It should be noted that more elaborate ornamentation can be produced in narrow than in broad cord effects.

1. 改变绒纬交织规律 /Changing the Interlacing of Pile Fillings

通过改变绒纬交织规律，不让绒纬显示在织物表面。主要有两种思路：

（1）织入提花法。让绒纬与

By changing the interlacing of pile fillings, the pile weft is prevented from showing on the surface. There are two chief methods:

(1) Interlacing the pile wefts and ground ends. The

285

地经交织，成为地布的一部分，如图 7-2-16（a）所示。纬浮长缩短，纬纱高度变低。在割绒时，导针越过这一区域，该区域纬纱不被割断。由于绒纬组织断裂，形成水平横条，织物表面呈现方格状外观，如图 7-2-17 所示。为了便于织入纬纱，一般用经重平组织切断绒纬浮长。

（2）飞毛提花法。取消部分绒根，使绒纬浮长跨越两个条子，在割绒时，两导针中间的一段绒纬被两端剪断并由吸绒装置吸去，该部分全部露底，形成花纹，如图 7-2-16（b）所示。

surplus pile weft is bound in on the underside in the same manner as on the face as shown in Fig. 7-2-16(a). Due to the shortening of the weft floats, the pile wefts are lowered. The guide wire will overpass the weft and leave the weft uncut. Check effect appears on the surface as shown in Fig. 7-2-17 because the horizontal lines can be formed simply by discontinuing the pile weave and inserting the required number of ground picks consecutively. In order that more fillings can be inserted, the warp rib weave is used.

(2) Discarding tufts. Some binding points are removed, therefore, the pile weft is floated loosely over two cords. After the cutting operation, the pile between two guide wires is brushed away as waste. The bare ground forms the figure as shown in Fig. 7-2-16(b).

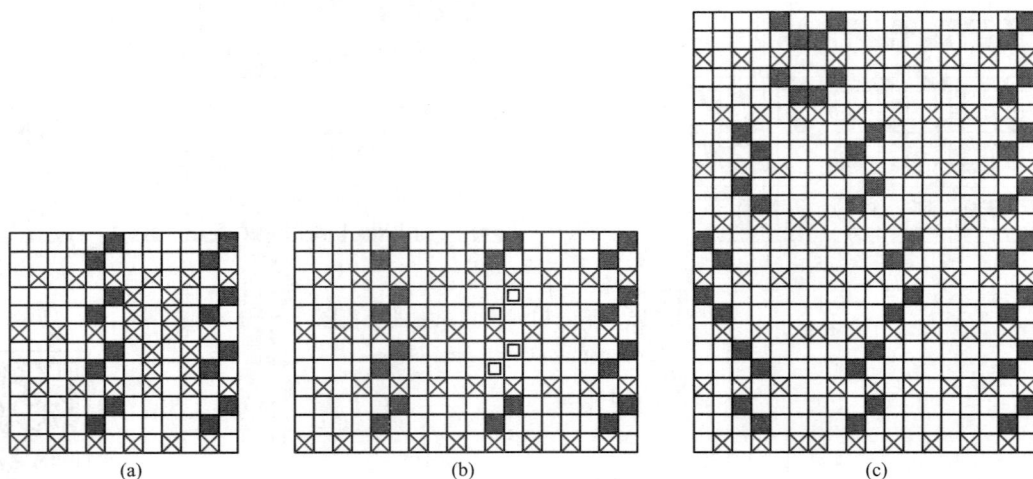

(a)   (b)   (c)

图7-2-16　花式灯芯绒组织
Figured corduroy weaves

图7-2-17　花式灯芯绒织物
The figured corduroy fabric

2. 改变绒根分布 /Changing the Arrangement of the Knuckle of the Tufts

改变绒根的布局，使绒毛长短发生变化，高度不齐，绒条按绒根分布成花型，如图7-2-16（c）所示。

还有多种方式生产花式灯芯绒。比如从不同的位置进刀割绒，形成高低交替、宽窄不同的条绒。或者采用图7-2-18所示的组织，形成双层织物，在正反两面两次割绒，形成双面灯芯绒。用超细条绒的仿平绒织物，还有色织泡泡灯芯绒、提花灯芯绒等。有些条子不割，形成浮长线和条绒交替条带结构。这种变化专门用于提高绒毛耐磨效果。

The positions of the pile wefts and pile ends are arranged so that the float lengths of the pile wefts vary, which makes the height of the piles change accordingly. The cords are formed with the binding weave as shown in Fig. 7-2-16(c).

There are other ways can be used to produce figured corduroys. The corduroys can be cut at different positions on the pile weft. The cords become "high-low", i.e. wide, high ribs alternating with smaller low ones. The double face corduroy weave can be produced by a two-layer fabric and cut twice on both the face and the back as shown in Fig. 7-2-18. Ultra-fine corduroys can be used to imitate velveteen, and seersucker corduroys and Jacquard corduroys are also available. Sometimes, alternate cords are left uncut so that a stripe of tufted cord alternates with a stripe of float construction. This modification is useful in fabrics intended for heavy wear as it results in an improved tuft anchorage.

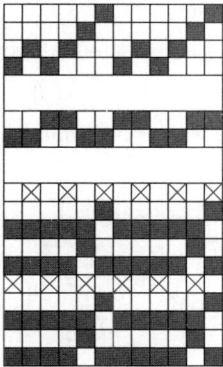

图7-2-18　双面灯芯绒组织
A double face corduroy weave

第三节　经起毛组织 /Warp Pile Weaves

经起毛组织是由两个系统的经纱与同一个系统的纬纱交织而成。地组织是地经纱和纬纱的交织规律，绒经组织是毛经纱和纬纱的

Warp pile weaves are composed of two series of the ends interlacing with the filling yarns which may be one series. The ground is interlaced by ground ends and the fillings while the piles are formed by the pile ends and the

交织规律。经起毛织物的生产方法有三种：①杆织法；②经浮长通割法；③双层织制法。

fillings. Three methods are used to manufacture warp pile fabrics, viz, ① with the aid of wires; ② by cutting the warp floats, and ③ on the principle of face to face.

一、杆织法 /With the Aid of Wires

在生产经起毛织物时，在织物幅宽方向，将起毛杆插入由绒经提起和织物布面形成的梭口中，绒经绕过起毛杆成圈。起毛杆若在一端上方留有刀片，则在抽出时，将线圈割断起绒；若没有刀片，则抽出后形成毛圈。起毛杆有圆柱状、扁平状和开槽等几种形式。绒毛高度由起毛杆的粗细决定，圆柱状适用于平绒，扁平状适用于仿皮毛和地毯织物。绒毛高度 1.5 ～ 25mm 之间。

杆织法经起毛组织的经纱由两个系统组成，即绒经和地经。每织几根地纬，织入一根起毛杆。地经与地纬交织形成地组织；绒经与地纬交织形成固结方式；起毛杆处，形成经浮长。常用地组织有平纹、$\frac{2}{1}$ 经重平、$\frac{2}{2}$ 经重平组织等。固结为 V 型、U 型时，地纬与起毛杆比例为 2：1；W 型时，地纬与起毛杆比例为 3：1。地经与绒经排列比一般为 1：1、2：1。图 7-3-1 是在平纹基础上的杆织法经起毛组织及其结构图，a 纱是绒经，每 2 纬织入一根起毛杆，故箭头指向的纬纱 3 就是起毛杆。可知，该图中，地经：绒经 =2：1，地纬：起毛杆 =2：1，绒经采用 V 型固结。图 7-3-2 是以 $\frac{2}{2}$ 经重平为地组织

To produce the warp pile fabrics a wire is inserted across the width of the warp into a shed formed by the raised pile ends and the foundation of the fabric. Loops are formed when the pile ends wrap around the wire. The wire is either bladed on the upper edge of the strip at one end so that when it is withdrawn the loops are severed to form a cut pile, or unbladed, when a loop pile is left on withdrawal. It can be round rod, flat strip or slotted, etc. The height of the pile is controlled by the thickness of the wire. The round rod is suitable for the production of short pile while the flat strip is for imitation of fur fabrics and carpets. The height of the piles ranges from 1.5mm to 25mm.

The cloth has two series of ends, pile ends and ground ends. A wire and several picks will be alternately inserted across the fabric. The ground weave is formed by ground ends and ground picks while the binding weave is formed by pile ends and ground picks. Warp floats appear where the wire is inserted. The common ground weaves include plain weave, $\frac{2}{1}$ irregular warp rib, $\frac{2}{2}$ warp rib, etc. When V-form pile and U-form pile are applied, the ground pick and the wire are arranged in the ratio of 2 to 1, for W-pile, 3 to 1. Usually, the ground ends and pile ends are arranged in the proportion of 1:1 or 2:2. In the warp a pile weave with a plain ground as shown in Fig. 7-3-1, a denotes the pile end, and a wire is inserted every two ground picks. The pick No. 3 is just the wire as shown by the arrow. It should be observed that the ground ends and the pile ends are arranged in the proportion of 2 to 1, and the ratio of ground wefts and wires is 2:1. The pile ends are V-form pile. The warp pile weave in Fig. 7-3-2 has almost the same

的杆织法经起毛组织，其余参数配置与图 7-3-1 所示的组织相同。

杆织法占用空间大，生产效率低。生产过程中，经常由于起绒杆抽出偏斜，导致割绒偏斜，可能造成起绒杆印等疵点。

parameters except for with a $\dfrac{2}{2}$ warp rib weave as the ground weave.

The method of producing the warp pile with the aid of wires is of low production efficiency, and more floor space is required. In addition, the cloth might have the defects such as wire marks due to bad cutting and the distinct diagonal alignment of pile in the direction of wire withdrawal.

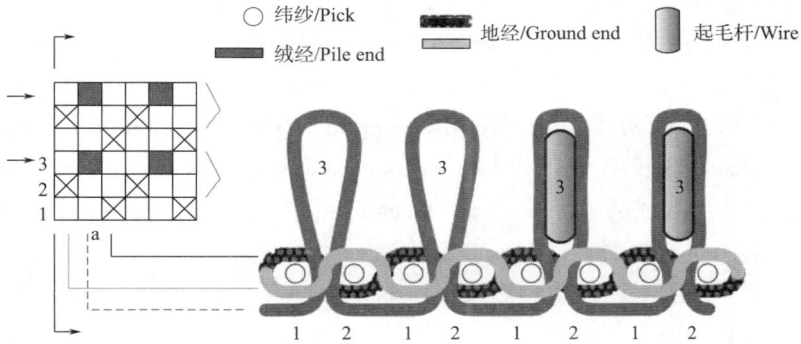

图7-3-1　杆织法原理图
The principle of making a warp pile fabric with the aid of wires

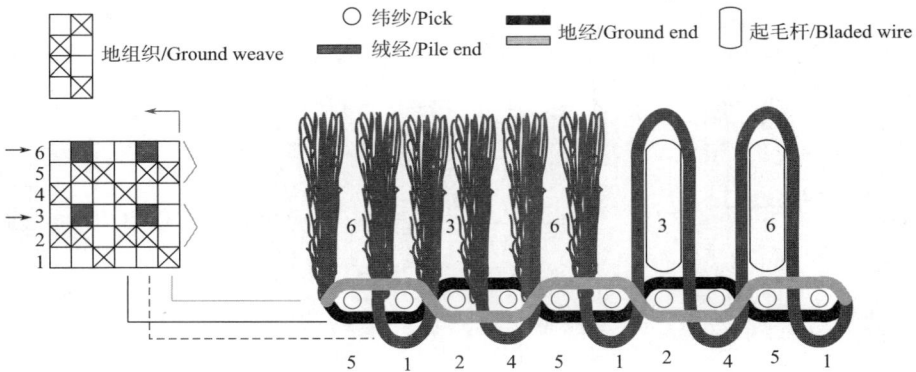

图7-3-2　杆织法经起毛组织
The pile fabric weaving by inserting a wire

二、经浮长通割法 /By Cutting the Warp Floats

经浮长通割法的原理与灯芯绒相同，但要割绒的是经浮长线。经纱由地经与绒经两个系统构成，纬

The principle of producing the warp pile fibric by cutting the warp floats is similar to for corduroys. There are two series of ends—ground ends and pile ends but only one

纱只有一个系统。整个组织由地组织与绒经组织构成，其中地经与纬纱形成地组织，绒经组织由经浮长与绒根固结组织构成。

图7-3-3（a）是一经浮长割绒法的经起毛组织图。地组织是平纹组织［图7-3-3（b）］，经纱Ⅰ、Ⅱ是绒经，绒根采用V型固结。地经与绒经排列比是1:1，经浮长长度是5。将绒根所在纬纱去除后，剩下的组织实际是经二重组织［图7-3-3（d）］，经向可密性大。经纱浮长重叠凸起，在浮长中间沿箭头方向割绒，刷毛后形成绒毛。

因为经浮长采用横向割绒，对机器连续化生产不利，绒毛密度也不如杆织法高，因此用该法起绒并不常见。

series of weft threads. The whole weave is composed of the ground weave and pile end weave, in which ground ends and weft threads form the ground weave, and the pile end weave is composed of warp floats and binding weave.

A warp pile weave by cutting warp floats is shown in Fig. 7–3–3(a). The ground weave as shown in Fig.7–3–3(b) is a plain weave and pile ends are located at Ⅰ and Ⅱ. V–form pile is used according to the binding weave. The ground ends and pile ends are arranged in the ratio of 1:1 while the length of the warp float to cut is 5. If the wefts of the binding weave are removed, then the rest part is virtually a backed warp weave[Fig.7–3–3(d)], which means that it is possible to allow more warps to be woven in the structure. The long warp floats overlap the ground weave and are ready to cut along the direction of the arrow at the center. The piles are then brushed and tufts project on the ground.

The warp floats are cut transversely, which is not good for continuous machine production, and the density of tufts is not as high as by inserting wires. Therefore, the method is not common.

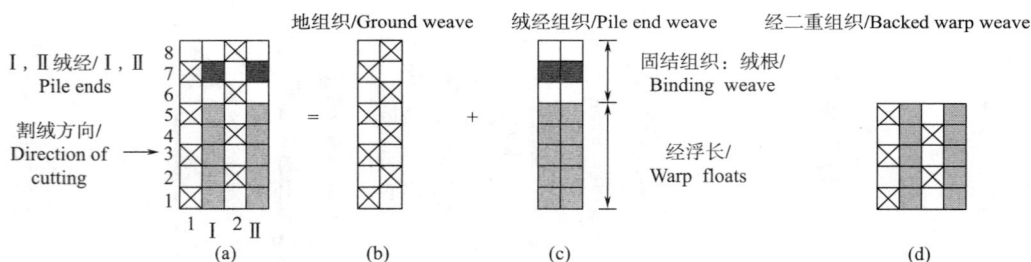

图7-3-3 经浮长通割法经起毛组织构成
Constitute of a warp pile weave by cutting the warping floats

三、双层织制法 /On the Principle of Face to Face

用绒经与上、下两层两块织物交织在一起，在织造过程中，用往复运动刀片将绒经割断，形成两块绒布的方法称双层织制法。下层织

Based on the double–layer method, two fabrics are woven face to face with the pile ends interlocking. During the weaving process, a reciprocating knife cuts through these pile ends to produce two separate pieces of velvet—

物绒毛向上，上层织物绒毛朝下。根据绒毛长度的需要，可以调整两层织物的间距。与起毛杆法、经浮长通割法相比，双层织制法生产效率大大提高。

这种双层织制法的经起毛织物原理看起来与接结经双层组织类似，都需要上下两层组织（地组织）、经纱排列比（表里绒）、纬纱排列比、接结组织（绒经组织）等设计参数，连接两层织物的是绒经或接结经。区别在于这两层织物并不是紧密地接结在一起，而是中间留有一定的距离，用来割绒经。绒根组织不需要考虑接结点遮盖关系，但要考虑绒毛固结牢度与绒毛密度。正因为要求不同，与接结经组织织物在设计和织造中差异很大。根据织机类型，有两种方法生产经起毛（绒）织物，即单梭口法和双梭口法。

the bottom cloth with the pile facing up, and the top cloth with a similar pile facing down. The distance between the ground fabrics is regulated according to the required length of piles. By comparison with the wire insertion system or cutting warp floats there is greatly increased production.

The principle of weaving the warp pile fabrics seems to be similar to that for center-warp stitching double fabrics. Both require the following parameters: upper-layer weave and bottom-layer weave(ground weaves), arrangement of warp threads (top:bottom:pile end or binding end), arrangement of weft threads, and binding weave or pile weave,etc. It is the pile ends or stitching ends that interlace between the two layers of the fabric. The difference lies in two aspects: The two layers are not closely connected, instead, two layers are separated at a certain distance for cutting pile ends. It is unnecessary to consider the binding points in pile weave to be covered by their neighbors as in center-warp stitching, however, the fastness of the tufts of pile and the density of the tufts are the key points to design. For those very reasons, there are a huge difference in designing and weaving between them. Actually, there are two ways to weave the pile fabrics: single-shuttle weaving (or single-shedding weaving), and double-shuttle weaving (or double-shedding weaving).

（一）双层织物织制经起毛的原理 /Formation of the Warp Pile Fabrics

Formation of the Warp Pile Fabrics

采用单梭口法织造时，与普通织法一样，主轴每回转一周，要么形成上层梭口，要么形成下层梭口，织入一根纬纱。单梭口法疵点少，织造可靠，品种适用性强。图 7-3-4 是单梭口织造经起毛织物的原理图，当织物前行，绒经被刀割断后，上下两层织物分别同步卷绕在不同的卷布辊上。

采用双梭口法织造时，主轴每回转一周，同时形成上、下两层

As ordinary manner, in single-shedding weaving, the main shaft rotates once, forming either an upper layer shed or a lower layer shed, and a pick is inserted. The method is preferred for various cloth as its use renders the production of defective cloth less liable. Fig. 7-3-4 shows the schematic of single-shedding weaving. As the fabric proceeds, the pile end is cut, and the top cloth and the bottom cloth are synchronously wound onto separated cloth roller.

In double-shedding weaving as shown in Fig. 7-3-5, when the main shaft rotates once, two sheds are formed,

梭口，同时各投入一根纬纱，如图7-3-5所示。下层梭子沿走梭板投纬，上层梭子在上层梭口的下层经纱之上投梭，其位置比下层梭口的上层经纱略高。剑杆织机则采用双层剑杆取代梭子引纬。

one above the other, and two shuttles are thrown across simultaneously so that one pick is inserted in both the top and the bottom fabric at the same time. The lower shuttle runs on the warp on the race-board in the ordinary way, while the upper shuttle runs on the lower line of the top shed, which is usually higher than the upper line of the bottom shed. In a rapier loom, a twinned (two-tier) rapier is used to replace the shuttle for insertion.

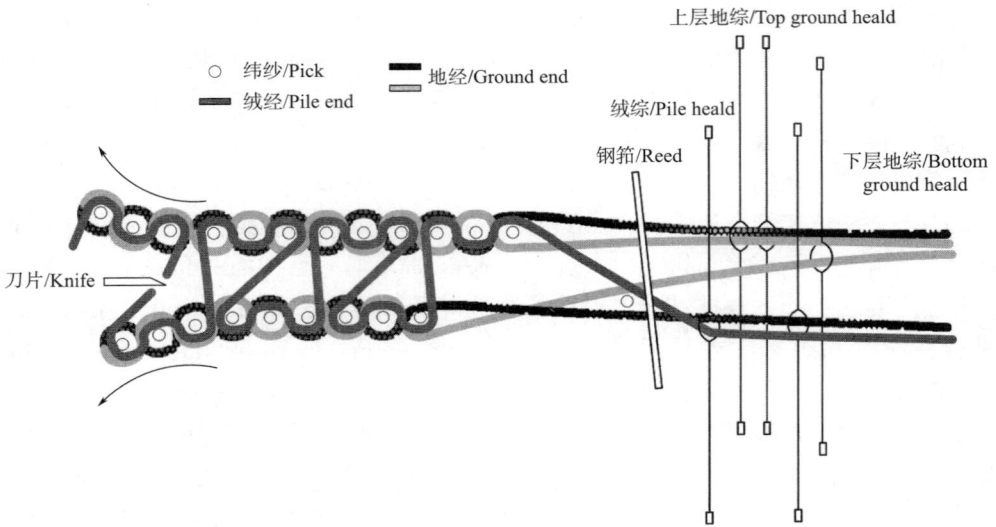

图7-3-4　单梭口织造法
Single-shedding (single-shuttle) weaving

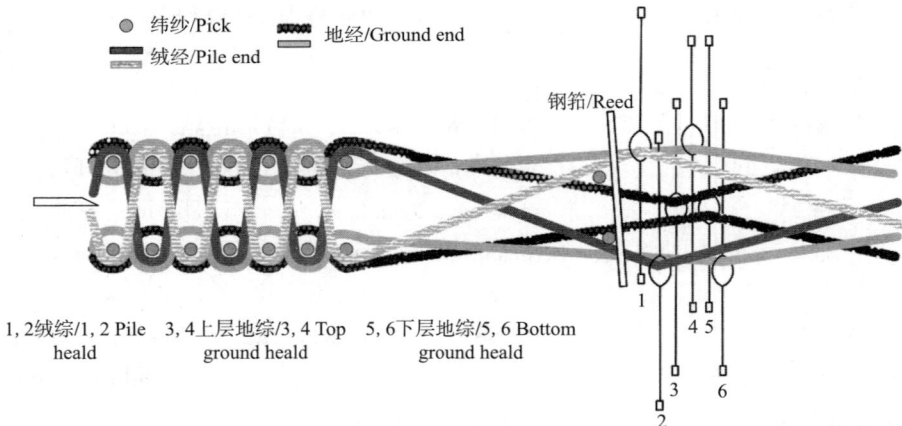

图7-3-5　双梭口织造法
Double-shedding(double-shuttle) weaving

显然，该法生产效率高，但仅适合 1∶1 的纬纱排列，绒经只能与一半的地纬交织，且需要特殊提综装置和图 7-3-6 所示的特殊综丝。绒综的综丝眼在综丝中间位置，上下综丝的综眼大致在 1/3 处，便于开口并减少摩擦。上、下层地经可以分别穿入上层综丝和下层综丝中，也可以同时穿入一根双眼综丝中。上、下综丝的动程与普通织造大致相同，对照图 7-3-5 可知，绒综的升降动程大致是地综的 2 倍，因此，普通提综装置需要改造。

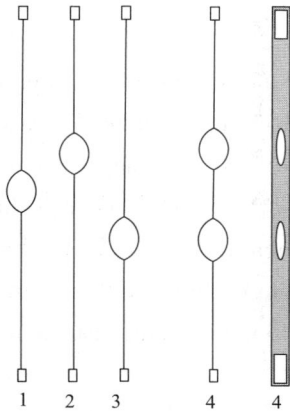

Obviously, the weaving by double-shedding achieves high production. However, the manner only applies to the ground arrangement of 1 top weft and 1bottom weft, and the pile ends can only interweave with half of the ground weft threads. In addition, special shedding mechanism and special heddles as shown in Fig. 7-3-6 are required. The mails of the pile heddles are midway between the shafts but the ground heddles are constructed with the mail eyes one-third of the distance between the top and bottom shaft in order to assist the shedding and to reduce friction. The ground ends in top cloth and bottom cloth are drafted on the top heddle and bottom heddle respectively, or drafted on the same twin heddle. As illustrated in Fig. 7-3-5, the ground ends are moved the usual distance between the top and bottom of their respective sheds but the pile threads move about twice that distance oscillating between the top line of the upper shed and the bottom line of the lower shed. Consequently, the ordinary shedding mechanism requires adjustment.

图7-3-6　双梭口织造法所用综丝
Heddles for double shedding weaving
1—绒综综丝/Pile heddle　2—上层综丝/Top heddle　3—下层综丝/Bottom heddle
4—双眼综丝/Twin heddle

（二）设计要点 /Key Points to Design

1. 绒毛类型 /Types of Tufts

绒毛高度与上、下层地经综眼间的垂直距离和绒经的送经量有关。根据绒毛在织物表面的高度，经起毛织物可分为短毛绒和

The height of tufts is related to the amount of letting-off and the vertical distance between the mails of top heddle and bottom heddle. According to the height of the tufts on the surface of the fabric, the fabric can be divided into plain

长毛绒两种。短毛绒高度一般为
1～3mm，以棉纤维和蚕丝起绒
较多，如棉平绒、乔其绒，用作
坐垫、军服领章、女士高档服装、
帷幕、沙发套和首饰盒垫等。长
毛绒高度7～10mm，一般用羊
毛、马海毛和化纤起绒，如提花长
毛绒，仿兽皮的长毛绒长度可达
15～20mm，可用作长毛大衣呢、
童装、冬季服装衬里等。

　　绒毛高度影响绒毛的手感、色
泽和绒毛的整齐度。绒毛短，则容
易直立，光泽均匀，绒面平整，弹
性足；绒毛长，则容易倒伏，不易
平整，绒面不够光亮，手感粗糙。

2. 绒经固结方式 /The Manner of Binding the Pile Ends

　　绒经的固结方式实际就是固结
组织，影响绒毛的牢固程度和绒
毛密度，用截面图可以精确表达。
图 7-3-7 所示为多种固结方式，每
根绒经的组织规律可用绒根固结形
态、固结梭数、绒头纬纱循环数、
起毛配置、绒经根数等表示。

　　（1）固结形态。按照绒根固结
形态来分，有 V 形固结［图 7-3-7
（a）（b）（g）（h）］、U 形固结［图 7-3-7
（b）］和 W 形固结［图 7-3-7（c）
（e）（f）］。U 形固结增加了固结的
纬纱根数，要比 V 形固结牢固。W
形固结绒经非常牢靠，而图 7-3-7
（i）所示的固结方式是 W 形的扩
展，更为坚固、复杂。V 形固结绒
毛密度最大，W 形固结绒毛密度相
对小，绒毛短、结实耐用，常用于
轻薄织物起绒。此外，还有 V 型和
W 型固结联合形式。

　　（2）固结纬纱根数 (N_{pb})。固结
纬纱根数指绒经割绒后单块织物上

velvet and plush. The height of plain velvet is generally between 1~3mm, by cotton fiber or silk, for the fabrics such as cotton velvet, georgette velvet, used as cushion, military uniform collar badge, dressing clothing, curtain, sofa cover and jewelry box cushion, etc. The height of tufts on plush is 7~10mm, generally with wool, mohair and chemical fiber for the fabrics such as Jacquard plush. The height of the tufts for imitated fur plush can be as high as 15~20mm, and the longer tufts are used as overcoat, children's clothing, winter clothing linings.

　　The height of the tufts affects its feel, luster and uniformity. Short tufts tend to be upright and resist pressure, uniform in luster, smooth and springy; longer tufts are liable to flatten under pressure, which makes the fabric uneven, dim and rough.

　　The manner of binding the pile ends or binding weave affects the fastness and density of the tufts. The manners of the binding can be accurately represented by cross-sectional diagram, as shown in Fig. 7-3-7. Four parameters are used to describe the manner of binding: shape of binding, number picks to bind a tuft, number of picks to complete tufting, mode of lifting pile ends, and the number of pile end series.

　　(1) Shape of binding. By the shape of the bend or knuckle of the pile, piles are referred to as V-form binding pile [Fig.7-3-7(a)(d)(g)(h)],U-form binding pile [Fig.7-3-7(b)] and W-form binding pile [Fig.7-3-7(c)(e)(f)]. Due to being bound by more picks, U-form binding pile is firmer than V-form binding pile. W-form binding gives a short, very firm and durable binding of pile. The binding as shown in Fig. 7-3-7(i) is the extension of W-form pile and is firmer and more complicate. W-form pile is extensively used for light textures. The density of tufts is the highest with V-form pile and the lowest with W-form pile. In addition, V-form pile can be combined with W-form pile.

　　(2) Number of picks (N_{pb}) to bind a tuft. The number picks to bind a tuft is the number of the picks to encircle

经纱 1 个绒根绕过纬纱的数量。当绒根固结形态确定后，绕过绒根的固结纱线纬纱根数就确定了，一般来说，V 形固结纬纱数量肯定是 1，U 形固结纬纱数量一般是 2，图 7-3-7（e）（f）（i）所示都是 W 形固结形态，但固结纬纱数量分别是 3、4 和 5。图 7-3-7（i）所示的是特殊 W 形固结，绕过纬纱数则是 5。经起毛织物中，二梭固结、三梭固结、四梭固结最常用。

（3）绒头循环纬纱数 (N_{pt})。按照绒经割绒后单块织物上 1 根经纱绒头循环的纬纱数量来分，有一梭（纬）绒头［图 7-3-7（a）］、二梭（纬）绒头［图 7-3-7（d）］、三梭（纬）绒头［图 7-3-7（c）（e）］、四梭（纬）绒头［图 7-3-7（b）(f)］，甚至更多，如图 7-3-7（i）所示。图 7-3-7（g）

the knuckle of the pile. When the shape of the bend of the pile is determined, the number of picks to bind the pile is also determined. Generally, 1 pick is employed to bind a V-form pile, 2 picks to bind a U-form pile and 3 picks to bind a W-form pile. For the special W-form pile as shown in Fig.7-3-7(i) 5 picks are required. The manners of 2-pick binding, 3-pick binding and 4-pick binding are most widely used.

(3) Number of picks (N_{pt}) to complete tufting. The number of picks to complete tufting is the number of the picks between the corresponding positions of the two neighboring tufts after cutting. There are 1-pick tufting [Fig.7-3-7(a)], 2-pick tufting [Fig.7-3-7(d)], 3-pick tufting [Fig.7-3-7(c)(e)], 4-pick tufting [Fig.7-3-7(b)(f)], and even more [Fig.7-3-7(i)]. In Fig.7-3-7(g), the tufting in each cloth is complete on two picks although each tuft is bound by 1 pick, therefore, it is actually a 2-pick tufting. H forms a tuft every 4 picks thereby makes a 4-pick tufting. As can be seen that the number of picks to complete tufting

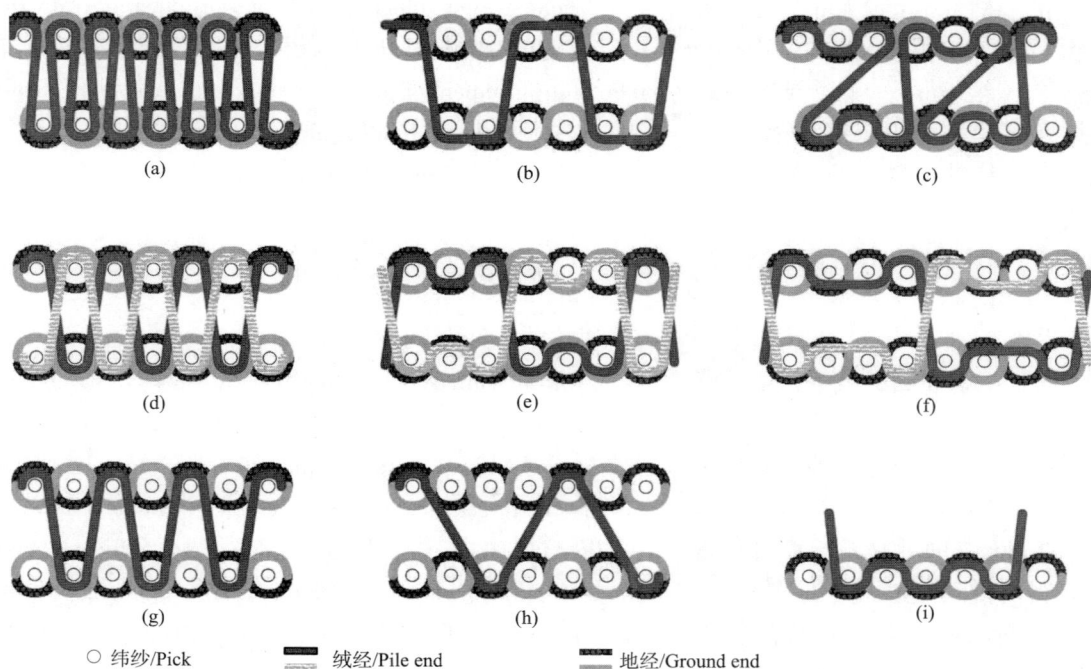

○ 纬纱/Pick　　　　绒经/Pile end　　　　地经/Ground end

图7-3-7　绒经固结方式
The manner of binding pile ends

中绒头为 1 根纬纱固结，但是在单层中一个循环是 2 根纬纱，故是两梭（纬）绒头。图 7-3-7（h）中每 4 纬一个循环，故是四梭（纬）绒头。可以看出，绒头纬纱循环数不小于固结纬纱循环数。

（4）起毛配置。如果绒经与上下两层的纬纱全部交织，则称全起毛配置，如图 7-3-7（a）（c）所示；如果仅与上下层一半的纬纱交织，则称半起毛配置，如图 7-3-7（b）（d）（e）（f）（g）所示；而图 7-3-7（h）则是 1/4 起毛配置，依次类推。

（5）绒经根数。在图（c）、（g）和（h）中，只有一根绒经，这将会导致绒经呈现条形分布。为了使得绒毛在织物表面均匀分布，可以采用 2 根绒经，如图（d）（e）和（f）。

3. 绒毛密度 /Density of Tufts

绒毛密度 D_t（根 /cm²）与单层织物经密 P_j（根 /cm）、单层织物纬密 P_w（根 /cm）、绒经与单层地经排列比 $n:m$、固结纬纱根数 N_{pb}、绒头循环纬纱根数 N_{pt} 有关。其计算式为：

通常织物表面的绒毛密度足够大，才能耸立在织物表面。为了使得绒根牢固，需要加大地经地纬密度，牢固夹持绒毛。若绒毛密度不足，通过后整理中的顺毛整理使得绒毛倒伏得到足够的绒毛覆盖率，并在背面涂胶以获得足够的绒毛牢度。

is no less than the number of binding picks.

(4) Mode of lifting pile ends. If a pile end interweaves with all the picks in both top cloth and bottom cloth, the pile end is completely interwoven or full lift as shown in Fig.7-3-7 (a)(c). The modes as shown Fig. 7-3-1 are called half interlacing or half lift because the pile end only interweaves with half of the wefts at either top cloth or bottom cloth. The modes as shown in Fig. 7-3-7(h) is a quarter lift, and so forth.

(5) Number of pile end series. In (c)(g) and (h), only one series of pile thread is used, so that the tufts of pile are distributed in horizontal lines or ridges. To make the tufts evenly distributed on the surface, two series of pile threads are used as shown at (d)(e) and (f).

Density of Tufts

The density of tufts D_t（tufts/cm²）is determined by the density of the single cloth P_j（ends/cm）, P_w（picks/cm）, ratio of arrangement of pile ends to ground ends of the single cloth $n:m$, number of binding picks N_{pb}, number of picks N_{pt} to complete a tuft. D_t is calculated by the formula:

$$D_t = \frac{N_{pb}P_jP_w n}{2N_{pt}m}$$

The pile surface of a cloth is usually satisfactory when the pile is dense and the tufts stand vertically from the foundation. The ground ends and picks require to be set close enough to nip the pile threads and hold the knuckles of the tufts firmly in position. If the density of the pile is deficient, a sufficient cover can be obtained by laying the pile over in the finishing process, and the required firmness is secured by treating the ground fabric with resin or latex on the underside.

4. 绒经交织截面图绘制 /Drawing of the Sectional Diagram of Pile End Weave

简单的固结方式用全起毛和半起毛方式、固结纬纱数、绒头循环纬纱数和固结形态即可表示，但是一些复杂的固结方式，如图 7-3-7（f）(h)(i) 所示的固结方式必须要用截面图才能正确表示。在全起毛配置时的截面图 7-3-8（a）中，实心圆圈处组织点不容易判断其沉浮属性。为了避免误判，全起毛配置时，绒经截面图中纬纱虽然分为上下两层，但前后位置按照投纬顺序或者在组织图中对应的顺序排列，如图 7-3-8（b）所示。也可以看出，全起毛配置时，地经纱和地纬纱仍然按照双层组织绘制，但绒经按照平纹规律与地纬交织。

在半起毛或者其他配置时，其绒经截面图绘制方法与接结经类似，仍然按照双层模式绘制，如图 7-3-7（d）(e)(f) 所示。组织图绘制方法也一样。

Simple pile end weaves can be represented by the arrangement of pile end lifting, number of binding picks, number of picks to complete tufting, and the shape of binding. Other complicated pile end weaves [Fig.7-3-8(f)(h)(i)] can only be accurately represented by the sectional diagram. However, for the sectional diagram as shown in Fig. 7-3-8(a) for a complete interlacing, it is not easy to distinguish the weaving point at the solid circles. To avoid the misjudgement in the sectional diagram for a complete interlacing or full lift of the pile end weave, all the picks are placed at the order in weave pattern or picking as shown in Fig. 7-3-8(b) although they are arranged in two layers. It can be observed that the pile end is actually interlaced with all the ground picks in plain order. The ground ends and ground fillings are still drawn in the way for two-layer weave.

When the pile end weave is half lift, the sectional diagram as shown in Fig. 7-3-7(d)(e)(f) is drawn in the way similar to that for center-warp stitching double weave, so does the drawing of the weave pattern.

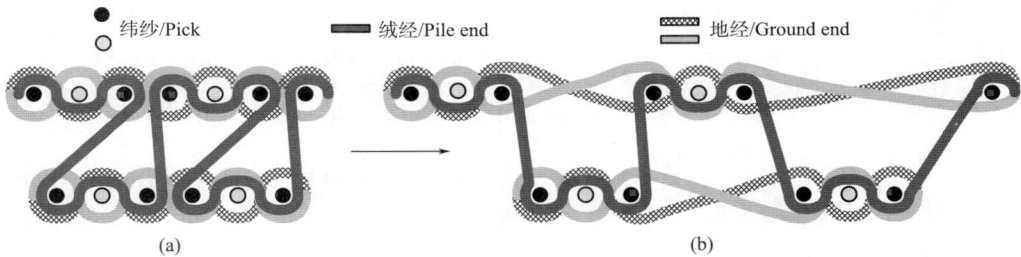

图7-3-8 全起毛配置的截面图

Sectional diagram of the pile end weave with full lift

5. 地组织选择 /Choices of Ground Weaves

地组织选取应根据织物品质确定。若要求织物手感柔软，则采用 $\frac{2}{1}$ 经重平、$\frac{2}{1}$ 纬重平、$\frac{2}{2}$ 方平；

The ground weave should be determined according to the requirements for the fabric quality. If the fabric is expected to be soft, the ground weave should be made

若要求织物挺括，则采用平纹地组织，且配捻度大的经纬纱。

of $\dfrac{2}{1}$ warp rib, $\dfrac{2}{1}$ weft rib, $\dfrac{2}{2}$ matt. If a crisp fabric is desired, the plain weave and the yarns (both the warps and wefts) with hard twist are preferred.

6. 纱线排列比 /Arrangement of Threads

（1）地经绒经排列比有 1∶1、2∶1、4∶1 等，常用 2∶1，即 1 根下层地经、1 根上层地经、1 根绒经。若采用半起毛配置，每层 1 根地经、1 根绒经，其起毛效果与全起毛时的地经∶绒经 =2∶1 相同。

（2）投纬比指纬纱轮流投入上梭口、下梭口的次数。双梭口织造时，只能采用 1∶1。单梭口织造时，与绒经的固结方式有关。V 型固结为 1∶1 或 2∶2；W 型固结为 3∶3 或 4∶4。

(1) The ratio of the ground ends and the pile ends can be set as 1∶1, 2∶1, 4∶1, and so on. If 2∶1 is used, namely 1 ground end at bottom layer, 1 ground end at top layer, and 1 pile end. If the half lift for pile end weave is adopted, and each layer is arranged as 1 ground end and 1 pile end, the effect of tufts is similar to that of the ratio of ground warp∶pile end as 2∶1 when the pile end weave is completely lift.

(2) The arrangement of the top pick and bottom pick to be inserted depends on the number of the sheds formed on a time and the number of the picks to complete a pile tuft. For the double-shedding weaving, the ratio can only be set as 1∶1. For the single-shedding weaving, the ratio is set as 1∶1 or 2∶2 for V-form pile, 3∶3 or 4∶4 for W- form pile.

7. 绒经组织与地组织的结合 /Combination of the Pile End Weave and the Ground Weave

为了达到一定的绒毛抱合和耸立效果，经起绒组织设计时，要考虑地组织与绒经组织的恰当配合。

图 7-3-9 是一种立绒经起绒织物的上机图，上、下层地组织分别如图 7-3-9（a）（b）所示，均为 $\dfrac{2}{1}$ 经重平组织。地经与绒经的排列比为 4∶1，上下层纬纱表里排列比为 3∶3，绒经为 W 型固结，全起毛配置。图 7-3-9（c）所示组织图中，a 为绒经，符号"▲"表示绒经组织点；1、2 等符号表示纱线为上层经纱、上层纬纱，符号"■"表示上层织物经组织点；Ⅰ、Ⅱ等符号表示纱线为下层经纱、下层纬纱，符号"×"表示下层织物经组织点；符号"○"表示投入下层纬

In order to obtain the desired effect of cover and upright, the pile end weave should be designed by considering the combination with the ground weave.

Fig. 7-3-9 shows the looming plans for a warp pile weave with upright tufts. The ground weaves for the top cloth and the bottom cloths are both $\dfrac{2}{1}$ warp rib weave [Fig. 7-3-9(a)(b)]. The ends are arranged 4 grounds, 1 pile, and the wefts 3 top, 3 bottom. Pile ends are W-bound and completely lift with ground wefts. In the weave pattern as shown in Fig. 7-3-9(c), an English character such as "a" represents a pile end, the mark "▲" represents a riser on a pile end, and an Arabic number denotes the thread in top cloth while a Roman character such as I or II for the thread in bottom cloth. The mark "■" represents for a riser at top weave while "×" for a riser at bottom weave. The mark "○" is a lifter, or the crossing point of the top end and bottom weft. The longitudinal cross-section diagram of the weave

纱时，上层经纱提起。截面示意图如图 7-3-9（d）所示。当织物纬向打紧后，上层纬纱 3、4 之间空隙消失，两侧地组织的组织点全部相同，在同一梭口形成浮长线，聚集收拢，紧紧抱合绒经，使绒经直立。在上层纬纱 1 和 6 之间、下层纬纱 Ⅰ 和 Ⅵ 之间、Ⅲ 和 Ⅳ 之间也是如此。为了使绒毛分布更均匀，还使用了另外一根绒经 b［图 7-3-9（e）］。

is shown in Fig. 7-3-9(d). After the wefts are beaten-up, the gap between the wefts No. 3 and No. 4 vanishes. Since the weave points at the ground weave on wefts No. 3 and No. 4 are all at same shed, the float will aggregate and hold the pile end making the pile tuft projecting upright. It is the same for the interweaving points between wefts No. 1 and No. 6, I and VI, III and IV. To make the tufts evenly distributed on the surface, a second series of pile end b is applied as shown in Fig. 7-3-9(e).

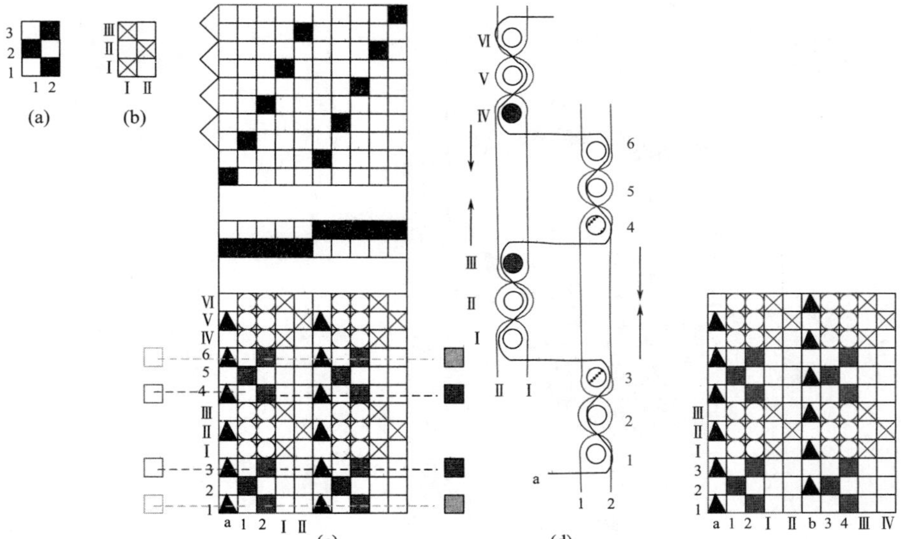

图7-3-9 立绒经起毛组织上机图
The looming plans for a warp pile weave with upright tufts

Fig. 7-3-10
show the looming

图 7-3-10 是一种伏绒（顺毛绒）经起绒织物的上机图，其中上下层地组织均为平纹组织［图 7-3-10（a）（b）］。地经与绒经的排列比为 4:1，表里纬纱排列比为 3:3，绒经为 W 型固结，全起毛配置。所有的符号 a、▲、1、2、■、Ⅰ、Ⅱ、× 和 ○ 表示的意义与图 7-3-9 相同。截面示意图如

Fig. 7-3-10 shows the looming plans for a warp pile weave with laying-down tufts. The ground weaves for the top cloth and the bottom cloths are both plain weave [Fig. 7-3-10(a)(b)]. The ends are arranged 4 grounds, 1 pile, and the wefts 3 top, 3 bottom. Pile ends are W-bound and completely lift with ground wefts. In the weave pattern as shown in Fig. 7-3-10(c), all the symbols—a, ▲, 1, 2, ■, I, II, ×, and ○ has the same meanings as in Fig. 7-3-9. The longitudinal cross-section diagram of the weave is shown in Fig. 7-3-10(d). After

图 7-3-10（d）所示。当织物纬向打紧后，上层纬纱 3、4 之间空隙消失。但地组织中两相邻组织点不同，纱线之间有交错，毛经绒头只能从交叉处倾斜伸出，因此适合制作倒伏的顺毛绒织物。在上层纬纱 1 和 6 之间，下层纬纱 I 和 Ⅵ、Ⅲ和 Ⅳ之间也是如此配置。

the wefts are beaten-up, the gap between the wefts No. 3 and No. 4 vanishes. Since the weave points at the ground weave on wefts No. 3 and No. 4 are different, there is the intersections between the neighboring points. The pile ends can only lean out from the crossing points, therefore, the weave is suitable for the fabric of laying down pile. The situations are the same for the interweaving points between wefts No. 1 and No. 6, I and Ⅵ, Ⅲ and Ⅳ.

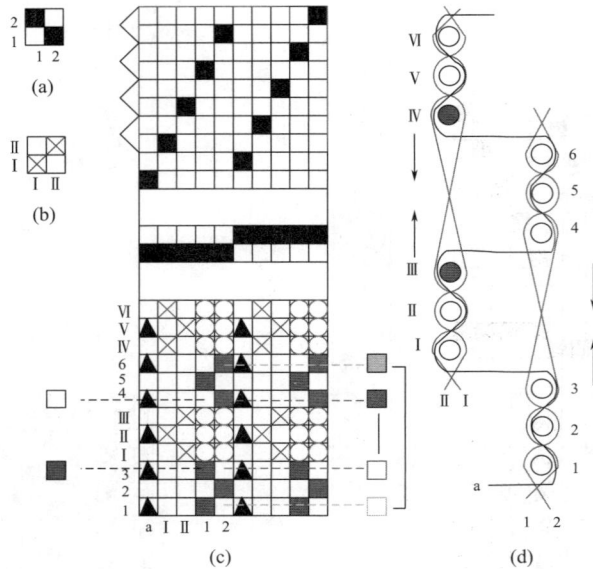

图7-3-10 伏绒经起毛组织上机图
The looming for a warp pile weave with laying-down tufts

图 7-3-11 是地组织采用 $\frac{2}{2}$ 经重平组织时，绒经固结方式对绒毛耸立影响程度的示意图。图 7-3-11（a）中采用 V 型单梭固结，绒毛都是从组织点交错处伸出，容易倒伏。图 7-3-11（b）中采用 W 型三纬固结，绒毛一半从组织点交叉处伸出，另一半从浮长中间伸出，绒毛部分耸立，部分倒伏。图 7-3-11（c）中采用 W 型四纬固结，毛绒头全部从地组织非交叉点处伸出，

Fig. 7-3-11 demonstrates how the pile end weave affects the status of the tufts of pile based on a $\frac{2}{2}$ warp rib. In Fig.7-3-11(a), pile ends are V-bound by one pick and extended out from the intersection of the weaving points thereby are liable to get flatten. In Fig.7-3-11(b), pile ends are W-bound by three picks, and half of the tufts extend from the intersections while the other project from the center of the floats. Therefore, half of the tufts tend to upright, half of the tufts tend to get flatten. In Fig.7-3-11(c), pile ends are W-bound by four picks and project out from the non-intersection point of the ground weave. Consequently, the

地经浮长线可以夹紧毛绒，增加毛绒的抱合和耸立。

tufts are gripped by the floats and are liable to stand upright.

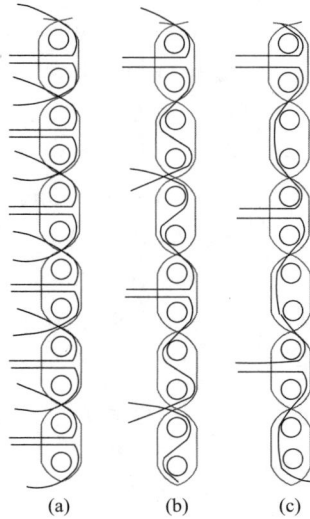

图7-3-11 绒经固结方式与地组织的配合
Combination of the pile end weave and the ground weave

（三）单梭口法经起毛组织图的绘制 /Drawing of the Design with Single-Shedding Weaving

采用单梭口法织制时，经起毛组织的设计步骤如下：

（1）计算 R_j。根据双层组织循环计算方式，得到一个循环内地经总根数 R_{jg}。根据地经、绒经排列比计算绒经数量 R_{jp}，则 $R_j=R_{jg}+R_{jp}$。

（2）计算 R_w。根据上、下层地组织，考虑投纬比、绒毛固结方式，绘制织物的纵向截面图，得到 R_w。在绒经组织采用全起毛、半起毛配置时，R_w 就是双层地组织的纬纱循环根数 R_{wg}。

（3）在意匠纸上框出范围，在经纱方向标出表经、里经、绒经排列序号，在纬纱方向标出上层纬纱、下层纬纱排列序号。

（4）填绘组织点。用多种符号

The steps of designing a warp pile weave for single-shedding weaving are illustrated as following:

(1) Calculate R_j. The number of ground ends R_{jg} in a weave repeat is calculated based on the principle of calculating the size of repeat for a double-layer weave, then the number of the pile ends R_{jp} is calculated according to the arrangement of ground ends and pile ends. $R_j=R_{jg}+R_{jp}$.

(2) Calculate R_w. Based on the ground weaves, wefting, manner of binding the pile ends, the longitudinal sectional diagram can be drafted to determine R_w. If the pile end weave is completely lift or half lift, R_w can be calculated based on the principle of calculating the size of repeat for a double-layer weave.

(3) Determine the range of the weave on the design paper, then the pile ends, top ends, bottom ends, top picks and bottom picks are numbered in order with different characters respectively.

填绘组织点，便于理解。地经地纬交织处方格按双层组织绘制方法填绘。绒经地纬交织处，根据截面图确定组织点。全起毛时用平纹方式填绘；半起毛及更少的交织时，用双层组织方式绘制组织点。

下面设计一种单梭口法织制经起绒织物的上机图。要求如下：上下两层地布均为平纹组织，地经与绒经的排列比为 2:1，表里纬纱排列比为 2:2，绒经为 V 形固结，绒经组织为半起毛配置。

首先，计算双层地组织的 R_{jg} 和 R_{wg}。$R_{jg}=$ lcm［lcm(2,1)/1，lcm(2,1)/1］$\times$(1+1)=4。地经:绒经 = 2:1，故绒经根数为 2，所以 $R_j=$ 4+2=6，$R_w=R_{wg}=$lcm[lcm(2,2)/2，lcm(2,2)/2]$\times$(2+2)=4。因为半起绒方式，故按照双层组织方式绘制组织图。在图 7-3-12 中，（a）（b）是地组织,（c）是绘制的组织图,（d）为修订后的组织图，（e）是经纱纵截面图，（f）是上机图。a、b 为绒经，符号■表示上层织物经组织点；1、2 为上层经、纬纱，符号"×"表示下层织物经组织点；Ⅰ、Ⅱ为下层经、纬纱，符号○表示投入里纬时，上层经纱提起；符号▲表示绒经组织点。

若地组织为 $\frac{2}{2}$ 纬重平，地经与绒经的排列比为 4:1，表里纬纱排列比为 4:4，绒经采用四梭绒头、W 型三纬固结，全起毛配置。为了使得绒毛更平整，一个循环采用 2 根不同交织的绒经。故地组织 $R_{jg}=$ lcm[lcm(2,1)/1，lcm(2,1)/1]$\times$(1+1)=4。地经:绒经

(4) Fill different symbols in the squares for risers for the convenience of understanding. Fill the marks according to the principle of drawing the double weave where the ground ends interweave the ground weft threads. At the square where a pile end interweaves with a ground pick, the property of the points is determined by the longitudinal section diagram. Fill the squares in plain order if the pile end weave is completely lift. However, for half lift or less lift, the squares are still filled based on the principle of drawing a double-layer weave.

Example: Design a warp pile weave on single-shedding weaving. The requirements are as following: the ground weave are both plain weaves, the arrangement of ends are 2 grounds and 1 pile, the order of wefting is set as 2:2. The pile ends are W-form bound and the pile end weave is half lift.

First, the size of the ground double-layer weave is calculated. $R_{jg}=$ lcm[lcm(2,1)/1, lcm(2,1)/1]$\times$(1+1)=4. Due to the ratio of number of ground ends to pile ends is 2:1, the number of the pile ends in a repeat is 2. Thus, R_j=4+2=6 and $R_w=R_{wg}=$ lcm(lcm(2,2)/2, lcm(2,2)/2]$\times$(2+2)=4. Because the pile end weave is half lift, the warp pile weave can be drawn based on the principle of drawing a double-layer weave as shown in Fig. 7-3-12 where the ground weaves are shown in(a) and(b), the warp pile weave is shown in(c) and then changed to (d). (e) and (f) are the longitudinal section diagram and looming plans respectively. In Fig. 7-3-12, a and b represent pile ends, the Arabic number such as 1, 2 denotes the threads in top cloth while a Roma character such as Ⅰ or Ⅱ for the thread in bottom cloth. The symbol "■" represents for a riser at top cloth , "×" for a riser at bottom cloth and "▲" for a riser on a pile end. The symbol " ○ " represents for a lifter, crossing of a top end and a bottom weft thread.

If the ground weaves are both $\frac{2}{2}$ weft rib weave, the arrangement of ends are 4 grounds and 1 pile, and the order of wefting is set as 4:4. The pile ends are W-form bound, the number of binding picks is 3 and tufts are completed on

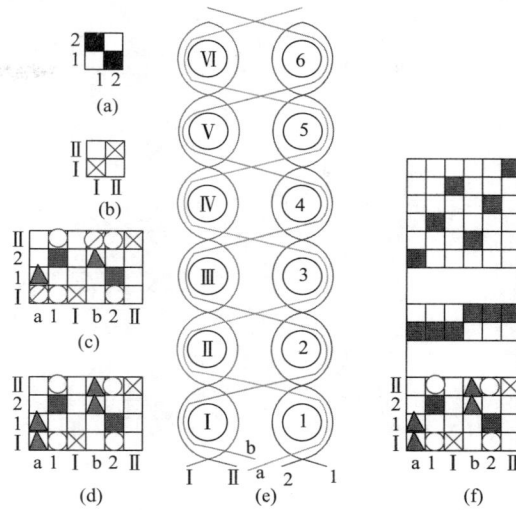

图7-3-12　经起毛组织上机图

The looming plans for a warp pile weave

=4∶1，因绒经根数 R_{jp} 为 2，所以 R_j=8+2=10，R_w=R_{wg}= lcm[lcm(4,4)/4, lcm(4,4)/4]×(4+4)=8。因为全起绒方式，故按照平纹方式绘经向截面图。在图 7-3-13 中，（a）和（b）是地组织，（c）是经纱纵截面图，（d）是上机图。其余符号代表意义与前面的组织相同。

4 picks, and the pile end weave is half lift. 2 pile ends are employed in a weave repeat unit to get uniform tufts. R_{jg}= lcm [lcm(2,1)/1, lcm(2,1)/1]×(1+1)=4. Due to the ratio of number of ground ends to pile ends is 2∶1, the number of the pile ends in a repeat is 2. Thus, R_j=8+2=10 and R_w= R_{wg}= lcm [lcm(4,4)/4, lcm(4,4)/4]×(4+4)=8. Because the pile end weave is completely lift, the warp pile weave can be drawn based on the plain weave as shown in Fig. 7-3-13 where the ground weaves are shown in(a) and(b), the warp pile weave is shown in(c). (d) is the longitudinal section diagram and looming plans respectively. The symbols in the figures have the same meaning as the previous diagrams.

（四）单梭口织造经起毛组织上机 /Looming for Single-Shedding Weaving

1. 织造要点 /Key Points to Weaving

织制经起毛织物时，穿综采用分区穿法。绒经要求张力小，故需穿在前区，上层地经穿在中区，下层经地经穿在后区。

地经张力比绒经大很多，地经、绒经原料不同，织造缩率相差

The warp pile fabrics are divided drafted. The pile ends require small tension, so they are drawn upon the front healds. Top ground ends are drawn upon the central healds while the back ground ends are drawn upon the rear healds.

The tension of ground ends is much higher than of pile

Looming for Single Shedding Weaving

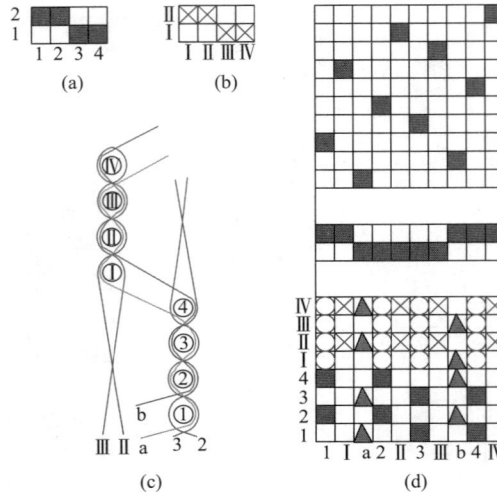

图7-3-13　W型四纬绒头经起毛组织上机图（全起毛法）
The looming plans for a 4-pick tufting warp pile weave (full lift，W-form pile)

悬殊，需要卷绕在不同的轴上织造。绒经需要积极送经，送经量是地经的 5 倍到 10 倍以上。织造棉绒、丝绒等短毛绒织物时，上、下层地经在一个轴上，绒经在另外一个轴上；织造长毛绒织物时，上、下层地经各在一个轴上，绒经另卷一轴，共三个轴。

有梭织机上一般用两把梭子织造，分别织上下层，否则割绒后会造成毛边。为了避免复杂的梭箱编链，表里纬纱排列比出现 3∶3 或 4∶4。

同组绒经、地经必须穿在同一筘齿中；绒经与地经在筘齿中的排列位置有两种情况：①绒经在筘齿中的位置以靠筘齿边。绒经的张力小，地经张力大，如果绒经在筘齿中被夹在地经中间，那么很容易被地经夹住而影响正常的开口运动，造成绒面不良；②绒经仍然放置在地经中间。蚕丝光滑，作为绒经不受影响；织物经密稀疏，为了绒经

ends, and different materials for warp threads are used, so there is a great difference of contraction of pile ends and ground ends in weaving, therefore, the pile ends and the ground ends are wound on the separated beams. Pile ends need to be let off positively and delivered at the ranges from five to ten times or more the length of the ground ends. For velvet of cotton or silk, the top ends and bottom ends are wound on one beam, and pile ends on another. For plush fabrics, the top ends and bottom ends are wound on different beams, and the pile ends are placed on a third beam.

Generally, two shuttles are used to weave the top cloth and the bottom cloth respectively in case of fringe formed in cutting. In order to avoid the complicated shuttle box chain, the arrangement of wefting is 3 top picks, 3 bottom picks, or 4 top picks, 4 bottom picks.

The pile ends and the ground ends of a pair must be placed in the same split. There are two cases for the pile ends to be put in the split. ① The pile end must be arranged at the edge of the dent. If sandwiched between the ground ends in the dent, the pile end of the low tension is likely clamped by the ground ends of the high tension, therefore, the shedding will be affected and a bad pile surface might appear. ② The pile end is placed in the center of the split

能很好地耸立在织物表面。

for velvet because smooth silk acting as the pile end is not affected by the ground ends. When the fabric is open, in order that the pile end to stand upright well on the surface of the fabric, the pile end is placed at the center of the split.

2. 单梭口法织造问题 /Problems in Single-Shedding Weaving

单梭口织造可能存在如下问题：①上、下两层距离过大，织造困难；②上下层纬纱一般是相同的，若用1把梭子，双层割开后，采用1:1投纬，双侧毛边；若采用2:2投纬，单侧毛边。为了避免毛边，用2把梭子，但需要多梭箱装置。如果纬纱排列比为3:3或者4:4，在上、下层梭口连续轮流投纬，某层连投几纬后，织口不齐，难以打纬紧密；若1:1投纬，梭箱编链非常复杂。

There are still some problems encountered in single-shedding weaving: ① the long distance between the two layers makes weaving difficult; ② If a single shuttle is used and the arrangement for wefting is one top, one bottom, fringe selvage is formed at both side of the fabric after cutting. If the weft yarns are arranged in proportion of 2 top to 2 bottom, fringe edge is formed at one side after cutting. Two shuttles can be used to avoid fringe edge. However, box motion is needed. Moreover, due to the arrangement of wefting of 3:3 or 4:4, after continuous insertions of fillings in top cloth or bottom cloth, the fells of the top cloth and the bottom are not at the same level, and resulting in difficulty to beat the weft tightly. If the weft yarns are arranged in the ratio of 1:1, it is too complex in designing the shuttle box chain.

（五）双梭口织造经起毛组织上机 /Looming for Double-Shedding Weaves

Designing for Double-Shedding Weaving

1. 特点 /Characteristic

双梭口织造经起毛织物不存在毛边和打纬织口不齐的问题，一次引入两纬，织造效率提高。

In double-shedding weaving, the problem of forming a fringe selvage or uneven fell does not exit. Two fillings are inserted at a time, and weaving efficiency is improved.

2. 上机图设计 /Designing for Double-Shedding Weaving

双梭口织造中，绒经至多与一半地纬交织。因为双梭口的上、下层经纱同时运动，所以提综图是以组织图上、下层各一纬为提综图的一横行（相当于一纬），地纬上、下层排列比只能是1:1。

图7-3-14所示为三纬绒头长毛绒织物上机图。地组织为$\frac{2}{2}$纬重平，W型固结，半起毛配置，地

In double-shedding weaving, the pile ends can only interweave with at most half of the ground picks. Because the top ends and bottom ends in the double shed move simultaneously, 2 horizontal spaces in the lifting plan just control the shedding of one filling yarn. The top weft and bottom weft can only be arranged at 1:1.

Fig. 7-3-14 shows the looming plans for a 3-pick tufting flush fabric. The ground weaves for the top cloth and the bottom cloth are both $\frac{2}{2}$ weft rib weave[Fig. 7-3-

经与毛经的排列比为 4:1。图 7-3-14 中，（a）和（b）是地组织，（c）是经纱纵截面图，（d）和（e）分别是单梭口织造与双梭口织造的上机图。组织图中符号代表意义与前面的组织相同，即符号"■"表示上层经纱或毛经在上层纬纱之上；符号 × 表示下层经纱在下层纬纱之上；符号"▲"表示毛经在下层纬纱之上，即处于双梭口的中间位置；符号"○"表示上层经纱在下层纬纱之上；符号"□"表示各种经纱在纬纱之下。

双梭口织造不能采用纹钉控制综框提升，而是采用凸轮开口装置，故上机图的右侧是提综图，不是纹板图。在提综图上，每 2 横行规律完全相同，标明上下梭口同时形成。此时，符号"■"表示上、下层地经及毛经在各自梭口的上方位置（上层地经在上、下层纬纱之上；下层地经在上层纬纱之下，下层纬纱之上；毛经在上、下层纬纱之上）；符号"日"表示上、下层地经及毛经在各自梭口的下方位置（上层地经在上层纬纱之下，下层纬纱之上；下层地经在上、下层纬纱之下；毛经在上、下层纬纱之下）；符号"◆"表示毛经在上、下层纬纱之间的中间位置。

图 7-3-14（e）所示为采用单梭口织造法的上机图。可以看出，除了提综图不同外，单梭口与双梭口两种方式的各自组织图的不同之处仅仅在于纬纱的排列顺序，相同的纬纱上对应的组织点完全相同。

14(a)(b)]. The ends are arranged as 4 grounds, 1 pile, and pile ends are W-bound and half-lift with ground picks. Fig. 7-3-14(c) is the longitudinal cross-section diagram of the weave. The weave patterns for the double-shedding weaving and single-shedding weaving are shown in Fig. 7-3-14(d)(e) respectively. In Fig. 7-3-14(d), an English character such as "a" represents for a pile end, the mark "■" for a top end or plie end passing over a top weft; the mark "×" for a bottom end passing over a bottom weft; the mark "▲" for a pile end passing over the bottom weft; the mark "○" for a lifter, or a top end passing over a bottom weft; and a blank for any end under a ground weft.

Pegged lags can't be used in controlling the healds in the double-shedding weaving. Instead, the healds are controlled by a cam shedding system. Therefore, the plan beside the weave design is a lifting plan rather than a pegging plan. It is observed that every two horizontal spaces are identical, which means that two sheds are formed simultaneously. In a lifting plan for the double-shedding weaving, the mark "■" represents that all the ends are at the upper line of shedding of itself respectively. That is, the top ground end is above the top picks and bottom picks; the bottom end goes under the top picks but above the bottom picks; the pile end is above both the top picks and bottom picks. The mark "日" represents that all the ends are at the lower line of shedding of itself respectively. In other words, the top ground end is under the top picks but over the bottom picks; the bottom end goes under the top picks and the bottom picks; the pile end is under both the top picks and the bottom picks. The mark "◆" represents that the pile end is at the middle position of the top picks and bottom picks, or at the lower line of the top shed or the upper line of the bottom shed.

Fig. 7-3-14(e) shows the looming plans for the same fabric by single-shedding weaving. As will be seen, the difference between double-shedding weaving and single-shedding weaving only lies in the order of wefting except for the lifting plan, and the marks of interweaving for each pair of corresponding picks are just the same.

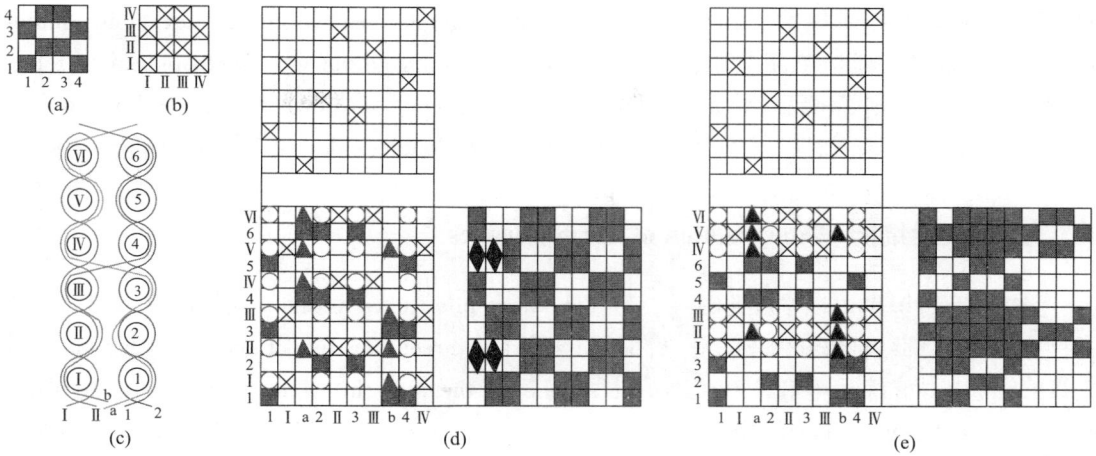

图7-3-14 三纬绒头长毛绒两种织造方法对比
Two manners of weaving a 3-pick tufting flush fabric

第四节 毛巾组织 /Terry Weaves

毛巾组织是经起毛组织，部分经纱在织物表面形成毛圈。毛巾组织是由两个系统的经纱（毛经纱和地经纱）和一个系统的纬纱构成。地经纱与纬纱形成地布，是毛圈的附着基础。毛圈是由相对松弛的毛经纱系统形成的，如图7-4-1所示。毛圈使织物具有良好的吸湿性、柔软性、蓬松性、保温或保暖性，非常适合用作浴巾、垫子、窗

Terry is a class of warp pile structure in which certain warp ends are made to form loops on the surface of the cloth. Only one series of weft threads is used but the warp consists of two series of threads, the ground and the pile. The former produces with the weft the ground cloth from which the loops are formed by the pile ends project. These loops as shown in Fig. 7-4-1 are formed by an extra series of comparatively slack pile warp threads. The loops result in the properties such as good moisture absorption, softness, considerable bulk, heat preservation or warmness, which

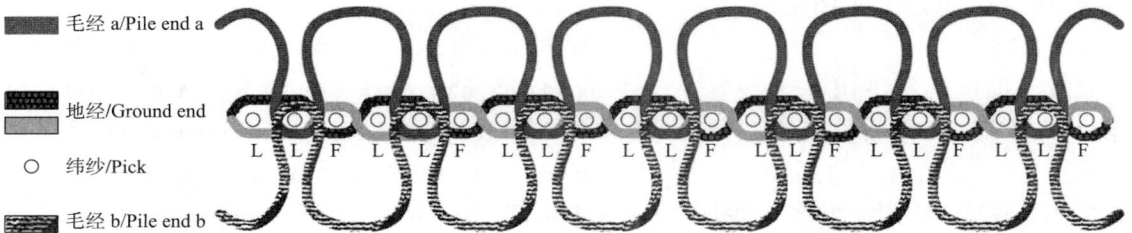

毛经 a/Pile end a
地经/Ground end
○ 纬纱/Pick
毛经 b/Pile end b

图7-4-1 毛巾织物结构示意图
Schematic of a terry structure

帘、女士长袍和睡袍等织物。如果仅在一面有毛圈，则是单面毛巾；若在双面有毛圈，则是双面毛巾。

are eminently suitable for towel, mats, curtainings, ladies' overcoats and dressing gowns, etc. The loops may be formed on one side only or on both sides of the cloth thus producing single-sided and double-sided structures respectively.

一、毛巾组织的结构 / Construction of Terry Weaves

毛巾起毛组织也可以根据形成一行毛圈所织入纬纱根数分为三纬毛巾、四纬毛巾、五纬毛巾和六纬毛巾，如图7-4-2所示。每行毛圈织入更多的纱线是为了更好地固定毛圈，但是会减少毛圈密度，且成本提高。如果六纬毛巾要达到三纬毛巾的毛圈密度，纬纱密度必须加倍。绝大多数毛巾织物使用三纬毛巾。

图7-4-2（a）～（d）所示的组织为三纬毛巾组织，使用非常广泛。每三根纬纱形成一个毛圈，地组织为$\dfrac{2}{1}$变化经重平组织。图7-4-2（e）（f）所示的组织为四纬毛巾组织，每四根纬纱形成一个线圈，地组织为$\dfrac{3}{1}$变化经重平组织。图7-4-2（g）是五纬毛巾组织，而图7-4-2（h）（i）则是六纬毛巾组织。

地经与毛经的排列比一般是1:1（单单经单单毛）、1:2（单单经双双毛）和2:2（双双经双双毛）。地经与毛经的排列比对毛圈结构基本没有影响。

由于毛巾的经纱由毛经纱与地经纱组成，可以将毛巾组织分成毛圈组织和地组织两部分，如图7-4-3所示。毛圈组织是毛经纱和纬纱之间的交织规律，地组织是地经纱和纬纱的交织规律。

The terry pile weave as shown in Fig. 7-4-2 can also be classified as three, four, five or six-pick terry fabrics by the number of the wefts inserted for successive horizontal rows of loops. The object of inserting a greater number of picks for each row of loops is to produce a superior fabric and to bind pile warp threads more securely to the foundation texture. However, the density of loops is reduced and the cost is increased. To produce the same pile coverage in a 6-pick, as in a 3-pick cloth twice as many picks per cm are required. Most of terry fabrics are constructed with three-pick.

The 3-pick terry structures as shown in Figs. 7-4-2(a) ~ (d) are employed most extensively. The loop is formed every 3 filling yarns. The ground weave is $\dfrac{2}{1}$ varied warp rib weave. In 4-pick terry structures as shown in Fig. 7-4-2 (e)(f), the loop is formed every 4 filling yarns. The ground weave is $\dfrac{3}{1}$ varied warp rib weave or $\dfrac{2}{2}$ warp rib weave. Fig. 7-4-2(g) shows a 5-pick terry while (h)and(i) are both 6-pick terry.

The ratio of the ground end and the pile end is usually arranged as 1:1, 1:2 or 2:2, which has little effect on the structure of the terry cloth.

Since the ends in a terry cloth comprise of the pile ends and ground ends, a terry weave can be broken into 2 parts: the pile weave and ground weave, which are shown in Fig. 7-4-3. The pile weave means the interlacing pattern formed by the pile end and the fillings. The ground weave means the interlacing pattern formed by the ground end and the fillings.

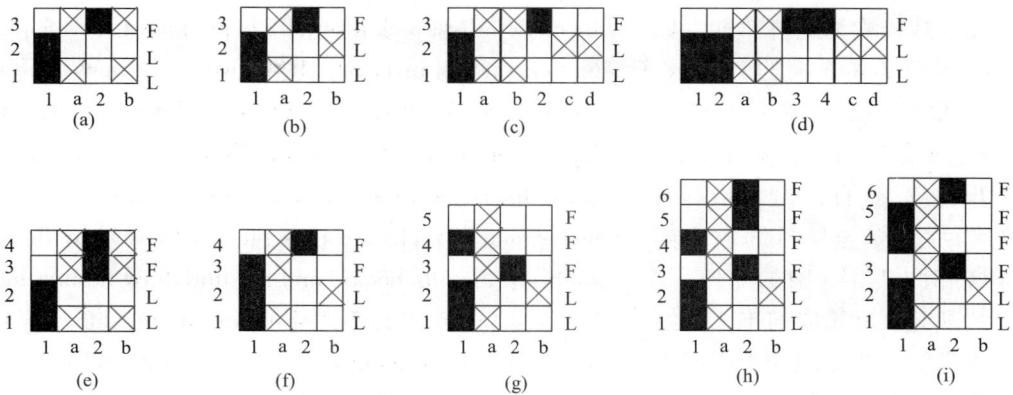

图7-4-2　常见毛巾组织
Various terry weaves

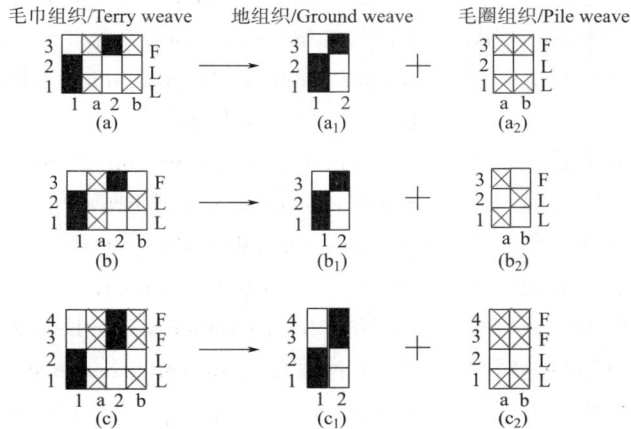

图7-4-3　毛巾组织分解图
Constitute of terry weaves

二、毛圈的形成原理 /Formation of Terry Weaves

（1）利用起毛杆（可认为是纬纱）起圈。起毛杆织入经纱梭口后，将其抽出，浮在该起毛杆的所有经纱都形成同等数量的毛圈。这种方法已在未割的丝绒中使用。

（2）利用毛巾打纬机构起圈。在织造时，将几根纬纱引纬离梭口（上一根纬纱织入处）一定距

(1) By means of wires (as if they were picks of weft) that are inserted in the warp sheds and subsequently with drawn, all warp threads pass over the wires to form a corresponding number of loops. The method has been introduced in the uncut velvet fabrics.

(2) The common way is by means of what are known as "terry pile motions" whereby, during weaving, several picks of weft (or loose wefts as shown at L in Fig. 7-4-2

离，产生一定空隙，即采用短打纬。这几根纱线称为短打纬纱，如图7-4-2和图7-4-3中的L所示。再织入一根纬纱后，这几根纬纱都被一起向前推到织物中的最终位置（长打纬）。直接由钢筘打向织口的纬纱，称为长打纬纱，如图7-4-2和图7-4-3中的F所示。每组纬纱都会被钢筘向前推，毛经纱按照预定的方式在织物单面或者双面弯曲，在织口处竖起成圈，如图7-4-4所示。这种成圈形式为毛巾线圈或非割绒毛圈，用于毛巾织物，仅靠改变梭口和钢筘的相对位置在经起毛组织上形成毛圈。很明显，毛圈的相对密度与每行毛圈所织入的纬纱数量成反比。

从原理上分析，毛巾起毛牢固度要比前面其他任何一种起绒组织都好。毛经纱与地组织交织更为紧密，减少了在使用时毛圈滑脱的可能性。无论是几纬毛巾组织，打纬时只有第一根和第二根纬纱是短打纬，其余纬纱都是长打纬。为了方便阐述，本书中的毛巾组织图的第一纬纱是第一短打纬的纬纱。

and Fig. 7–4–3) are inserted a short distance from the fell of the cloth (or last pick inserted), to produce a short gap, after another pick is inserted, all the wefts are pushed forward together to take their final place in the fabric. The wefts that are put forward to the fell directly by reed are called fast wefts as shown at F in Fig. 7–4–2 and Fig. 7–4–3. As each group of picks are thus pushed forward by the reed, pile warp threads buckle and are uprighted to form loop at fell as shown in Fig. 7–4–4 either on one side only or on both sides of the cloth as predetermined, and so develop the characteristic loops of pile known as "terry loop, or uncut pile". The loops in the warp–pile fabric are created by varying the relative positions of the fell and the reed. It should also be observed that the relative density of loops of pile is either greater or less in a measure which is inversely proportional to the number of picks inserted for each horizontal row of loops.

Terry pile fabrics constructed on this principle are relatively stronger and firmer texture than either of the previous examples; also pile warp threads are more firmly interwoven with the foundation texture, and therefore less liable to be accidentally or otherwise withdrawn when the fabric is in use. No matter how many picks are inserted for a horizontal row of loops, the first two fillings are of loosen picks while the remains are all fast picks. For a better illustration, the first loosen filling yarn is the first weft in the terry weaves of this book.

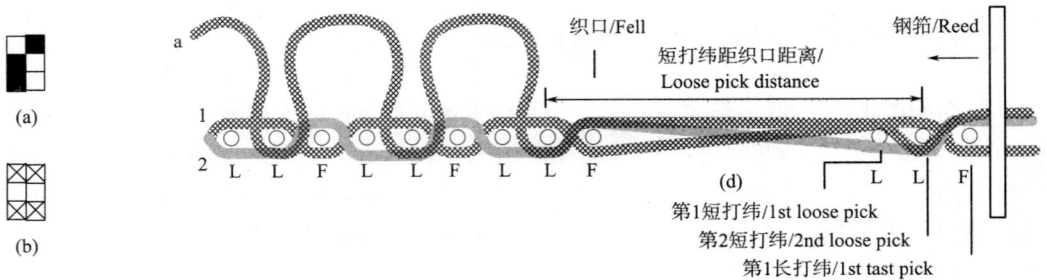

(a)

(b)

织口/Fell　　钢筘/Reed

短打纬距织口距离/Loose pick distance

a

1

2　L　L　F　L　L　F　L　L　F　　　　　　(d)　　　　L　L　F

第1短打纬/1st loose pick
第2短打纬/2nd loose pick
第1长打纬/1st tast pick

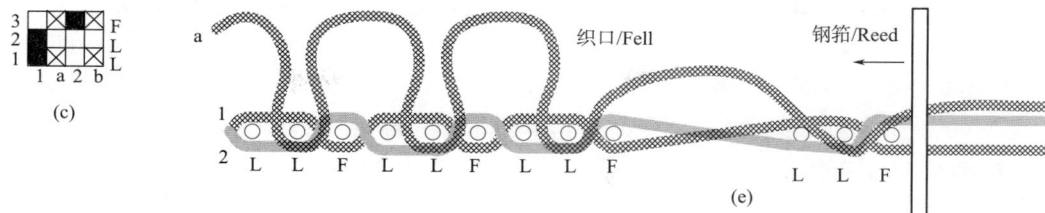

图7-4-4　三纬毛巾成圈原理
Mechanism of forming loops in a 3-pick terry weave

三、毛圈成形良好的关键 /The Conditions Required for Terry Loops

（一）特殊的打纬机构 /Special Beating-up Motion

Special Beating-up Motion

毛巾织物只能在配备特殊打纬机构的织机上生产。通常，钢箱有两个打纬位置，箱座上有专门机构控制毛圈形成过程中的不同打纬位置。不同的纬纱，箱座动程不同。短打纬时，箱座摆动到离织口一定距离后停下；长打纬时，箱座运动距离稍长些，在织口处停下。

在设计毛巾织机时，其箱座在前两个短打纬时，仅摆动到距离织口一定的距离处。在形成一行毛圈的第三纬和其后的纬纱打纬时，箱座摆动距离长。这意味着，只有前两纬是短打纬，纬纱被推到离织口一段距离停下，当第三纬投纬后，三根纬纱被一起向前推。在纬纱沿着张紧的地经从临时位置推到最终位置的同时，这三根纬纱与松弛的毛经纱之间的摩擦力已经足以使得所有的毛经纱也向前行并弯曲，形成织物表面的毛圈。如果毛经浮在上方，则在正面成圈；若毛经沉在下面，则在背面成圈。成圈原理示意图如图 7-4-4 所示。

The production of terry fabrics is only possible on the looms with special equipped beating-up motion. In general, the reed has two beat-up positions. The sley has a special mechanism built in which allows different beat-up positions for pile formation. At a different pick, the movement of the sley changes. For a loosen pick, the sley moves a shorter distance and stops a distance from the fell while for a fast pick the sley moves a longer distance and stops at the fell.

The terry pile motion is designed to oscillate the sley for a shorter distance for the two consecutive loose picks, and for a greater distance for the third and following fast picks inserted for each horizontal row of loops. This means, first two out of three or more picks are beaten up within a short distance from the fell of cloth, and then after the third pick is inserted in the warp shed, the three picks are thrust forward together. As the picks are pushed forward from their temporary positions to their final positions, they slide along the tense ground warp threads; but the degree of frictional resistance between the three picks and slack pile warp threads is sufficient to draw the latter forward, thereby causing them to bend and thus form a series of loops to constitute the pile surface. If the pile warp threads float, then loops are formed on the surface; otherwise, on the back. The principle is illustrated in Fig. 7-4-4.

（二）毛地组织的恰当配合 /The Match up of the Pile Weave and the Ground Weave

起毛组织和地组织都是经重平组织，但是毛、地组织的起始位置和配置方式对于成圈的形状十分重要。

Both the pile weave and the ground weave are derivative warp rib weaves. The suitable arrangement of the two weaves is crucial to the formation of loops.

1. 毛圈组织 /Pile Weaves

第1短打纬。第一次短打纬时，毛经纱上组织点，应该与前面的组织点（最后一次长打纬）的组织点相同（同一梭口），以保证毛圈达到一定的高度。当织入纬纱1后，毛圈的高度就几乎确定了。短打纬时，钢箝最前方与织口的距离的一半，就是毛圈的高度。

第2短打纬。纬纱2织入后，如果纬纱1和纬纱2与毛经纱有交织，则毛圈的结构被固定下来；否则，同一梭口中的毛经纱容易被后退的钢箝带回部分，使得将来形成的毛圈参差不齐。因此，在短打纬的两根纬纱中，毛圈组织的组织点是相反的。

第1长打纬。当第一次长打纬时，应该与前一组织点交错（相反），这有助于减少由于纬纱的反拨而导致的毛圈变形。

Loose pick 1. The interweaving points at the loose pick 1 should be designed the same as the previous ones (or on the last fast weft) to ensure the height of the loops. After inserting the loose pick 1, the height of the loop is almost determined since the gap of the temporary position to the fell of the cloth is almost 2 times of the loop height.

Loose pick 2. After inserting the loose pick 2, the structure of the loop is anchored if there is an intersection between the loose pick 1 and loose pick 2 to the pile end. However, if they were in the same warp shed, the pile warp threads would retreat when the sley moved backward, which caused the deformed loops in different height. Therefore, it will be observed that the two loose picks are always inserted in different pile warp sheds.

Fast pick1. When the first fast weft is beaten up, there should be intersection with the previous interweaving point, which is beneficial to relieve the rebound of the wefts.

2. 地组织 /Ground Weaves

在设计地组织的起始位置时，要考虑如下因素：①地组织交错次数要小。长打纬时，三根纬纱要同时被推向梭口，必须减少交错次数，以减小打纬阻力小。②地纬对毛经纱的夹持力要大，特别是前两根地纬纱对毛经纱的夹持力要大，否则造成毛圈高度不齐。③最后一根纬纱反拨要小。长打纬后，当钢箝回退时，最后一根纬纱也会随之

In arranging the ground weave, the following factors should be considered: ① Less intersections. When the reed pushes forward the 3 wefts to the fell, it is necessary to have less intersections of the ground weave to reduce the resistance to beating–up. ② High grip to the pile warp threads from the ground wefts is required, especially for the first 2 loose wefts. Otherwise, the loops aren't uniform in height. ③ Less rebound from the last weft. After beating–up, the newly inserted weft retreats with the reed, which might affect the grip of the pile loop and cause uneven height in loops.

回退一些，可能影响对毛圈的夹持能力，造成毛圈高度不齐。

在图7-4-5（a）中，地组织每经都交织了2次，长打纬阻力大；而在图（b）（c）中则是1次，长打纬阻力小。在图（a）所示组织中组织点之间均有交错，纬纱间距大，聚集和夹持能力弱；在图（b）所示组织中2、3纬间有浮长，有集聚能力，对毛经纱有夹持，但1、2纬间有交织，夹持力小，导致毛圈不齐；在图（c）所示组织中1、2间是浮长线，对前两根纬纱有收缩集聚拉拢作用，对毛经纱夹持力强。图7-4-5（a）中纬纱3与纬纱1在同一梭口中，长打纬后容易反拨；图7-4-5（b）中，纬纱3与纬纱1不在同一梭口中，长打纬后反拨不严重，但后退会减小纬纱2、3之间的夹持力；图7-4-5（c）中纬纱3即使反拨，对纬纱1、2之间的夹持力影响也不大。

For the ground weave as shown in Fig. 7-4-5(a), there are 2 interlacings for each ground ends whereas there is only 1 interlacing for the weaves as shown in Fig. 7-4-5(b) and (c), therefore, more obstruction is created in beating up the fast picks in (a). There is an intersection between each neighboring interweaving points on ground ends in (a), thereby produces a longer distance between the wefts, which results in a weak grip for the pile ends. The pick 2 and 3 in the weave as shown in Fig. 7-4-5(b) are at same warp shed and have some ability to grip the pile ends, however, there is less grip between the pick 1 and pick 2, which might lead to uneven loops in height. For the weave as shown in Fig. 7-4-5(c), the pick 1 and pick 2 are inserted the same ground warp shed, and the long float will contract and bunch the first 2 picks, giving strong grip to the pile end to ensure the loop of a uniform length. For the weave as shown in Fig. 7-4-5(a), the pick 3 and pick 1 are at the same shed, the pick 3 tends to rebound after beating-up. For the weave as shown in Fig. 7-4-5(b), there is intersection between the pick 3 and pick 1, and the retreat will decrease the grip between the pick 2 and pick 3. For the weave as shown in Fig. 7-4-5(c), there is little effect on the grip between the pick 1 and pick 2 even there is some retreat from the pick 3.

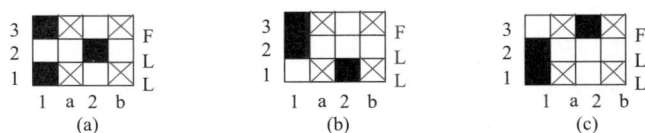

图7-4-5 三纬毛巾毛地组织的配合比较
Comparison of various ground and pile weaves for 3-pick terry

（三）织机特殊的送经运动 /Special Letting-off Motion of the Loom

毛圈只有在织机上进行恰当的送经运动才能形成。毛经纱应该松弛，以便弯曲形成毛圈。否则，张力将把毛圈拉掉。张紧的地经纱可以使得毛圈被推向织口。因此，需

Terry loops can only be formed with a proper letting-off motion of the loom. The pile ends must be slack so that they are easily buckled to form loops, otherwise the tension would pull the loops out. The tighten ground ends enable the terry loops to be pushed forward to the fell. Therefore,

要两根织轴，一个卷绕松弛的毛经纱，一个用于高张力的地经纱。

毛圈的高度 H 由长短打纬距离的差值，也就是短打纬到织口的距离（图7-4-4）决定。线圈的高度大约是其距离的一半。即 H = 短打纬到织口的距离 /2。在送经时，毛经纱的送经量大。毛经纱和地经纱送经量的差值大小和打纬距离的差值应该一致。毛经与地经的送经量之比称毛倍数 R，不同的毛巾具有不同的毛倍数。对于手帕来说，R 一般为 3∶1，面巾和浴巾为 4∶1，而螺纹毛巾为 5∶1 ~ 9∶1。

two beams are used. One is for the lightly tensioned pile ends, and another is for the tightly tensioned ground ends.

The height H of the loop is determined by the loose pick distance– or the distance to the loosen pick from the cloth fell. The height of the loop is about half the distance, or H = the loose pick distance/2. In letting–off, more pile end is forwarded. The amount difference between the letting–off of the pile ends and the ground ends must be coincide with the loose pick difference. The ratio R of the amount of the letting–off for pile ends and the ground ends is called multiple of pile length, which varies with the varieties of the terry fabrics. Different varieties require different R. R is usually 3∶1 for handkerchiefs, 4∶1 for bath towel and face towel, 5∶1~9∶1 for spiral towel.

四、毛巾组织设计 /Designing of Terry Weaves

1. 地组织 /Ground Weaves

地组织地经纱上前两根短打纬的组织点应该相同。因此，三纬毛巾地组织只能选 $\frac{2}{1}$ 变化经重平；四纬毛巾地组织可选 $\frac{3}{1}$ 变化经重平或 $\frac{2}{2}$ 经重平组织。

The first two loose picks are placed in the same warp shed. Thus, the ground weave for a 3–pick terry can only be a $\frac{2}{1}$ varied warp rib weave and can be either a $\frac{3}{1}$ varied warp rib or a $\frac{2}{2}$ warp rib weave for a 4–pick terry.

2. 毛圈组织 /Pile Weaves

毛圈组织的纬纱循环应与地组织相同。按照毛、地组织配合原则，确定毛组织的起点。故三纬毛巾选 $\frac{2}{1}$ 变化经重平，四纬毛巾只能选 $\frac{3}{1}$ 变化经重平。最后，毛圈组织也取决于是否可正反使用。单面毛巾所有毛经纱上组织规律相同，双面毛巾在毛地排列比为 1∶1 时相邻毛经纱组织规律相反。

The weft–wise size of the ground weave must be equal to that of the ground weave. The commencing point is determined by the combination of the pile weave and the ground weave. Therefore, $\frac{2}{1}$ varied warp rib weave is chosen for 3–pick terry and $\frac{3}{1}$ varied warp rib for 4–pick terry. Finally, pile weaves are also determined by the reversible or irreversible effects of the terry. All the pile ends have the same interlacings for single–sided terry weave while the neighboring pile ends have the opposite interlacings for double–sided terry weave if the pile ends and the ground ends are arranged in the proportion of one to one.

3. 确定地经与毛经排列比 /Ratio of Ground Ends and Pile Ends

地经纱与毛经纱的比例可以随意选择，通常为 1:1，也可以为 2:2。无论哪种情况，结果大致相同。偶尔也有 1:2 排列。

The particular disposition of warp threads, however, is quite optional. Some advocate an alternate distribution of ground and pile warp threads, whilst others prefer to dispose them in alternate pairs of each series of warp threads. In both cases the ultimate results are virtually alike. The ends are occasionally arranged in the proportion of one ground end to 2 pile ends.

4. 确定 R_j 和 R_w/Determine R_j and R_w

R_j= 地经纱和毛经纱的数量之和，R_w= 地组织纬纱循环根数。

R_j is the sum of the ground ends and pile ends, and R_w is the same as the size of the weft repeat of the ground weave.

5. 填绘组织点 /Fill the Marks

方格纸上画出组织范围，标出毛经与地经。根据地经和毛经位置，分别填绘对应地组织和起毛组织的经组织点的符号。

图 7-4-6 是按该步骤设计的三纬毛巾组织［图 7-4-6（a）（b）］、四纬毛巾组织［图 7-4-6（c）（d）］的上机图。图 7-4-6（b）所示组织是单面毛巾，其余均为双面毛巾组织。

Draw the square range on the design paper, and number the positions of the pile ends and the ground ends. Fill the marks on the ground weave and the pile weave respectively.

Two 3-pick terry weaves[Fig. 7-4-6(a)(b)]and two 4-pick terry weaves [Fig. 7-4-6(c)(d)]are designed based on the principle illustrated above and shown in Fig. 7-4-6. The weave as shown in Fig. 7-4-6(b) is a single-sided terry weave and the others are all double-sided terry weaves.

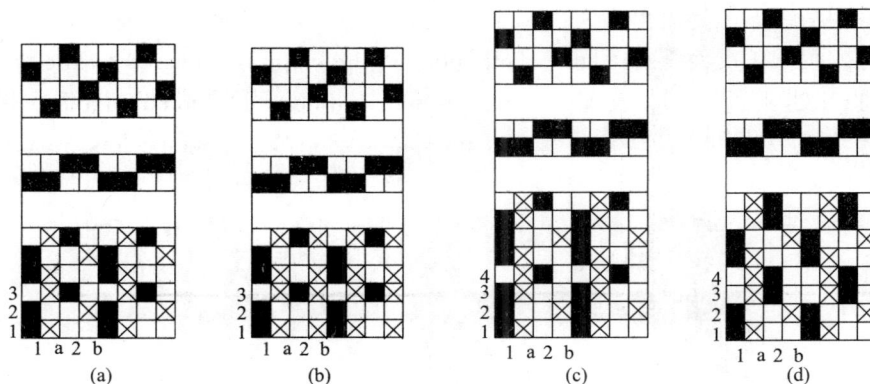

图7-4-6　两种三纬毛巾组织和两种四纬毛巾组织的设计
Designing of two 3-pick terry weaves and two 4-pick terry weaves

五、花式毛巾组织 /Fancy Terry Weaves

通过毛圈的割圈、正反面位置的分布、不同颜色毛纱的排列和毛圈高度的控制，可以得到各种花式毛巾。毛圈不仅可以均匀分布在表面，也可以同时在织物的正反两面，以形成非常平坦的表面；或者用毛圈在平纹或者简单地组织上织制花纹图案，有时线圈也可以割绒产生平绒效果。

The loops may be distributed uniformly either on the face side only, or on both the face and back of the fabric, to form a perfectly even surface; or the loops of pile may be developed in such manner as to create a figured design upon a plain or simple ground. Sometimes, the loops may be cut to form a velveteen effect.

1. 割圈毛巾 /Cut Loop Terry

在后整理工序中对线圈顶端进行割绒形成平绒织物的表观，如图 7-4-7 所示。通常仅在一面进行割绒，另一面仍保留普通毛圈。

Cut pile terry effects are sometimes produced by cropping, during a finishing operation, the tips of the loops in a terry cloth. Usually only one side of a fabric is so treated the other retaining the normal loop formation.

2. 花色效应 /Introducing Color Combining Placing the Loops

如果毛经纱采用不同的颜色，并且单面线圈和双面线圈交替使用，将会在织物表面形成特定图案。

If the pile ends of different colors are used, and the single-sided terry loop and double-sided terry loop are used alternately, a special pattern will be shown on the fabric face.

图 7-4-8 采用单花色毛纱且单面成圈的毛巾组织，交替在正面和反面成圈形成图案，因此，织物正反面交替呈现毛圈和地组织。

In Fig. 7-4-8, only one series of pile threads is used, and the color is introduced either in the pile or in the ground threads. The pattern is due to the pile threads forming loops on the face and back in turn, so that alternate sections of pile and ground are produced on both sides of the cloth.

图 7-4-9 采用双花色毛纱双面成圈的毛圈组织，但是表里交换颜色。此时，织物正反面都有毛圈覆盖，但是各处颜色不同，交替出现。

In Fig. 7-4-9, two series of pile ends are employed, and the pile ends interchange from face to back. In this case both sides of the cloth are covered by the loops, but one series of ends is differently colored from the other series, so that alternate sections in different colors are formed.

图7-4-7　割绒毛巾
Cut loop terry

图 7-4-10 中组织使用了 2 组毛经纱，单面毛圈和双面毛圈间隔配置。此时，织物某一面显示两种颜色的毛圈，但另一面显示地组织。

图 7-4-11（a）所示的毛巾组织中，地经与毛经排列比为 1:1，采用 a 和 b 两种色经纱。可知其正面图案如图 7-4-11（b）所示，背面图案如图 7-4-11（c）所示。

In Fig. 7-4-10, two series of pile ends are employed, and the single-sided loop and double-sided loop alternately placed. In this case one side of the cloth is covered by the loops of two colors, but the other side shows the ground.

In the weave as shown in Fig. 7-4-11(a), the ground ends and the pile ends are arranged 1:1, and two colored pile ends a and b are employed. Based on the above principle, the patterns of the face side and back side are shown in Fig. 7-4-11(b) and(c) respectively.

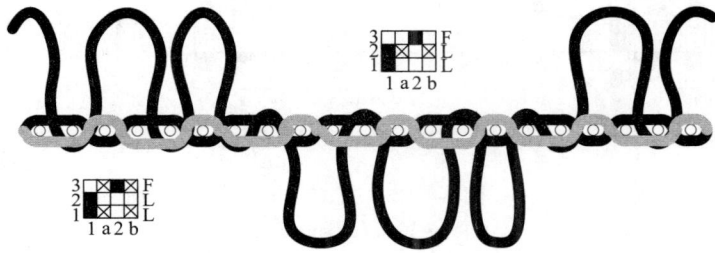

图7-4-8 单色花式毛巾
A figured terry of single color

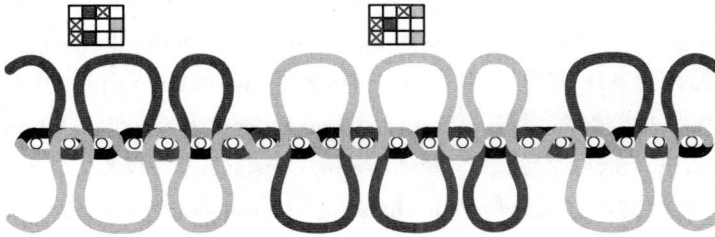

图7-4-9 双面换色花式毛巾
A interchanging figured terry

图7-4-10 双色花式毛巾
A figured terry of 2 colors

(b) 正面图案/Pattern of face side

(c) 背面图案/Pattern of back side

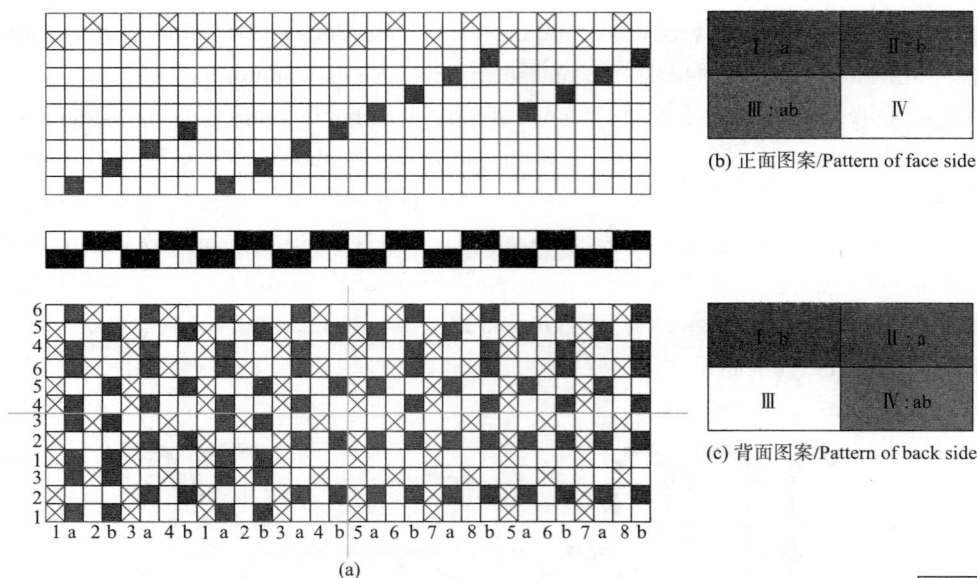

图7-4-11 花式毛巾设计
Designing of a figured terry

Changing the
height of loops

3. 改变毛圈高度 /Changing the Height of Loops

在现代毛巾织机上，可以通过电子装置可编程自由控制毛圈高度。毛圈高度可以在 1 ~ 10mm 之间以 0.1mm 的步长改变。织机需要两种装置：①任意改变短打纬到织口距离的装置；②电子送经装置。这样，每组纬纱毛圈的高度都可以变化，有 180 种短打纬到织口距离，也就有这么多的毛圈高度，毛圈高度的顺序可以根据需要任意设定。根据这个原理，可以生产立体浮雕毛巾。

With a modern terry loom, the height of the loop can be freely programmed from one pick to another pick on the loom by electronic devices. The height of the loop can be changed from 1mm to 10mm within 0.1mm in fabric length. Two mechanisms are required: ① a mechanism to change the loose pick distance freely; ② an electronical letting-off mechanism. The pile height can be freely programmed from one pick group to another. In this way, 180 different loose pick distances, and hence the same number of pile heights, can be programmed in any order desired. Sculptured terry has been developed by the principle.

六、毛巾织物上机 /Looming for Terry Fabrics

毛经与地经送经量差异很大，需要分别卷绕在不同经轴上，地经纱采用消极式送经，毛经纱采用积极式送经。毛经纱要求张力较低，

Due to the great difference in the amount of letting-off, the pile ends and the ground ends are wound on the separated beams. The pile ends are supplied by positive letting-off mode while the ground ends negative letting-

为保证开口清晰，穿在前面综框中，其位置比地经纱稍高，避免因地经张力大，升降时挂带松弛毛经的现象发生。

筘号不宜过高，因毛经松，会使织造困难。通常 2 纱 / 筘，若地经、毛经排列比为 1:1，则同组毛地经穿入同一筘齿。若地经、毛经排列比为 2:2，则同系列地经或毛经穿在同一筘齿中。因为同一筘齿中地经沉浮规律相反，地经与毛经有时在同一梭口中，因被筘齿隔开，开口更加清晰。

由于开口过程中承受更多的摩擦，地经纱上浆率高，约为 9.5%，毛经纱上浆率约为 1.5%。毛经纱选用低捻度，利于织物吸水、柔软。毛巾织物的布边组织宜选用 $\frac{2}{2}$ 经重平。

off. Pile ends are drawn upon the front healds due to the requirements of low tension for a clear shedding. The positions of the pile ends are set higher than the ground ends in order to avoid the slack pile ends are drawn by the tight ground ends in lifting.

The count of the reed should not be too high as the smaller split between the wires causes the difficulties in weaving due to the slack pile ends. Usually, two threads are placed in each split of the reed, and in the 1 ground, 1 pile order, and one of each series is placed in the same split. In the 2 ground, 2 pile order, however, two ends of the same series are placed together. The threads in each split work opposite to each other, and at the same time the pile and ground threads, which on some picks work alike, are separated by the wires of the reed, so that a clear shed is more readily obtained.

The ground ends should have a much higher sizing percentage due to more abrasion during shedding. Usually, the size taking is 9.5% for ground ends while 1.5% for pile ends. Soft twist yarns are preferred for the pile ends since the terry fabrics are expected to be absorbent and soft. A $\frac{2}{2}$ warp rib is suitable for the selvedge weave of the terry fabric.

习题 /Questions

1. 以 $\frac{2}{1}\nearrow$ 为地组织，地纬与绒纬之比为 1:2，R_j=6，绒根固结方式为 V 型，试作灯芯绒织物的组织图。

2. 已知地组织为 $\frac{2}{2}$ 纬重平，绒根用复式 V 型、W 型固结，地纬与绒纬之比为 1:2，绒根的固结位置自己决定，试作灯芯绒组织图，并标出割绒位置。

1. Draw the corduroy weave under the following conditions: ground weave $\frac{2}{1}\nearrow$, the ground weft and the pile weft are arranged at the ratio of 1:2; the size of repeat in warp direction is 6, and the pile is V–form pile.

2. Draw the corduroy weave and mark the position of pile cutting under the following conditions: ground weave $\frac{2}{2}$ weft rib; the ground weft and the pile weft are arranged at the ratio of 1:2; the location of binding is chosen by the designer; and the pile wefts are bound by the

3. 某长毛绒织物，以 $\frac{2}{2}$ 纬重平为地组织，经绒 W 型固结，绒经与地经排列比为 1:4，每个组织循环中有 2 根绒经纱，绒经均匀固结，上下层投梭比为 3:3，采用双层单梭口全起毛织造方法，试作该织物的上机图及经向截面图。

4. 以 $\frac{2}{2}$ 纬重平为地组织，绒经 W 型固结，半起毛单梭口，双层织造，上下层投纬比为 4:4，地经与绒经排列比为 2:2，试作长毛绒组织的上机图及经向截面图。

5. 已知某长毛绒织物的经向截面图如习题图 7-1 所示，试作出该组织合理的上机图。

6. 试作双双经双双毛四纬双面毛巾的上机图。

7. 毛巾纹样如 $\begin{array}{|c|c|} \hline A & B \\ \hline B & A \\ \hline \end{array}$ 所示，其中 A、B 各代表不同的色区，每区由 4 根毛经纱、4 根地经纱和 2 个全打纬（2 个打纬循环）组成，地经:毛经 =1:1，试作该表里换层的异色花式四纬毛巾的组织图和经向截面图（毛、地组织自选）。

8. 判断习题图 7-2 所示的组织类别，说明其在恰当工艺条件下的外观特点。

9. 习题图 7-3 所示的组织都是灯芯绒组织。请判断哪个是细条灯芯绒，哪个是阔条灯芯绒，哪个是中条灯芯绒。请说明原因？

combination of V–form pile and W–form pile.

3. Draw the looming plans for a plush weave and its longitudinal section diagram under the following conditions: ground weave $\frac{2}{2}$ weft rib; W–form pile; the order of warping is 1pile, 4 grounds; 2 pile ends that are uniformly placed in each repeat; the order of wefting is 3 top picks, 3 bottom picks; the single–shed weaving and pile end weave is completely lift.

4. Draw the looming plans for a plush weave and its longitudinal section diagram under the following conditions: ground weave $\frac{2}{2}$ weft rib; W–form pile; the order of warping is 2 piles, 2 grounds; the order of wefting is 4 top picks, 4 bottom picks; the single–shed weaving and pile end weave is half lift.

5. The longitudinal section diagram of a plush fabric is shown as below. Try to draw the reasonable looming plans for the fabric.

6. Draw the looming plans for a 4–pick terry weave with the order of warping: 2 pile ends, 2 ground ends.

7. The design pattern of a terry fabric is shown as $\begin{array}{|c|c|} \hline A & B \\ \hline B & A \\ \hline \end{array}$, in which A and B represent different color. Each square consists of 4 pile ends and 4 ground ends, and includes 2 picking cycles. The order of warping is 1 ground end, 1 pile end. Try to draw the interchanging multi–color 4–pick terry weave and its longitudinal sectional diagram (Note: the pile weave and ground weave are determined by the designer).

8. Judge the following weaves and describe the appearance of the weave at the reasonable looming conditions.

9. There are 3 corduroy weaves as below. Which one can be used to produce fine cord, medium cord and broad cord respectively?

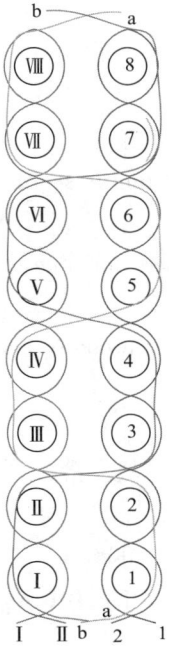

习题图7-1

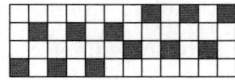

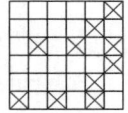

(a)　　　　　　(b)　　　　　　(c)　　　　　　(d)

习题图7-2

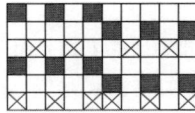

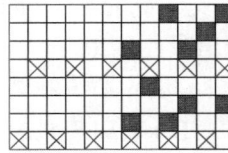

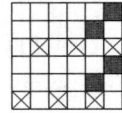

(a)　　　　　　(b)　　　　　　(c)

习题图7-3

参考文献 /References

［1］GROSICKI Z J．Watson's Advanced Textile Design–Compound Woven Structures ［M］. Cambridge: Woodhead Publishing limited, 2004.

［2］GROSICKI Z J．Watson's Textile Design and Colour–Elementary Weaves and Figured Fabrics ［M］. 7th ed．Cambridge: Woodhead Publishing limited, 2004.

［3］NISBET H．Grammar of Textile Design ［M］. 2nd ed．London: Scott, Greenwood & Son, 1919.

［4］ADANUR S．Handbook of Weaving ［M］. London: CRC Press, 2001.

［5］（美）托托拉,（美）默克尔. 仙童英汉双解纺织词典 ［M］. 7 版. 黄故，等，译. 北京：中国纺织出版社，2004.

［6］ROUETTE H K．Springer 纺织百科全书 ［M］. 北京：中国纺织出版社，2008.

［7］MCINTYRE J E, DANIELS P N．Textile Terms and Definitions ［M］. 10th ed. Manchester: The Textile Institute, 1995

［8］蔡陛霞. 织物结构与设计 ［M］. 2 版. 北京：中国纺织出版社，1988.

［9］顾平. 织物结构与设计 ［M］. 2 版. 上海：东华大学出版社，2004.

［10］郑秀芝，刘培民. 机织物结构与设计 ［M］. 北京：中国纺织出版社，1992.

［11］吴汉金，郑佩芳. 机织物结构设计原理 ［M］. 上海：同济大学出版社，1990.

［12］李栋高. 纺织品设计学 ［M］. 北京：中国纺织出版社，2006.

［13］谢光银，张萍. 纺织品设计 ［M］. 北京：中国纺织出版社，2007.

［14］吴坚，李淳. 纺织品功能性设计 ［M］. 北京：中国纺织出版社，2007.

［15］姚穆. 纺织材料学 ［M］. 2 版. 北京：中国纺织出版社，1990.

［16］沈兰萍，白燕，陈益人. 织物组织与设计 ［M］. 北京：化学工业出版社，2014.

［17］PEIRCE F T．The geometry of cloth structure ［J］. Journal of the Textile Institute, 1937, 28 （3）: 45–96.

［18］SEYAM A M．Structure design of woven fabrics ［J］. Textile Progress, 2002, 31 （3）: 1–36.

［19］郑天勇，黄故. 机织物结构外观分析及三维模拟显示 ［J］. 纺织学报，2001 （2）: 40–43.

［20］郑天勇，王跃存，郭新生. 判断单层、多重、多层织物层数的关键 ［J］. 纺织学报，1997 （6）: 40–42.

［21］郑天勇. 几种变化组织织物的计算机三维模拟 ［J］. 纺织学报，2005, 26 （6）: 35–39.

［22］KEMP A．An extension of Peirce's cloth geometry to the treatment of non–circular threads ［J］. Journal of the Textile Institute, 1958, 49 （1）: 45–48.

［23］HU J．Theories of woven fabric geometry ［J］. Textile Asia, 1995, 26 （1）: 58–60.

［24］HAMILTON J B. A central system of woven-fabric geometry ［J］. Journal of the Textile Institute, 1964, 55（1）: 67-82.

［25］ZHENG T, ZHANG X, ZHAO Z, et al. Geometric structure model of plain woven fabric based on progressive spring-slide mechanics ［J］. Textile Research Journal, 2014, 84（17）: 1803-1819.